The American Robot

THE AMERICAN ROBOT

A Cultural History

DUSTIN A. ABNET

The University of Chicago Press
Chicago and London

The University of Chicago Press, Chicago 60637
The University of Chicago Press, Ltd., London

Published 2020
Printed in the United States of America

29 28 27 26 25 24 23 22 21 20   1 2 3 4 5

ISBN-13: 978-0-226-69271-5 (cloth)
ISBN-13: 978-0-226-69285-2 (e-book)
DOI: https://doi.org/10.7208/chicago/9780226692852.001.0001

Library of Congress Cataloging-in-Publication Data

Names: Abnet, Dustin A., author.
Title: The American robot : a cultural history / Dustin A. Abnet.
Description: Chicago : University of Chicago Press, 2020. | Includes bibliographical references and index.
Identifiers: LCCN 2019037835 | ISBN 9780226692715 (cloth) | ISBN 9780226692852 (ebook)
Subjects: LCSH: Robots—Social aspects. | Robotics—Social aspects.
Classification: LCC TJ211 .A34 2020 | DDC 303.48/3—dc23
LC record available at https://lccn.loc.gov/2019037835

♾ This paper meets the requirements of ANSI/NISO Z39.48-1992 (Permanence of Paper).

For Nicole

Contents

Introduction: An Intimate and Distant Machine *1*

Part 1 **God and Demon, 1790–1910** *13*

1 The Republican Automaton *19*

2 Humanizing the Industrial Machine *41*

3 Mechanizing Men *71*

Part 2 **Masters and Slaves, 1910–1945** *97*

4 Symbolizing the Machine Age *103*

5 Building the Slaves of Tomorrow *131*

6 Conditioning the Robot's Brain *163*

7 A War against the Machine Age *187*

Part 3 **Playfellow and Protector, 1945–2019** *209*

8 Preserving American Innocence *215*

9 The Postindustrial Gift *241*

10 Cheerful Robots *271*

Epilogue: The American Robot *295*

Acknowledgments *303*

Notes *309*

Index *365*

Introduction

An Intimate and Distant Machine

In March 1999, Matt Groening's animated sitcom *Futurama* introduced audiences to Philip J. Fry, an idiot who falls into a cryogenic tube on December 31, 1999, and awakens exactly one thousand years later. In the interim, humans have traveled through space and met hundreds of alien species. They have invented fantastic and horrifying new technologies. Earth has been invaded, destroyed, and rebuilt multiple times. Yet, this future maintains a curious nostalgia for elements of twentieth-century culture, including the Harlem Globetrotters, Richard Nixon, and 1950s-style hygiene films. When, in a season 3 episode, Fry dates a robot that physically and vocally duplicates the actress Lucy Liu while poorly performing a "personality mathematically derived" from her movies, his disgusted friends show him one of those hygiene films, *I Dated a Robot!*[1]

As a teenage boy and girl sit at a diner, a middle-aged white male narrator walks to their table and lectures, "Ordinary human dating. It's enjoyable and it serves an important purpose," before flipping their table over to reveal a screaming baby. "But when a human dates an artificial mate, there is no purpose. Only enjoyment. And that leads to tragedy." As "Billy Everyteen" succumbs to the temptations of his "Marilyn Monroebot," he grows too lazy to walk his dog, deliver newspapers, or make out with a girl from across the street. "In

Fig. I.1. "Billy Everyteen" making out with his "Marilyn Monroebot" instead of walking his dog, finishing his paper route, or even making out with a human girl in the *Futurama* episode "I Dated a Robot." Animators added a (barely visible) aura around the robot to reinforce the distinction between authentic and inauthentic that the episode satirizes.

a world where teens can date robots," the narrator asks, "why should he bother? Why should anyone bother?" Skipping eighty years into the future, viewers see Billy die alone with his robot just before Earth suffers the consequences of his unrestrained pursuit of pleasure: annihilation.

Despite its satire, the episode raises numerous questions of increasing importance in an era of advanced automation, artificial intelligence, and digital personas. If a technology can simulate human affection and fulfill individual desires on command, what is the point of risking rejection or laboring to build relationships? Do we lose a sense of purpose by devoting ourselves to easy pleasures offered by technology rather than more difficult tasks? Are we, like the episode's robots, mere algorithms of the data we present to the public, or is there something deeper that defines us that no machine can duplicate? Will the technologies of the twenty-first century enable people to experience the bliss of individual fulfillment, or will the pursuit of our desires destroy society and, with it, ourselves? The episode answers none of these questions; it treats them as jokes because they are as unanswerable as they are fundamental to life in the digital age. Collectively, the questions raised by technology in the twenty-first century are overwhelming; but our culture, like the *Futurama* episode, has a way of grounding them in an icon that is simultaneously literary and technological, humorous and horrifying, intimate and distant: the robot. A concrete yet symbolic anchor in this maelstrom of questions, the robot tethers our conversations about science, technology, identity, purpose, and power like few other icons or devices—and it has done so, at least in some form, for hundreds of years.

The American Robot is a history of the idea of the robot in American culture. It explores how robots and their kin—automatons, mechanical men and women, androids, artificial intelligence, and cyborgs—have embodied and conceptually linked some of the most critical questions of modern culture: What is a human and what is a machine? Does free will exist, or are people merely programmed by internal or external forces? Is the machine a model of human identity and behavior, or its antithesis? What in modern life might make certain people appear machinelike, and what might enable them to maintain their humanity? Which tasks in the home, workplace, or military should be mechanized, and which should be left for people to perform? Does the pursuit of science and technology need to be controlled, and if so, who should have that power? The answers people have given to these and similar questions have rarely been absolute or universal; frequently, they have not even been completely coherent. But in the struggle to answer them, people have imagined, depicted, viewed, and, occasionally, built robots. Robots have been important not just because they have raised questions but because they have spawned fantasies that people have tried to make real.[2] A creature of both fiction and fact, the robot has been a particularly potent force in the cultural history of the United States.

Of course, the robot is neither exclusively American nor modern. The term itself is Czech. Clockwork, steam, and water-powered automatons date to the ancient world, as do stories of robotlike creatures, such as golems and the statue come to life in *Pygmalion*. In medieval and early modern Europe, automatons were a fixture of both religious and monarchical life, and the devices would remain predominantly European through the nineteenth century.[3] The classic story of robot rebellion duplicates much of the British writer Mary Shelley's *Frankenstein*. The most significant American writer of robot fiction, Isaac Asimov, emigrated from Russia. Since the proliferation of robotic toys in the 1950s, the figure has been associated with Japanese culture. An ancient and global icon, the robot apparently transcends both geographic and temporal boundaries to address a question that seems universal: what does it mean to be human?

But the deceptively universal nature of the robot obscures the ways that its meanings are shaped by context, that different times and places inspire different visions, interpretations, and, ultimately, consequences. The earliest American device discussed in this book is an automaton Indian from the late eighteenth century; the most recent is the indigenous "host" Akecheta from HBO's twenty-first-century television show *Westworld*. Though each would appear in the form of a Native warrior, their vastly different contexts give them distinct meanings. Appearing in Philadelphia near the dawn of efforts to expand a new nation and construct new racial hierarchies, the automaton Indian offered a fantasy of control that reduced its Native subject to a mindless body that needed to be tamed by the audience. Appearing during an era far more critical of racial stereotyping and the violence associated with imperial expansion, Akecheta challenges audiences to empathize with the device, question the stereotypes that shaped its creation, and root for its efforts to escape to a digital space that is truly—unlike the myths that white settlers told of the real West—untouched by any, save its creator.[4] The robot might be universal, but its meanings shift to fit particular beliefs, ideals, hopes, fears, and longings.

In imagining robots, Americans have adjusted ideas and iconography created abroad to fit the specific cultural concerns and tensions of their society. Since the eighteenth century, their visions of robots have developed in conversation with the violence of slavery and Western expansion; the theology of Puritanism and evangelical Protestantism; the individualism of republican, liberal, and democratic traditions; the expansion of an industrialized and commercialized economy; and the continual efforts of marginalized groups to secure freedom and equality. None of these aspects has been unique to American history, but the combination has created a distinct cultural environment that has shaped how people have understood the robot's meaning. The tendency to see the robot as a universal phenomenon across both time and space ignores such particularities and thereby conceals the degree to which its primary technology is not electromechanical but ideological; that, rather than a piece of machinery, it is a flexible concept deeply rooted in the power relations of a society.[5]

The robot's enduring ideological power comes from its ambiguity. People never have agreed on a single meaning for the term. A derivative of the Czech word *robota* meaning "drudgery" or "servitude," *robot* comes from Karel Čapek's 1921 play *R.U.R. (Rossum's Universal Robots)*, which depicts the devices as biological but artificial humans rather than as metallic inventions. In the context of the Russian Revolution, people initially understood Čapek's rebellious robot as a metaphor for alienated workers, particularly those who toiled along Henry Ford's assembly line. Yet, others soon connected the term to the emergence of automated machinery in the same automobile industry.[6] From nearly the moment the term arrived on American shores, *robot* referred to both workers and the machines that could replace them, a duality that has persisted even as it has been transformed. In World War II, people applied the term both to fascist soldiers who seemed to lack willpower and to remotely-guided technologies such as the V1 and V2 rockets. During the Cold War, sociologist C. Wright Mills critiqued white-collar workers as "cheerful robots" because they seemed to lack independence, while others used the term to explain electronic computers and automation technologies that enabled machinery to operate more independently. Even the *Futurama* episode's use of celebrity robots satirized both technology and the commodification of actresses' identities. Understood as both a *humanized machine* and a *mechanized human*, the robot has linked two core themes of modern life: the actual replacement of people with machines and the metaphorical transformation of people into machines. Its significance, historically and today, derives from the relationships that its creators and consumers have posited between those two trends—how it has symbolically linked the advances of science, technology, and industrial capitalism with the transformation of the individual soul.

As a *humanized machine*, the robot has helped Americans understand and relate to powers and processes that seem beyond the control of any individual. While not entirely linear, interest in robots and related devices has grown steadily in American culture since the end of the eighteenth century, largely due to the industrialization, commercialization, rationalization, bureaucratization, and democratization of the nation. Seemingly possessing an energy independent of human actors, these interrelated processes have had concrete effects on the lives of Americans that few could control and none could stop. In the nineteenth century, critics such as British writer Thomas Carlyle employed the metaphor of "the machine" to describe this apparent loss of human agency.[7] By anthropomorphizing the machine, people who have imagined robots brought such large-scale developments into the more intimate space of human identity, where people could understand and relate to them. Through humanization, people could talk, fantasize, and even joke about processes and powers beyond their control in a much more relatable and concrete fashion. Sometimes such humanization resulted in nightmares of machines stealing jobs, enslaving individuals, or exterminating humanity. More frequently, though, they satirized such fears, offered fantasies of reclaiming power over machinery, and spread dreams of a postindustrial world free of degrading work where each person could possess all that he or she desired without any cost. Deeply tied to the longings of an individualized, commercial society, the humanized machine has been an important ideological weapon for the legitimization of industrial capitalism.[8]

As a *mechanized human*, the robot has symbolized both the inherently mechanical nature of all people and the supposedly mechanical nature of specific types of people. Numerous intellectuals, in denying the existence of a supernatural soul and occasionally free will, have thought that people are machines crafted by comprehensible processes rather than divine or metaphysical energies. In such an interpretation, people are like automatons or robots because they follow predictable and potentially controllable patterns. The scientist and inventor Nicola Tesla wrote in 1900, "I am an automaton endowed with power of movement, which merely responds to external stimuli beating upon my sense organs, and thinks and acts and moves accordingly." Yet, most forms of American culture have professed the exceptional, spiritual, and vitalist nature of humanity. Few myths have been more potent in American cultural history than the assumption that people—especially the powerful—should be more than "mere machinery," that they should possess qualities that machines, made entirely of matter, seem to lack: reason, sentiment, consciousness, and, above all, willpower and agency.[9]

Though unwilling to acknowledge that they themselves might be machines, many American elites have been willing to accept that others might or should be. The most common forms mimicked by performing automatons in the nineteenth century were nonwhites, women, and children—precisely the groups whom powerful men deemed to lack the qualities they believed necessary to exercise self-control and the rights of citizenship. While stage automatons satirically insinuated that such people were nothing more than machinery, metaphorical uses of the word *automaton* indicated that even people born fully human could acquire a mechanical nature by losing self-control. In the nineteenth century, writers applied *automaton* to gamblers, fashion devotees, and rival politicians who seemed particularly beholden to party. Soon, critics of the emerging factory system adopted the term to explain the destructive effects of the deskilling and repetitive motions of industrial labor. During and after the Civil War, however, Americans applied the term to soldiers to both critique and praise military discipline. Though the central term shifted to *robot* in the 1920s, the metaphor has remained consistent: the mechanized human is one who seems to lack a unique, independent soul capable of demonstrating willpower. In their more self-conscious moments, people have applied these terms to themselves to signify how modern life turns all people into machines; but more commonly they have applied them to others—especially those they deem inherently inferior or who seem to have had their individuality crushed by the modern workplace, totalitarian governments, or mass culture. Never self-made, the robot has been a metaphor for a person who lacks agency and authenticity, who is either intrinsically a machine, who has let himself or herself become a machine, or who has been forced to become one.

Such a theme has proven crucial within American cultural history because separating human from machinelike persons has been one of the key ways of rationalizing efforts to restrict the rights, freedoms, and powers of others. When attached to privileged individuals, robotic metaphors can pointedly and humorously critique power; when applied to entire groups, they dehumanize and delegitimize claims to basic rights. In the 2012 presidential election, opponents caricatured Mitt Romney as the "Romney-bot" because his carefully tailored campaign personality appeared to lack human authenticity. Barack Obama's opponents, however, caricatured the president's supporters as "Obamabots" because they seemed uncritically beholden to him.[10] Though each usage dehumanized opponents, one targeted the authenticity of a powerful politician while the other targeted the rationality of an entire group of people, many of whom came from communities whose rights (including the right to vote), remain threatened.[11] Such rhetoric may have little power today, but historically identifying groups as machinelike has often been used to justify discrimination and even violence.

It should come as no surprise, then, that the character of the robot has predominantly been the creation of a small subset of Americans: middle- and upper-class white men. In a mass consumer society, anyone can encounter a robot in a magazine, theater, or store, just as anyone can imagine one. Yet the subset of people who have publicly envisioned, displayed, and discussed robots is strikingly small. With only a few exceptions before the late twentieth century, robots have been the expressions of men of power and privilege. Part of this narrowness stems from the limited access other groups have had to engineering, intellectual, and even popular culture. But even within women's magazines or the black press, discussions of robots and automatons have rarely appeared.[12] While working-class publications used the term and iconography, they, unlike their middle- and upper-class brethren, defined it almost exclusively as a machine, not as the people who tended those machines.[13] Even ostensibly feminist stories criticizing men who lust after "female" androids generally have come from men such as Ira Levin, the author of *The Stepford Wives*, Joss Whedon, the producer of *Buffy the Vampire Slayer*, and the almost entirely male *Futurama* staff.[14] It is revealing that the most prominent female and African American writers of science fiction from the second half of the twentieth century—Ursula Le Guin, Octavia Butler, Samuel Delaney—rarely, if ever, depicted robots. Fundamentally slaves in both humanized machine and mechanized human form, robots have been primarily imagined and built by men whose gender, whiteness, training, or wealth has taught them that they were entitled to privilege.

But slaves, as Americans know well, might rebel. In the real world, robots never revolt; they break down and make mistakes, but revolution requires consciousness of social position (something that, as far as I know, machines do not have—yet). But in fiction, the robot rebellion is so prevalent it has become cliché. In Čapek's original story of robotic revolution, the creatures win; but American stories of mechanical rebellion have typically resulted in human victory, usually through the strength, intelligence, violence, or virtue of white men. Fusing with one of the country's most prominent contributions to global culture—the Western—the American story of robotic revolt has offered a vision of reclaiming white, male authority from out-of-control peoples and technologies. Unlike its twenty-first century descendent, Michael Crichton's original *Westworld* film tells a story of the restoration of manhood through having sex with robotic prostitutes and killing rebellious robotic gunslingers. Linda Hamilton's Sarah Connor might have become an icon of empowered womanhood for facing down Arnold Schwarzenegger's murderous T-800 in *The Terminator*, but it is her son, John, who is destined to save humanity from the power of the artificial intelligence Skynet. The sexual allure of the enslaved Marilyn Monrocbot threatens to destroy humanity in *Futurama*, but if men like Billy would just tame their own desires, then humanity might be saved. While real robots offer the possibility of slavery without risk of rebellion or moral guilt, fictional robots offer fantasies of control and empowerment.

Robots have remained potent fantasies for such men because they promise to resolve the fundamental tensions between American myths and American realties that have dominated the country's culture since the very beginning. Since the eighteenth century, American development has been shaped by tensions between freedom and slavery, equality and hierarchy, inclusion and exclusion, peace and violence, social responsibilities and individual desires. Existing both among people and within individual consciences, such tensions demand resolution, often because people dedicated to creating a more inclusive and just society demand it. As mechanized humans, robots rationalize efforts to control, exclude, and wage war against other groups of people; as humanized machines they offer a "techno-fix" that promises to quickly and permanently settle social issues without much difficulty.[15] It is no coincidence that dreams of automatons performing degrading labor emerged in the urban North prior to the

Civil War as an alternative to both slavery and wage labor, each of which challenged visions of America as a nation of relatively equal small farmers and craftsmen. Similarly, domestic robots were often promoted by men as an alternative not to housework itself but as a response to women's requests that men help with housework. The same has been true of military robots who were imagined as an alternative to jeopardizing the lives and moral innocence of human soldiers. To white upper- and middle-class men, robots offer the fantasy of maintaining the benefits of power and privilege without having to deal seriously with the demands of other people and any accompanying moral guilt.

When Philip Fry dates the Lucy Liubot, he joins a long history of white American men embracing the robot as an individualistic fantasy of power over the body of someone they could not control—in this case, an Asian American actress. Yet Fry does not start with sex but with an inane conversation that is, for him, surprisingly deep. Like many real and fictional users of sex robots, he wants communication and companionship, not simply control.[16] The power fantasy offered by the robot is not enough; after all, almost any machine can offer a fantasy of control. The robot is unique because it mixes control with a simulation of connection; it offers not just a means of distancing the self from the machine but the fantasy of connecting with it on the most intimate of levels. This is the source of the robot's power and its greatest limitation. At its most utopian, the robot offers the possibility that everyone can satisfy even the most intimate of desires without having to deal with the competing wishes of others. Robots will, according to their advocates, free us from degrading work, liberate us to be ourselves, and ensure equality (at least among the people who matter). But, as critics have always maintained, that vision of equality is not social, not rooted in community; it denigrates and jettisons the other to allow one to focus exclusively on the self. It is revealing that, in pursuing his relationships with the Liubot, Fry—much like Billy Everyteen—ignores the obvious jealousy of a woman he loves, Leela, who longs to be with him. Abandoning a chance for human connection, he becomes consumed with a machine that is merely a reflection of himself.

Like the fictional Fry, Americans in the twenty-first century stand in a world beset by robots. In popular culture, humanized machines appear in novels, television shows, films, songs, video games, and YouTube videos. In the home, less-human-looking robots vacuum floors and pools and control lights and indoor climate; others entertain and provide companionship for adults, children, and even pets. In the factory, robots (often in the form of a mechanical arm) manufacture nearly everything with minimal aid from human laborers. In stores, robotic scanning machines replace the work of human cashiers in the name of efficiency for companies facing the nearly monopolistic power of Amazon.com and convenience for customers harried by the time constraints of modern life. Digital assistants—powered by artificial intelligence and given names such as Siri or Alexa, as if they were persons—gather, analyze, and redistribute data to create digital profiles of customers that ensure even greater convenience and efficiency in the pursuit of individualized desires.

Even as humanized machines dominate our lives, the plight of the mechanized human remains. On July 15 and 16, 2019, Amazon's "Prime Day" sales event, the company's workers went on strike to protest degrading and dehumanizing working conditions. As they protested, they held signs that echoed sentiments that unappreciated and dehumanized workers had articulated since the dawn of industrial capitalism in the nineteenth century: "We Are Humans, Not Robots!" On one level, the workers' actions were meant to pressure Amazon directly by hindering efforts to quickly distribute goods to demanding customers. But such actions also called on consumers to realize their complicity in a system that wrecks the bodies, minds, and maybe even souls of fellow human beings. In proclaiming themselves humans amid a system that treats them as robots, striking workers asked for empathy for those who suffer so that others might consume; they asked for human intimacy in an economic system that depends on creating distance, not just between workers and consumers but between everyone. Of course, while workers protested in the name of their humanity, Amazon's engineers were hard at work developing robotic carts, scanners, planes, and drones to replace them.[17] Such is the story of the American robot.

PART 1

God and Demon 1790–1910

In 1872, *Scribner's Monthly* magazine published a strange but illuminating tale with a horrifying ending. Late in 1867, the story told, the sweet and innocent Nellie Swansdown attended an automaton performance in the town of Mullenville with her would-be beau Sam Gumple. As the performance started, the automaton stepped out of a black box "with an air of jaunty assurance, with flaxen hair and whiskers, a nobby suit of clothes, an eyeglass, a cane, and patent leather boots . . . with a bow and a smirk, just as any human being might have done, only with infinitely more grace and ease." Most of the crowd gaped as the device "ogled the girls . . . cracked jokes with the men . . . danced a hornpipe, and whistled 'Hail Columbia' and 'Yankee Doodle.'" Yet Nellie did not share in the enthusiasm because of a nightmare the previous evening. In her dream, she wed an automaton, but her husband soon "turned into the tall old-fashioned clock which stood in the dining-room; and while she was standing looking up at it in dismay at the thought of being united to such a thing, and bound to honor, love, and obey it all her life, it fell over on to her and crushed her to pieces." Petrified by her vision of death by clockwork routine, Nellie sat silently through the performance. Her horror increased when the automaton produced a nosegay and threw it into her bosom before disappearing with an "infernal cachinnation, as though some devilish joke were in the wind." As Sam escorted Nellie home, "a phantom carriage with jet-black horses" approached. From the carriage "sprung a goblin" that Sam "recognized as the automaton." After knocking Sam out, the automaton "seized Nellie round the waste . . . leaped with her into the phantom coach," and "disappeared into the night with a rumble like an earthquake." The town found Sam the next morning raving like a lunatic. Nellie, however, was never seen again; presumably, she was dragged to hell to wed a demon automaton.[1]

It was a curious tale, made all the curiouser by its author: a former engineer named Julian Hawthorne, the only son of the acclaimed writer of *The Scarlet Letter*, Nathaniel Hawthorne. Born in 1846, Julian had been raised to appreciate nature, art, and imagination. His formative years had been spent in Concord, Massachusetts, with Nathaniel's fellow writers Ralph Waldo Emerson and Henry David Thoreau, both titans in the transcendentalist religious movement that sought to free the individual from the chains of society through the pursuit of a spiritual connection to nature and the rest of humanity. When Julian was a child, Emerson walked with him through "Fairy Land," their name for the Walden Woods that Thoreau would later make his home. Also while in Concord, Herman Melville—the fiercest literary critic of industrial capitalism in antebellum America—read Julian bedtime stories. Despite such a background, Julian initially exchanged the worlds of art and literature for the world of civil engineering. With the family financially strapped following his father's death, Julian chose the regular paycheck offered by a profession at the heart of economic life; further, he reasoned, the job would require him to spend time outdoors, which his father and mentors would appreciate. But Julian was horrible at mathematics, an increasing requirement in the profession, and when his nontechnical job failed to pay him enough to provide for his growing family, he turned to father's profession.[2]

This story, "The Mullenville Mystery" was one of his first. Highly experimental, it is told almost entirely in the past tense by an omniscient narrator. Starting the story two months before the fateful automaton performance, the narrator recounts how Nellie hated Sam's efforts to woo her, but soon met a man brought by the railroad who seemed much closer to her ideal: the "aristocratic" Ned Holland.

Unlike Sam, Ned presented an "enterprising countenance and easy confident bearing." He was, the narrator recalled, "a talented, audacious sort of chap, popular among both men and women,—for there is a large amount of pure romance mingled with his composition, and an impetuosity and fertility of thought and action such as girls always love, and men too, unless they happen to be jealous."[3] The courtship unfolds in a chaste manner appropriate for Victorian middle-class magazines, but then Julian abandons the narration of the past for the structure of a play—complete with parenthetical directions for actions and insights into the characters' emotional states.

As Nellie and Ned become actors performing roles for the narrator, the two fight over human authenticity. Ned asks, "What is it about me you like the most, Nellie?" To which she jokes about his "*rather large*" nose as the source of his "conceit." Offended, he chastises Nellie for her time with Sam. The frustrated Nellie exclaims, "Poor Sam! He's a human creature, at any rate,—not a machine." "Furious," per the stage directions, Ned replies, "Do I understand you to say *I'm* a machine, Miss Swansdown?" To this, Nellie unleashes a blistering combination of laughter and venom: "You always reminded me so much of a clock!—stuck up to be looked at; wound up to go, and always doing over the same things,—thinking yourself so clever, so accomplished, so knowing, and everybody else so vulgar, so stupid, so commonplace,—oh! You needn't say anything: one can always tell what a clock is going to say if one can see its face! But really now, Mr. Holland, if you only wouldn't pretend to be a man, you might be very interesting—as a machine." After he proposes marriage, she replies, "I'd rather marry an aw . . . awtomaton!" His manhood attacked, the infuriated Ned leaves Nellie with a prophetic: "I trust your wish may be gratified!!"[4]

This then was the source of Nellie's descent into an automaton's hell. Despite appearing amid the heightened industrialization of the post–Civil War years, Julian's robotic vision was not predominantly a humanized machine but a mechanized human, a man whose external perfection and regularity had made him seem to be performing humanity rather than possessing it. Instead of machines, Nellie was frightened by the regularity and artificiality of the age; that, not the railroad, telegraph, or any other mechanical improvement of the period, gave the automaton its demonic nature.

The contrast with Nathaniel Hawthorne is revealing because he also wrote about an automaton, in his case, a stunningly beautiful clockwork butterfly. Entitled "The Artist of the Beautiful," the 1844 story recounts the efforts of a skilled but self-absorbed and hubristic clockmaker named Owen Warland to imbue spirit into matter by building a beautiful, living automaton butterfly "that should attain to the ideal which Nature has proposed to herself in all her creatures, but has never taken the pains to realize." Guided by "Providence" and aided by long walks in nature that invoke the Transcendentalists, Warland eventually succeeds in fusing his "intellect . . . imagination . . . sensibility" and "soul" into the mechanical butterfly, despite the opposition of the town's utilitarian and materialistic men—a retired master clockmaker and a blacksmith—who see no financial or practical benefit to artistic machinery.[5] Though the butterfly is quickly destroyed by the blacksmith's child, the story suggests that natural and artificial, spirit and matter, beauty and machinery, are not entirely separate entities.[6] Fusion is possible, but only under the guidance of God.

Released just after the Civil War, "The Mullenville Mystery" captured subtle but significant transformations in literary critiques of industrial life. Writing near the dawn of American industrialization, Nathaniel Hawthorne, like Emerson and most men of his generation, retained faith in the ability of American institutions and culture to tame machinery in the name of beauty and freedom. As Emerson put it early in his career: "Machinery & Transcendentalism, agree well." Though he occasionally complained about the railroad, Emerson also believed that transportation and communications improvements could enhance people's connections to God, nature, and one another to build a harmonious community that still nurtured individual liberty.[7] Several decades of industrialization later, Julian Hawthorne told a less optimistic tale about the destruction of human relationships and loss of individual agency. The railroad still brought Ned to Nellie, but their relationship was doomed by his apparent lack of authenticity and their mutual inability to demonstrate agency beyond actions determined by the narrator. Similarly, while both Nathaniel and Julian recognized the magic of the automaton, for the former it remained a potential boon of God if it could be harnessed to the beauty of nature, while for the latter it was a demonic force enchanted by the artificiality of modern life. It is telling that Julian entered engineering believing, as his father did, in the coexistence of nature and mechanics; he left as it became clear that these were two very distinct things.

Between those gaps—natural to mechanical, real to authentic, freedom to determinism, Providence to the devil—lay several of the key tensions facing Americans as they sought to reconcile the development of industrial capitalism and modern science with their longings for a harmonious republican community that nurtured autonomous individuals capable of controlling their own destinies. Merging the worlds of engineering and literature, automatons and the like helped both critics and supporters grapple with the impact of material and intellectual transformations on the individual soul. As they debated the consequences of industrialization on human identity, Americans pondered whether the automaton was a gift of God or a demon that threatened to drag the country's innocence to hell.

1

The Republican Automaton

In 1788, newspapers in Philadelphia—the epicenter of the new country's economic and political life—printed numerous advertisements for an exhibition by "Signor Falconi" that included "INGENIOUS EXPERIMENTS" in catoptrics and electricity, along with an intriguing contraption: an automaton Indian that "will float an arrow into any number (which will be fixed in a frame) he is desired by any of the company, placed at the distance of twelve feet. In the same manner, he will answer any question that may be made him." To aid comprehension, Falconi included a sketch of the Indian adorned with a feathery headdress aiming an arrow at a circular frame with thirteen numbered targets. "Any Lady," a later advertisement announced, "may desire the figure to float at a particular number, which it will instantly do with the greatest of exactness. Any person may likewise write one of the numbers painted on the board, which may be folded up, and before it is seen by any one present the figure will strike the number written. And what is more surprising, any of the company may draw two or three dice under a hat, the amount of which the Automaton will strike before they are seen by any person."[1] Here was a performance like few others in the new nation, for it offered an opportunity to indulge, if only for a brief afternoon, in the

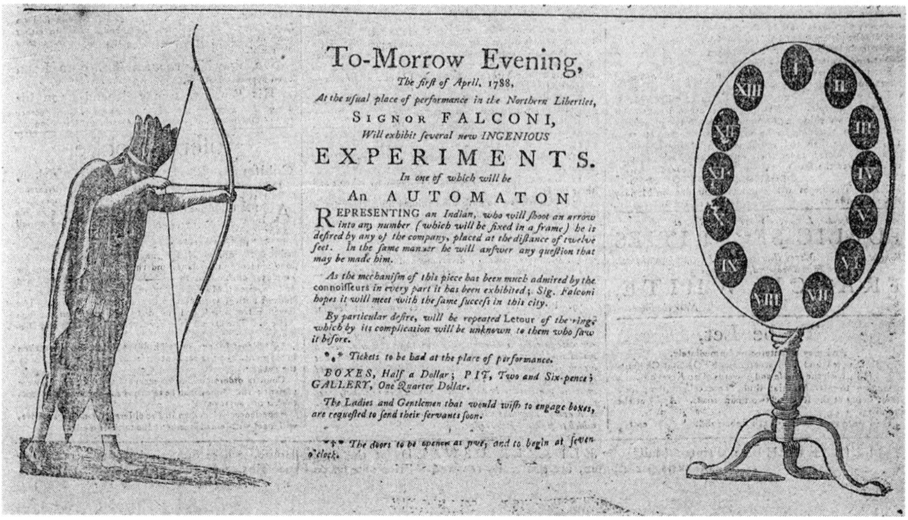

Fig. 1.1. A 1788 advertisement for Falconi's automaton Indian in Philadelphia's *Independent Gazetteer* that lays out the performance routine and a two-tiered ticket-pricing scheme that suggests the device attracted multiple social classes. Courtesy of American Antiquarian Society.

fantasy of transforming the body of a "savage" Native warrior into a weapon under the command of the audience, especially white women.

The fantasy offered by the automaton inverted the structure of early American captivity narratives in which white settlers, frequently women, were kidnapped and raised by natives. This performance brought the native body into a space where it could be contained by the eyes and voices of Falconi and the audience. Removed from its environment, the Indian did not degenerate into savagery—as many captivity narratives assumed would happen to white settlers—but grew civilized by following the orders of his social superiors. The audience became the rational mind of a creature assumed to be driven by mechanical reflex and violent passions. Though the act acknowledged a degree of contingency inherent in savagery, even rolling the dice left power in the hands of the audience members themselves; they could decide whether to allow the native to indulge in random violence or to force it to follow orders.[2]

By itself, Falconi's Indian offered audiences a humorous but powerful fantasy. When connected to the scientific work of the era, however, it became an ideological statement about the nature of people

Euro-Americans simultaneously feared and mocked. Several years earlier, one of the most popular European naturalists, Georges-Louis Leclerc, Comte de Buffon, published a series of natural history books on the flora and fauna of the Americas. In one volume, he claimed that the Native American male was "a kind of weak automaton" because of his inability to control himself or nature. For Buffon, who embraced vitalism when discussing Europeans, American Indian men were "mere automatons incapable of correcting Nature or seconding her intention" because "they had no control over either animals or elements; they had neither subjected the waves nor directed the motions of rivers, nor even cultivated the earth around them." Similarly, because they possessed only small and feeble "organs of generation," displayed "no ardour for the female," and were "possessed of less sensibility" than Europeans, they had "no vivacity, no activity of soul and that of body is less a voluntary exercise than a necessary action occasioned by want. Satisfy his hunger and thirst and you annihilate the active principle of all his motions; and he will remain for days together in a state of stupid inaction."[3] For Buffon, Native American men were deficient mentally, sentimentally, and sexually because they had failed to exert authority over nature, their bodies, and the bodies of women in the ways that European men had.[4]

Few Americans read Buffon's treatise, and even those who did may have rejected his claims, as future President Thomas Jefferson did in his *Notes on the State of Virginia*.[5] Yet, Falconi's automaton Indian translated Buffon's naturalism into a material form that entertained urban citizens for over a decade. In making his automaton a warrior, the exhibitor reduced the humanity of his subject to the pursuit of food—one of the most fundamental of necessities—and the impulse to violence, the basest of the passions that reason and sensibility were supposed to cure in "civilized" men. In his advertisement, Falconi stressed the immediacy of the connection between order and behavior. Like a beast, Falconi's Indian lacked the capacity for reason, morality, and self-control; it needed, as Buffon stressed, to be harnessed to the desires of civilized men and women.

Falconi's automaton and Buffon's theories came at a critical moment in the development of ideas about the relationship between machinery and human identity. Even before the codification of Isaac Newton's laws of motion in 1687, philosophers and scientists had debated which human behaviors were governed by similarly inviolable laws. Traditionally, writers interested in human identity understood the self as existing on three distinct levels. At the base of the self were the thoroughly mechanical reflexes and processes that functioned without any conscious thought. Next were the animalistic passions that fueled violence and desires. At the pinnacle of the human self was the mind, the basis of morality and reason, which came from God and gave people a limited ability to control their passions. Everyone agreed that the reflexes were mechanical, but they differed on the degree to which the passions, reason, and morality mimicked machinery.[6]

When Falconi introduced his automaton, two strains of thought on these issues battled in Euro-American culture. In the strain prominent in the late seventeenth and early eighteenth centuries, human beings were completely and totally mechanical. Slaves to their reflexes, necessities, and passions, they lacked the willpower to act morally and rationally in society. Others in this mechanistic strain of thought—especially some Puritans and deists—believed that people possessed the capacity to reason and act morally, but only because they acted out the will of God. During the second half of the eighteenth century, however, philosophers grew skeptical of the mechanistic view. Encouraged by the development of liberalism, unexplainable discoveries about the human body, growing interest in sensibility, and new theologies that stressed the role of choice, this strain emphasized that people's unique souls gave them the capacity to make independent, rational judgments. Associated with Romanticism in the nineteenth century, this strain rejected the idea that people were "mere machines," since they possessed reason, sentiment, and consciousness.[7]

Introduced to America only after the Revolution, automatons became a cultural means by which the country's white citizens attempted to resolve the inherent tensions between these two strains, between a need to see citizens as more than "mere machines" and a lingering suspicion that they might be. Tension could be resolved by using fantasy and humor to imply, as Falconi's automaton did, that some people are machines while others are not. The automaton Indian's performance told the audience of presumably white Philadelphians that they could reason while it could not; that they could make moral judgments while it could not; that they could control themselves while it could not, and, therefore, that they should have power while it should not.[8]

Matter, Mind, Soul

Before the 1780s, automatons were exclusively an Old World form of entertainment in a monarchical society. Part of a long tradition that stretched to the ancient world, these "self-moving" machines played a prominent role in early modern Europe as it began its slow transition from an agrarian society into an urban, industrial one. In the seventeenth century, artisans devoted significant time and effort to inscribing, with astonishingly accurate detail, the image and behaviors of animals and human beings into metal, wood, and clockwork. By the eighteenth century, two automaton manufacturers—Swiss clockmakers Pierre and Henri-Louis Jaquet-Droz and French physiologist and inventor Jacques de Vaucanson—built machines that duplicated human and animal behaviors so well that audiences frequently could not tell if the device was living. Watching Droz's musician or Vaucanson's flute player play beautifully while replicating the movements of a white aristocratic woman and shepherd astounded audiences while challenging them to contemplate the differences between people and machines. Droz's *The Draughtsman* and *The Writer*, both modeled on aristocratic children, similarly fascinated audiences by writing prearranged words and phrases. Mixing humor and precision, European automatons entertained and astonished the aristocracy with elaborate performances of music, dancing, writing and, in the case of Vaucanson's duck, defecation.[9]

Initially, many of the era's intellectuals thought automatons provided a useful metaphor for explaining the world and its inhabitants. The publication of Newton's laws of motion had popularized the possibility that all matter followed strict patterns and laws as if it were a living clockwork automaton crafted by God. Identifying the automaton as a "self-directed device" because it appeared to possess vitality and operate independently of external control, scientists applied the term as a metaphor for the orderly but active nature of reality.[10] To early chemist and physicist Robert Boyle, the world was God's "great automaton" and a "pregnant automaton . . . bound and maintained by God's laws of motion."[11] Most famously, philosopher Rene Descartes identified animal and human bodies as automatons but differentiated humans by their rational willpower. As he wrote in the early seventeenth century, "Apart from the guidance of the will," the human body did not "appear at all strange to those who are acquainted with the variety of movements performed by the different automatons, or moving machines fabricated by human industry."[12]

The most extensive expressions of the mechanistic view came in the middle of the eighteenth century. In 1748, French physiologist Julien Offray de La Mettrie published *Man a Machine*, which contended that people were entirely mechanical, even in their reason. Just two years later, Scottish philosopher Henry Homes, Lord Kames—a philosopher whose thinking heavily influenced Americans such as Thomas Jefferson and Benjamin Franklin—claimed that "the Deity is the primary cause of all things. In his infinite mind he formed the great plan of government, which is carried on by laws fixed and immutable." Such laws, he continued, "produce a regular train of causes and effects in the moral as well as material world, bringing about those events which are comprehended in the original plan, and admitting the possibility of none other." Ultimately, Kames reached for the clockwork metaphor: "This universe is a vast machine, winded up and set a-going: the several springs and wheels operate unerringly one upon another: the hand advanceth and the clock strikes, precisely as the artist had determined."[13]

In a culture grappling with the discovery of laws of matter and challenges to the authority of the monarch, the automaton linked the individual to the state, the universe, and God. In 1651 during the English Civil Wars, Thomas Hobbes opened his *Leviathan* by invoking the automaton to connect the human body to the absolutist state: "Why may we not say that all *Automata* . . . have an artificial life? For what is the *Heart* but a *Spring*; and the *Nerves* but so many *Strings*; and the *Joynts*, but so many *Wheels*, giving motion to the whole Body." Hobbes deployed the automaton as a metaphor because he denied free will and believed that passions and desires caused socially destructive acts. "Every act of mans will, and every desire, and inclination," he claimed, "proceedeth from some cause, and that from another cause, in a continuall chaine, (whose first link in the hand of God the first of all causes,) proceed from *necessity*." For Hobbes, the most prominent desire was one that guaranteed conflict: lust for power. Even the rational mind was, for Hobbes, a servant of desire. As he wrote, "The Thoughts, are to Desires, as Scouts, and Spies, to range abroad, and find the way to the things Desired: All steadiness of the minds motion, and all quickness of the same, proceeding from thence." If reason could not restrain passion and desire, then the only way to preserve peace was to ensure that people feared authority. For Hobbes, an automaton state was necessary to contain the human automatons bound to violence by necessity and desire.[14]

In a monarchical world, even religion embraced mechanism. For Anglo-Americans, the most powerful faith to employ mechanistic imagery was Puritanism due to its reliance on the Calvinist doctrines of predestination and inherent depravity. Writing in 1721, New England scientist and minister Cotton Mather wrote that the body of man "is a *Machine* of a most astonishing Workmanship and Contrivance! *My God, I will praise Thee, for I am strangely and wonderfully made!*" Mather's emphasis on the body suggested a Cartesian distinction between the mechanical body and the nonmechanical mind; the theologian, however, rejected such thinking. For Mather, reason was also mechanical—though tied to the spirit, it followed the laws set forth by God as logically as a mathematical equation. "Whenever any reasonable thing is offered," he wrote, "I have God speaking to me." For Mather, every part of the universe served as God's mechanical puppet: "The Great God not only has the springs of this immense machine, and all the several parts of it, in his own hand, and is the first mover; but that without his continual influence the whole movement would soon fall to pieces." Yet Mather rejected the clockwork metaphor. If a dog's motion were "caused by the striking of a spring," he proposed, "there can be no reason imagined why the spring . . . should not carry the machine in a direct line towards the object that put it in motion." But dogs rarely traveled the most direct route. God may have dictated the destination, but, unlike a "mere piece of clock-work," living creatures could choose their own paths.[15] The problem with clockwork as a metaphor was that clocks were never arbitrary; they slowed down but they never altered their patterns; no one, not even the mechanic, could alter the routine once the spring was wound.

The other major faith to endorse mechanistic thought was deism. Aside from a few years between the American and French Revolutions, deism was never a potent religion in America. Proposing that

God created the universe to follow set mechanical laws, deism appealed to the era's intellectuals who embraced science but were unwilling to deny the existence of God.[16] While materialistic, deism did not require an automaton view of humanity; however, its emphasis on mechanical laws encouraged people to adopt one. In 1725, Benjamin Franklin, born a Puritan but receptive to deist arguments, published *A Dissertation on Liberty and Necessity, Pleasure and Pain*, in which he speculated that God created human beings as machines. For the young Franklin, the idea of free will was ludicrous because it made God into a powerless clockmaker who designed a perfect machine but nevertheless "placed in it several other wheels endowed with an independent *self-motion*, but ignorant of the general interest of the clock" that would "every now and then be moving wrong, disordering the true movement, and making continual work of the mender."[17] God, he suggested, would not have designed a system with such an inherent flaw as free will.

Even as mechanistic philosophy appeared ascendant, alternative voices challenged its vision of humanity. To many, the problem with analogizing humanity to clockwork was obvious: while automatons may have looked and behaved similarly to living beings, they demonstrated no consciousness, willpower, reason, or sentiment; they possessed no internal source of power and could not vary their movements. By the 1740s, numerous European philosophers, doctors, scientists, and theologians had retreated from mechanism to adopt more vitalistic notions of identity that stressed the inherently unmechanical—though not necessarily immaterial—nature of the individual.[18]

In Britain and its colonies, the most important opposition came from religion. During the early seventeenth century, Dutch theologian Jacobus Arminius directly challenged Calvin's doctrines of predestination and inherent depravity by proposing that individuals could choose to follow God, act morally, and attain eternal salvation. In the English-speaking world, the most influential Arminians were the followers of Methodist cleric John Wesley who explicitly challenged Calvinism for its susceptibility to mechanistic metaphors. As Wesley wrote, "If all the passions, the tempers, the actions of men, are wholly independent on their own choice, are governed by a principle exterior to themselves, then there can be no moral good or evil;

there can be neither virtue nor vice, neither good nor bad actions, neither good nor bad passions or tempers." For Wesley, "the doctrine of necessity, as taught either by ancient Heathens, or by the moderns, (whether Deists or Christians,) destroys both [will and liberty], leaves not a shadow of either, in any soul of man; consequently, it destroys all the morality of human actions, making man a mere machine; and leaves no room for any judgment to come, or for either rewards or punishments."[19] Because they were made of unconscious matter that could not choose between morality and immorality, machines could never serve as an adequate analogy for human beings.

The Methodist challenge found support in new claims about the human body. While dissecting the bodies of living creatures, European physiologists discovered much—including the nerves—that science had not explained. Dissatisfied with mechanistic metaphors, scientists such as German philosopher Georg Ernst Stahl and Scottish physician William Cockburn studied the body as an organic entity powered by a vital, nonmechanical life force which most dubbed "the soul" or "anima." Scottish physician and vitalist Robert Whytt warned readers that "the human body ought not be regarded . . . as a mechanical machine, so exquisitely formed, as by the mere force of its construction, to be able to perform, and continue the several vital functions." Instead of mechanism, Whytt recommended that physicians examine the body as "a system, framed indeed with the greatest art and contrivance. . . . In which the peculiar structure of each part is not more to be admired than the wise and beautiful arrangement of the whole; nevertheless, as a system whose motions are all owing to the active power, and energy, of an immaterial sentient principle, to which it is united, and be which every fibre of it is enlivened and actuated."[20] Life, for Whytt and the new vitalists, did not come from the mechanical movements of the body but from a spirit that united all living things. Such a spirit could be rooted in a material substance—electricity or fluids predominantly—but it was not fundamentally mechanical.

Interest in such a spirit was part of an emerging European and American interest in "sensibility" as a distinguishing feature of humanity. Developed as lineage claims to authority disintegrated and the century's expansion of commerce required less violent forms of interaction, sensibility proposed new forms of social bonds based on feelings of sympathy and love and a shared human identity. Sensibility was revolutionary because it jettisoned the conventional three-part division of the human self into the reflexes, passions, and reason/morality in favor of a much more unified structure. Neither a form of passion or reason, sentiment relied on the physical body for its sensory organs but connected them to a moral and rational mind. Such an understanding was not necessarily vitalist, but it could encourage a conclusion that people shared a common spirit with all human beings. When it defined sensibility, Denis Diderot's *Encyclopédie* tied it to Whytt's conception of the human body: "the sentient principle, or the feeling of all the parts, the basis and preserver of life."[21] Denying other people possessed sentiment, as Buffon had done with indigenous Americans, denied those people humanity while justifying inhumane and violent actions against them.

In the Anglo-American world, much of this interest in the senses and sentiment came from the writings of John Locke, whose 1689 treatise *An Essay Concerning Human Understanding* popularized the idea that identity was not preordained but made by an individual's consciousness of experiences and sensations. In Locke's vision of the human individual, the "self is that conscious thinking thing (whatever Substance made up of whether Spiritual or Material, Simple or Compounded, it matters not), which is sensible, or conscious of Pleasure and Pain, capable of Happiness and Misery and so is concerned for itself, as far as that consciousness extends." Rejecting debates over matter and spirit, Locke defined the self by its ability to sense. He theorized, "That with which the consciousness of this present thinking thing can join itself makes the same Person, and is one self with it, and with nothing else; and so attributes to itself, and owns all the Actions of that thing as its own, as far as that consciousness reaches, and no farther; as every one who reflects will perceive."[22] In placing consciousness of the self and its relationship to the rest of the world at the heart of human identity, Locke established one of the central ideas that would distinguish human beings from machines through the twentieth century.

Though his vision of consciousness was materialistic, Locke rejected machinery as a metaphor. "By denying freedom to Mankind," Locke wrote, men such as Hobbes made "Men no other than bare Machines, they take away not only innate, but all moral Rules whatsoever." Such a choice led men to "reject all Principles of Virtue." People, he later wrote, "are endow'd with a power to suspend any particular Desire, and keep it from determining the *Will*, and engaging us in Action." Such an ability to willfully choose to resist desires, was, for Locke, the essence of liberty. "He that has a power to act or not to act, according to such Determination directs," the philosopher concluded, "is a free Agent; such Determination abridges not that Power wherein Liberty consists. He that has his Chains knock'd off, and the Prison-doors set open to him, is perfectly at liberty, because he may either go or stay, as he best likes." Repeated references to "Determination" avoided the debate between vitalism and mechanicalism by emphasizing a new ground on which the question of liberty should be debated. Human beings were not different from machines because they possessed free will, but because they were "free agents" whose consciousness and sentiments could teach them to become virtuous individuals.[23] Not bound to a clockwork fate by God, necessity, or desire, people could learn and thereby improve.

Over the second half of the eighteenth century, vitalism and mechanicalism vied for supremacy as thinkers remained fundamentally uncertain regarding the differences between human beings and machines. Yet, the very act of the Revolution suggested the power of free will and the possibility of change. If the entire world was God's great automaton, then efforts to disrupt its clockwork were pointless, foolhardy, and destructive. But that is precisely what colonial Americans did when they severed their fates from England. By the 1780s, one of the most defining assumptions of American culture was that citizens were—or at least, should be—more than mere clockwork. As Revolutionary War hero Ethan Allen wrote in his 1785 pamphlet *Reason: The Only Oracle of Man*, people are more than "mere machines" because they are endowed with reason and free will.[24] A deist and self-described "free-thinker," Allen was an anomaly in early American society; indeed, his *Reason* was widely panned by the late eighteenth century.[25] However, in embracing this mythical concept of the free agent, Allen reflected a much deeper transformation enshrined by the Revolution: the ideal voting citizen, a man who owned property, was to put aside his baser instincts and desires in the service of all. This man—the vitalistic individual of reason and sentiment—controlled himself while feeling sympathy for others. While his reason made him independent, his sentiment allowed him to forge bonds with others, and his consciousness enabled him to tame those wants that threatened the larger community. Freed from necessity and the will of others through the ownership of land and his own mechanical and animalistic passions, the virtuous republican citizen was to be the antithesis of the clockwork automaton.

Fantasies of Amusement and Power

The instability of power after the American Revolution encouraged changes in the meaning of the word *automaton*. The Revolution, as an expression of a much larger challenge to authority, encouraged marginalized people to assert authority over their own lives, identities, and government. Though those assertions of equality went largely unheeded, they prompted the powerful—landholding white men—to formulate new justifications for their authority. To do so, they built or reinforced social hierarchies on racial, gender, class, and age distinctions. The explicit goal of the framers of the Constitution was to locate power in the hands of those individuals who seemed to demonstrate the greatest capacity for independent reasoning and sentiment. In an environment where power was to reside in those who demonstrated virtue by being free of both their harmful "passions" and dependence on other people, stage automatons increasingly mimicked the identities of people denied the full rights of citizenship: women, children, and nonwhites.[26]

In Europe, automatons could safely exist as amusements, but Falconi was an entertainer in a middling, bourgeois republic that contrasted itself with European aristocrats by celebrating hard work, self-denial, and rational recreation. The virtuous and vitalistic citizen did not spend his leisure time in idle or selfish pursuits but in activities that could benefit all.[27] Quoting his Poor Richard character, Franklin wrote, "Leisure, is time for doing something useful; this leisure the diligent Man will obtain, but the lazy Man never; for 'a life of leisure and a life of laziness are two things.'"[28] To satisfy his audience's desires for productive leisure, Falconi presented his exhibitions as forms of amusement that also taught practical applications of science and mechanics.[29] By the nineteenth century, almost everyone in America saw science and industry as a means of securing the country's independence from Europe. People celebrated inventors such as industrialist Samuel Slater and cotton-gin inventor Eli Whitney as national heroes for enabling American independence of foreign-made goods.[30] Though Falconi adopted the title of "magician," he attempted to fit into this world by naming his displays "Natural and Philosophical Experiments."[31]

Falconi was also part of a new breed of entertainers who combined science with fraud.[32] His use of the word *automaton* implied that his device was made of clockwork, yet his advertisements avoided mechanical words such as *gear* and *spring*—such choices drew on a growing ambiguity in the meaning of *automaton*. Through the mid-eighteenth century, Europeans had used it to refer to self-moving machines that operated independently of human control. That meaning, however, changed in the late eighteenth century after the debut of the Hungarian aristocrat Wolfgang von Kempelen's automaton chess player. The chess player—which some Americans, including Franklin, had seen but most had only read about in newspapers—ostensibly played chess better than almost any human. In doing so, it exhibited precisely the characteristics of reason and consciousness of the world around it—sensibility—that seemingly defined humanity. Like Falconi, Kempelen never claimed to have invented a real, fully mechanical automaton, but he used the term for publicity. Widely popular across Europe, the chess player attracted audiences not because people believed it was a machine—the idea that a clock could win a game of chess was ludicrous—but because people wanted to decipher the nature of the fraud.[33]

Americans, however, found the automaton amusing. Falconi's device was, of course, an illusion, and at least some people knew it. Clockwork manufacturers had made automatons that wrote prearranged and even requested phrases, played music, and danced, but no one had invented a machine capable of reading a target, adjusting its aim, and successfully firing an arrow. Real automatons could sense pressure or shape, but sensing the world through sight, hearing, smell, or even touch was far beyond the capabilities of clockwork. Commentators, however, enjoyed the fraud. One reviewer observed that "the learned have derived much pleasure from his philosophical experiments; and the astonishing performances of his automaton and other figures, give them more the appearance of intellectual beings to the eyes of the spectator, than pieces of mechanism. Besides which, the exquisitely nice *deceptions* of signior Falconi, appear, many of them, to be preternatural."[34] The central joke of the automaton depended on people understanding that the device was an illusion, that an actor performed the role. As audience members laughed and watched in amazement at this actor pretending to be an automaton pretending to be an Indian, they could identify themselves as uniquely human because they, unlike both natives and machines, possessed self-control.[35]

This shift in the meaning of automatons reinforced a far more consequential transformation: as the devices became more fraudulent and less serious, the subjects of their performances replicated the emerging gender, class, racial, and age hierarchies of the modern world. The typical subjects of the most popular eighteenth-century European automatons had been figures who resembled their audiences: male and female elite. The Drozs' creations were dressed in the latest French fashions and even mimicked the routinized movements of the royal court.[36] Usually, the closest such automatons came to representing the form of an "other" were classically inspired pastoral figures such as Vaucanson's *Flute Player*. In their efforts to entertain, however, many late eighteenth- and nineteenth-century automaton manufacturers built their devices in the forms of groups of people left out of the emerging political structure, especially women, children, and, especially, nonwhites. Part of the chess player's success stemmed from its replication of the appearance of a "Turk"—a particularly symbolic identity in the Hapsburg city of Vienna because of the ongoing conflict with the Ottoman Empire. By the nineteenth century, various Middle Eastern–, Chinese-, African-, and Native American–themed automatons performed on stages as audiences looked for entertainment in forms that mechanized the bodies of those peoples Euro-Americans simultaneously romanticized as exotic and sought to control through imperial expansion.[37]

In community halls, taverns, and museums throughout America, entertainers exhibited real, working automatons with lifelike wooden bodies that danced, played instruments, performed along a slack rope, answered questions, or demonstrated magic tricks that could be inscribed into clockwork. Though not all automatons represented marginalized groups of people, many did. Falconi's first automaton was dressed in a "Turkish Dress" and would "answer by Signs any

Question put to it" and "guess the Number of Dice drawn by any One of the Company." A mechanical museum in Philadelphia advertised several automatons, including one that tumbled, an "Indian Automaton Sorcerer" that could tell time and perform card tricks, an astrologer named "The Little Magician" that read thoughts, "The Mysterious Penman, a small boy that would write the name of anyone in the audience," and "The Mameluke Cobler" with a "comic" appearance.[38] Another Philadelphia entertainer advertised "The Fair Arithmetician, or Writing Automaton . . . The Industrious Lilliputian Cobler at work . . . the Indian Oracle . . . the Miraculous Barrel" and "the tumbling vaulting and dancing figure."[39] Though only one of these devices, the "Indian Oracle" obviously represented a disempowered group of people, the use of the term *fair* suggested that the arithmetician replicated the appearance of a beautiful woman. Charles Willson Peale's American Museum in Philadelphia included a female piano playing automaton "dressed in a rick blue satin frock" with "her head covered with artificial wreaths and a few pearl beads," as well as a Chinese slack-rope dancer.[40] In New York, another automaton could "hold conversation the same as any human person, even to a whisper." Sophisticated Europeans, the exhibit's advertisement claimed, believed that this "beautiful Young Lady" automaton was equal to "any thing they have ever seen, for proportion of figure, elegance of form, beauty of complexion, and modest peasantry of countenance; she stands erect on a richly carved and gold pedestal, adorned with sundry ornaments—with an elegant green Parrott at her feet."[41] Apparently, a woman who engaged in conversations was just parroting the words of someone else.

Such caricatures reduced the identities of their subjects to easily controllable material bodies that remained still for most of their existence but could be unleashed by the winding of a spring. Like Falconi's fake Indian, these real machines often responded to their audiences' demands and offered the chance for observors to laugh at their ridiculousness while fantasizing about controlling such bodies. By the end of the eighteenth century, automaton manufacturers appear to have grown less interested in astounding audiences with mechanism and more interested in making audiences laugh in recognition of their superior humanity to the mechanical peoples depicted on stage. Working in an absolutist state, Vaucanson and the Drozses were skilled artisans who sought and gained respectability and patronage; working in a commercial republic, Falconi and his competitors were entertainers who sought money from larger audiences. Like Kempelen, they catered to their audiences' cravings for amusement rather than manufacturing a mechanical piece of art.[42]

As such automatons grew more common, educated citizens employed the label as a metaphor for people enslaved to the will of another or to their animal urges. In 1806, a lawyer told a jury that they should not simply follow the wishes of the prosecutor because they "would become as mere puppets in the hands of the juggler—they might be made the automata of government, to act their parts, according as the hidden mechanist puts them in motion." Addressing problems of the public debt, Thomas Jefferson complained that "we must live, as they [the English] do now, on oatmeal and potatoes; have no time to think . . . but be glad to obtain subsistence by hiring ourselves to rivet their chains on the necks of our fellow-suffers." The failure to live frugally, he feared, would make people into "mere automatons of misery" with "no sensibilities left but for sinning and suffering."[43] Instead of self-operating, people imagined automatons as enslaved.

Sometimes the automaton's enslavement originated from within the individual. "The minister who does not believe in God but still voices the doctrines," one minster preached, will "be merely a substitute for a Christian Pastor; an automaton, in the place, and dress, and business of a living man."[44] Other ministers invoked the automaton to critique sinful forms of leisure. The Presbyterian minister and president of Union College Eliphalet Nott told graduates that gambling destroyed the intellect and converted children "into automata; living mummies; the mere mechanical members of a domestic gambling machine, which, though but little soul is necessary, requires a number of human hands to work it."[45] For Nott, automatons were not just sinners who repeatedly gave into desires; they were ambulating corpses or disembodied "hands" serving a compulsion to sin.

Writers most frequently compared women to automatons. One male critic identified women "of fashion" as performing "graceful actions and other automaton accomplishments" without "mind" or "soul" because they used consumer goods to define themselves.[46] Most who used the term to refer to women, however, criticized women's subordinate status in the nation. For the Reverend Joseph Emerson, uneducated women were "without reasoning" and "little more than an automaton, or a domestic animal."[47] In such a state, he claimed, women "will be able to make improvement in nothing; she will drudge through the same dull round of operations, like a beast of burden."[48] Another writer lambasted critics who were against women's education by claiming that they "would convert 'heaven's last best gift' to man into a sort of automaton, with the body of a doll and the head of a parrot."[49] Still another advised men to no longer "value women exclusively for their external beauties and unmeaning accomplishments . . . [to] no longer believe that she is a mere automaton, possessing the power of motion, or like the chronometer, regulated at pleasure. . . . Let us then resolve to believe that their power of mind are equal with our own."[50] Regardless of whom they blamed, these admonishments suggested that American women were automatons because they lacked the faculty of reason.

Few Americans made the distinction between citizen and machine as well as Alexander Hamilton. In a 1777 letter to John Jay, the president of the Continental Congress, Hamilton recommended the enlistment of black Americans as soldiers. "I have not the least doubt," Hamilton wrote, "that the negroes will make very excellent soldiers, with proper management. . . . It is a maxim with some great military judges, that, with sensible officers, soldiers can hardly be too stupid." Realizing that Jay might think the notion absurd, Hamilton explained his reasoning: "I think their want of cultivation (for their natural faculties are as good as ours), joined to that habit of subordination which they acquire from a life of servitude, will enable them sooner to become soldiers than our white inhabitants." He ended with a generalization that encapsulated the emerging distinction between humans and machines: "Let officers be men of sense and sentiment; and the nearer the soldiers approach to machines, perhaps the better."[51] For Hamilton, the machine was not an analogy for human identity; it was an ideal that those below him in status should emulate.

The contrast with Jefferson was telling. Nominally a deist influenced by Kames, Jefferson embraced materialism.[52] Writing in 1786 on the contradiction between slavery and liberty, he exclaimed, "What a stupendous, what an incomprehensible machine is man! Who can endure toil, famine, stripes, imprisonment & death itself in vindication of his own liberty, and the next moment . . . inflict on his fellow men a bondage, one hour of which is fraught with more misery than ages of that which he rose in rebellion to oppose." Jefferson accepted that he—along with his fellow elite gentlemen, the rabble, and even slaves—was a machine. For all of Jefferson's sense that Africans were mentally inferior to whites, he accepted the universal mechanical nature of human life emblematic of the early Enlightenment. Jefferson then turned to his vision of how slavery would end: "But we must await, with patience, the workings of an overruling Providence, and hope that he is preparing the deliverance of these, our suffering brethren."[53] Hamilton imagined himself as a man of reason and sentiment and fought for much of his life for the abolition of slavery. Jefferson imagined himself as a machine waiting for God to compel the liberation of slaves. For the vitalist Hamilton, the machine was an ideal for the rabble to emulate; for the materialist Jefferson, the machine analogy was an excuse to enslave.

Early Industrial Automatons

It was fitting that the American who first envisioned the industrial automaton was Hamilton's coauthor on some of the earliest calls to introduce manufacturing to the new nation: Tench Coxe. A Philadelphian merchant, Coxe spent most of his life justifying the introduction of machinery to a country that most believed would remain agrarian. Several innovations convinced Coxe that the United States could enjoy independence from British manufactured goods but still ensure that most of its citizens remained farmers and independent craftsmen. Over the preceding century, a series of shuttling, spinning, and carding devices had been introduced to British textile factories that dramatically increased productivity, especially when harnessed to water power. In the 1780s, Oliver Evans built an automatic flour mill along Red Clay Creek in Delaware that used several conveyers to move grain throughout the entire enterprise without the need for human hands. Even greater productivity seemed feasible if such devices could be harnessed to the energy generated by James Watt's improved steam engine. "Factories, which can be carried on by water-mills, wind-mills, fire, horses, and machines ingeniously contrived," Coxe concluded, "are not burdened with any heavy expence [*sic*] of boarding, lodging, clothing, and paying workmen; and they multiply the force of hands to a great extent, without taking our people from agriculture."[54]

Gentlemen such as Coxe feared the movement of workers from farms to factories because they believed that a life permanently devoted to repetitive physical toil for others was ill suited to the cultivation of the mind and sentiments necessary to become a virtuous citizen. For America's elites, the ideal citizen was an independent farmer, artisan, or businessman—a man who did not work for others but instead worked for himself while supervising the labor of others so that he might have leisure to serve the community rather than his own desires. The introduction of manufacturing threatened this vision because in England factories had destroyed the environment while creating a class of men who toiled all day for others and spent their limited leisure time on mindless pursuits. American advocates of manufacturing overcame this skepticism by stressing that factory labor would raise productivity and allow more men to become independent. Limited factory labor, they predicted, could teach the habits of hard work and self-control necessary for virtue. As Coxe argued, manufacturing would "lead us once more into the paths of virtue by restoring frugality and industry, those potent antidotes to the vices of mankind." Industrial labor would not only manufacture native goods; it also would manufacture good citizens.[55]

In the 1780s, Coxe had linked his vision of machinery to the Philadelphia orrery of David Rittenhouse, a famed mechanical representation of the clockwork nature of the universe.[56] In 1810, Coxe instead invoked the magic of enlivened, humanized machines. In a statement for Congress, Coxe conjured an image of "*wonderful machines*" as "*animated beings, endowed with all the talents* of their *inventors*, laboring *with organs that never tire* and subject to no expense of food, or bed, or raiment, or dwelling." These machines would be "*justly considered as equivalent to an immense body of manufacturing recruits, suddenly enlisted in the service of the country*."[57] Eight years later, Coxe directly referenced the automaton: "The whole system and power of labour-saving machinery . . . may be considered as forming, by steam, by water, and by wooden and metallic machinery, a vast body of gigantic automatons in aid of the labour of our people and the draughts of our cattle." This "Herculean corps of automatons" could not operate on its own; it required the assistance of "some of our women, our children, and our acquired artists (with a very few of our native men) . . . as the *little* fingers" to guide and control the machines.[58]

With the phrase "little fingers," Coxe reduced the workers who toiled at machinery to their animal bodies in much the same way that Nott reduced gamblers to their hands and Falconi reduced the American Indian to an automaton. The vision of an automatic factory depended on articulating a class identity that stressed the innate differences between those who labored with their hands and those who labored with their minds and, perhaps, felt sentiment with their hearts. In early America, the people who were the least like automatons would farm or, increasingly, serve in white-collar professions, performing the work of the mind; those who were the most like automatons would serve as the hands of empowered machines.

Coxe's vision of empowered automatons aided by a small number of machinelike women, children, and "very few of our native men" reverberated through the first half century of American industrial development as people debated the merits of mechanization. Beginning in the late eighteenth century, local, national, and state governments standardized land sizes and parceled them to individuals. Soon a national system of roads and canals and later railroads and telegraph lines standardized and sped up the movement of people, goods, and information between communities. By the 1820s, in places such as Lowell, Massachusetts, manufacturers brought single-purpose machines together with low-wage female labor from the countryside in water-mill and, later, steam-powered factories that transformed cotton raised in the slave South into textiles for the nation. Manufacturers such as Eli Whitney and, later, Samuel Colt discussed creating standardized parts manufactured by machines that would accelerate production, lower costs, save labor, and make repairs easier. At federal armories in Springfield, Massachusetts, and Harpers Ferry, West Virginia, as well as Colt's private factory in Hartford, Connecticut, specialized machines and a small labor force of skilled craftsmen manufactured nearly identical weapons that popularized the idea of interchangeable, machine-made parts.[59]

As the country became industrialized, America's political elite continued to envision humanized machines performing jobs they saw as degrading in a way that identified some kinds of labor as fit only for machines. On his 1839 tour of England, the famed orator and senator from Massachusetts Daniel Webster visited Charles Babbage, the inventor of a mathematical engine and industrial theorist who forcefully advocated for the division of labor, and marveled at his automaton which mimicked the appearance of a dancing woman.[60] Not long thereafter, Webster connected automatons to industrialization while reevaluating the meaning of labor. Like most economists of the time, Webster accepted that labor was the source of all wealth. For Webster though, the laborers of the future were machines. "We commonly speak of mechanic inventions as labor-saving machines" he wrote, "but it would be more philosophical to speak of them as labor-doing machines; because they, in fact, are laborers. They are made to be active agents, to have motion, and effect, and though without intelligence, they are guided by those laws of science which are exact and perfect, and they produce results, therefore, in general, more exact and accurate than the human hand is capable of producing." Webster then alluded to a recent visitor to the United States: the automaton chess player, now owned by Johann Nepomuk Maelzel: "These automata in the factories and the workshops are as much our fellow laborers, as if they were automata wrought by some Maelzel into the form of men, and made capable of walking, moving, and working, of felling the forest or cultivating the fields."[61] In a country obsessed with business and work, with turning wilderness into civilization, machines could be made "active agents," philosophically equivalent to human workers—and perhaps even superior, because they produced better results.

Twenty years later, US Senator William Seward articulated similar thoughts but as a fantasy of slavery. In the senator's closing statement in a patent infringement case against Cyrus McCormick, he connected the inventor's reaper to automatons. "Mr. McCormick," Stewart claimed, "could not make a breathing man to perform the labor that was thus required, and no living man could perform it." Therefore, he "invented a mechanical man to perform that labor, and attached him to the machine. Wherever the machine goes now, there that mechanical man goes before it, always stooping and lifting up, and disentangling and dividing the grain." The mechanical man, the abolitionist concluded, was "the slave of Cyrus H. McCormick." The inventor "created the automaton, and the law made it his slave for fourteen years. . . . The defendants have appropriated that slave to their own use. I appeal to you as just and magnanimous citizens to restore it to its owner."[62] For Seward, the automaton offered the possibility of slavery without guilt.

Webster's vision of automaton factory workers and Seward's vision of automaton farmers suggested a central and enduring tension in the meaning and significance of work and its connection to freedom. Since the Revolution, the farmer's labor had been celebrated as the key to republican independence and virtue. By the 1820s, ordinary mechanics in places such as New York had claimed their professions as similarly worthy of adulation and praise.[63] Though immersed in a northern middle-class industrial culture that warned of the dangers of leisure and celebrated work as central to identity and purpose, both Webster and Seward identified those types of labor as unworthy of human effort. That Seward invoked slavery was revealing. Typically, abolitionists criticized slavery for impeding the development of a work ethic and tended to see southerners—white and black alike—as inherently lazy.[64] Yet, they never criticized automatons or even less fantastical forms of machinery for destroying the national work ethic. Instead they celebrated them as essential to American development. But Seward and Webster did not merely invoke machinery as sources of labor; they invoked humanoid automatons, devices that linked the dream of freedom from certain types of labor to a dream of mastery over the bodies of others. By the start of the Civil War, industrialization had transformed the fantasy of control initially offered by Falconi's Indian into a utopian vision appropriate for a country dominated by the tension between slavery and freedom: every man a master, no man a slave.

2

Humanizing the Industrial Machine

In 1868, newspaper editors imagined they had discovered the perfect symbol for the burgeoning industrial age: a "man" made of iron and powered by steam. Invented by Zadoc Dederick, a twenty-two-year-old Newark machinist, this "Steam Man" stood seven foot seven, weighed five hundred pounds, and carried a coal-burning steam engine in its chest. With a steam-engine heart providing the "vitality" to move its attached mechanical arms and legs, the machine could reportedly accomplish feats both human and superhuman. Per one account, it could perform "some of the most important functions of humanity," including "standing, walking, and running, as he is bidden." In another, it could pull a weight that "would tax the strength of three stout draught horses" and tow a wagon to any destination at speeds up to thirty miles an hour.[1] With such a device, writers implied, Americans no longer would be bound to a fixed path; instead, they would be free to venture wherever they wished at far faster speeds. Yet, in a culture that deified the power of machinery, the steam man offered something far more intimate and fun: a chance to identify with the machine, to acknowledge its superhuman capabilities while humanizing it with laughter—for, as amazing as the invention sounded, it was largely a fraud. As one

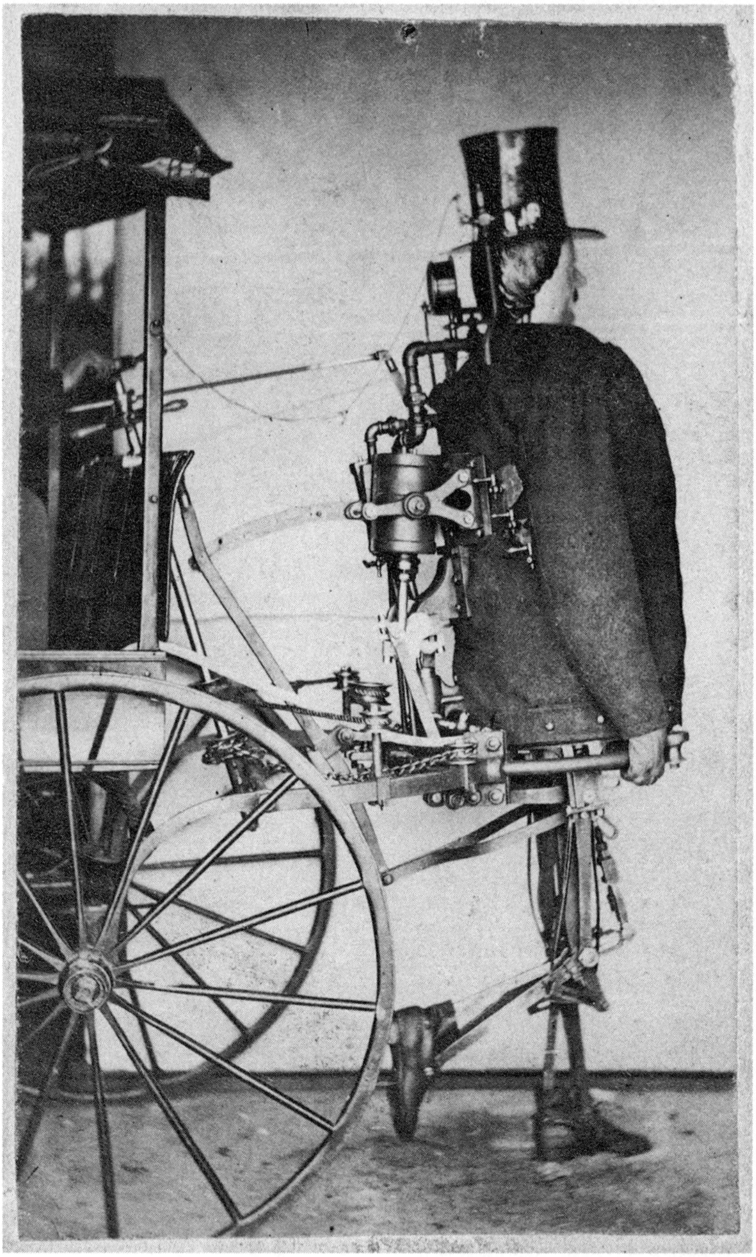

Fig. 2.1. An early photograph of Drederick's steam man that shows both the mechanisms that gave it vitality and the white gentlemanly facade that Drederick chose for its figure. From the Miriam and Ira D. Wallach Division of Art, Prints and Photographs: Photography Collection, New York Public Library, http://digitalcollections.nypl.org/items/2a1096e0-2289-0132-b3b6-58d385a7bbd0.

critic concluded, the device was "an unmitigated humbug. It never has walked a step since it was made. . . . It will tumble over as fast as it can be set up."[2]

Editors from Maine to the Kingdom of Hawai'i reprinted amusing accounts of the steam man to attract readers intrigued by the fantastical potential of the era's mechanical developments. Since the 1820s, Americans had practically worshipped the sublime power of machinery to guarantee their national independence and individual freedom, create a harmonious republican community, and civilize the Western frontier. Through the force of their minds and labor, Americans believed they could, under the guidance of God, transform the scarcity of the wilderness into an abundant cornucopia, a "second creation" that could peacefully coexist with the original. In sermons and speeches, general and specialized periodicals, Fourth of July parades, and tours of factories in cities such as Lowell, Massachusetts, practically all nineteenth-century white Americans lauded mechanical improvements for bringing material and moral progress. As one advocate of industrialization summarized in 1836, improvements created "a moral machine which . . . tends most effectually to perfect the civilization and elevate the moral character of the people."[3]

Nothing epitomized the moral, nearly divine power of machinery and its inventors as well as the steam engine. Unlike even the most amazing automatons, the steam engine appeared to possess its own source of vitality, its own metaphorical soul. According to one mechanic of that era, inventors had "created both god and armor—a Titan with body of iron and soul of steam."[4] While the "soul" metaphor suggested that American inventors had duplicated the life-giving power of God to become gods themselves, the term *titan* implied that men had not recreated themselves; they had created gods far more potent than any person. As rail lines spread across the nation in the mid-nineteenth century, the engines grew more mundane; but Americans continued to see something superhuman in their capabilities. Commenting on the thirty-nine-foot-tall, 680-ton Corliss steam engine at Philadelphia's Centennial Exposition, the editor of the *Atlantic Monthly* employed the metaphor of a magical djinn. The device's engineer, he claimed, was "like some potent enchanter . . . and this prodigious Afreet . . . his slave who could crush him past all semblance of humanity with his lightest touch."[5] Though not quite divine, the power of both the engineer and his machine seemed to far surpass that of ordinary people.

Debuting just three years after the Civil War and eight years before the Centennial Exposition, Drederick's "mechanical Frankenstein" emerged in a culture in which the steam engine was progressive but dangerous, powerful but mundane.[6] Drederick and his partner Isaac Grass patented the device to indicate their seriousness, but their claims for the device were undoubtedly hyperbole. The steam man was not an obvious fake like a chess-playing automaton; it could have worked. By all accounts, the steam man could move its legs; it just likely could not move without falling. The divide between its potential and the ridiculousness of the performance inspired laughter that humanized the machine while lampooning the hubris of inventors and the power of their devices. The *Newark Advertiser* joked, "The wonderful piece of mechanism, if it could but speak, would undoubtedly exclaim, 'Homo Sum!'"[7] Another reported that one businessman tried to order several to work on the prairies, while someone else asked for "a *pair* to repopulate his place. Five women write, ordering cast-iron husbands, and one gentleman sends for a wife. As the machine is speechless, the inventor replies to the the last he was doubtful whether a *woman* could be made a success."[8] The *New York Express* mocked plans to sell the steam man to consumers with the suggestion that Drederick was "ready to procreate—without any regard to the conventional idea of nine months parturition—steam men at a cost of $300 apiece."[9] Such jokes, spread nationally through press syndicates, acknowledged a tendency to identify with the machine, to see the colossal, potentially frightening steam engine in human rather than divine terms. Yet the absurdity of their suggestions also provided a chance to mock the era's celebration of the power of engineers and inventors to mimic the divine; Americans might be able to imitate nature, they intimated, but creating life remained a power unique to God.

Aside from a handful of commentators, most did not identify Drederick's invention with the automaton. Fundamentally preindustrial, automatons were objects of art created by craftsmen and entertainers and powered by human force. Spring-powered automatons had no vitality, free will, or, to use a phrase common at the time, "motive power." Set on a fixed path by their inventors, they could not move at will—in contrast to the most attractive and potentially dangerous element of Drederick's device. Prewar writers and contemporary exhibitors of chess-playing automatons credited magical or divine power for their devices' more lifelike qualities, but Drederick's invention offered a new breed of humanized machine whose body provided its own source of vitality. As knowledge of steam engines, gas-powered internal combustion engines, and electric "dynamo machines" seeped into American culture in the waning decades of the nineteenth century, both inventors and fiction writers imagined humanized machines powered by forms of energy that could be explained by science. Far more suited to a disenchanted world skeptical of the supernatural, such machines were fantastical but not magical. Instead of unique cultural objects that duplicated human appearances, these were potentially mass-produced consumer goods that humorously humanized the machine in ways that preserved vitalism.

The Magic of Capitalism

Even as science and industrialization disenchanted American culture in the second half of the nineteenth century, magic persisted in places such as New York's famous hall of amusement, the Eden Musee. Across the hall from a "Turkish" smoking room in the museum, a "Moorish" automaton named Ajeeb sat on a cushion atop a cabinet. Doors in the cabinet as well as in the device's chest lay open to offer visitors a glimpse into the intricate machinery that appeared to give it motion and sentience. By inserting ten cents into a slot, visitors could start a game of checkers; twenty-five cents yielded a game of chess. Regardless of the game, it usually won, leaving visitors to either try again or wander deeper into the museum. Some, however, likely lingered to ponder the magic by which the device had won. After a few minutes of thought, perhaps they would have asked the attendant, there to ensure that no one discovered the secret behind Ajeeb's mystery and, likely, to help the man hidden inside the machine escape once visitors had left.[10]

Built in England in 1868, Ajeeb debuted at the Eden Musee in 1885. For two years, it toured American cities before returning to the museum until the attraction's doors shut in 1915. After that, Ajeeb moved to Coney Island and was finally dismantled during World

Fig. 2.2. An 1886 photograph of "Ajeeb the Wonderful Chess Automaton" at the Eden Musee. As with the original automaton chess player, a human player operated the device from within the cabinet. Its exotic appearance was typical for performing automata of the time. TCS 1.183, Houghton Library, Harvard University.

War II. Throughout his nearly eighty-year existence, "Ajeeb, the mystifying Chess Automaton" was the most famous American automaton. While the original remained in New York, other versions appeared in cities including Chicago, Cincinnati, and Boston. Periodicals reported its activities; when the original's human operator died, apparently from natural causes, they printed obituaries that included such horrifying but humorous anecdotes as the times when a Texan shot him or a woman stabbed him—both for winning a match.[11]

Ajeeb's exhibitors used the term *automaton* to create a sense of mystery. In its first advertisement for the device, Austin and Stone's Dime Museum claimed that "a Genuine Surprise is in store for students, scientists and everybody else. We intended exhibiting this wonder a week ago, but in the process of unpacking 'AJEEB' after his ocean voyage it was found that some very important parts of his intricate internal machinery were seriously damaged." It then called on academic expertise: "Professor Daniels has been hard at work day and night to get ready for exhibition, and we are fully as pleased as he is himself to be able to say that 'AJEEB' will surely make his debut tomorrow!"[12] In such advertisements, exhibitors suggested Ajeeb's scientific merit while avoiding stating that it was a machine. Despite such efforts, no one who wrote about Ajeeb thought that it was mechanical, and no writer was invested in maintaining the illusion. The setting of the device in a hall of amusement indicated that this was an artifact of artifice and humor, not science. As one article debunking several automatons stated, "Every prestidigitateur, necromancer, magician or fakir of note in ancient and modern times has placed almost his chief dependence on automata, illusions or mechanical means for the entertainment of patrons."[13] Though exhibitors emphasized the "mysterious" nature of the device, Ajeeb was, at least for adults, a joke.[14]

Adults laughed at Ajeeb because they were in on the joke. Fifty years earlier, American writers had obsessed over discovering how Maelzel's automaton chess player worked.[15] For the *New York Evening Post*, that automaton was the "greatest piece of mechanism that the human mind has ever invented . . . no person who pays Mr. Maelzel a visit, will come away unremunerated for the time and trouble. Verbal description can convey no adequate idea of this magical exhibition. It has hitherto baffled all attempts of the most keen sighted to penetrate its mysterious movements."[16] Most accounts assumed that a person controlled its movements, whether by magnets or concealment in the bowels of the machine. As Edgar Allan Poe wrote in the most extensive analysis, "It is quite certain that the operations of the Automaton are regulated by *mind*, and by nothing else. . . . The only question is the *manner* in which human agency is brought to bear."[17] Unable to definitively prove the device a fraud, Poe resorted to logic: the device was too irregular to be a "pure machine." Once human beings discovered "the principle . . . by which a machine can be made to play a game of chess," he deduced, "an extension of the same principle would enable it to win a game—a farther extension would enable it to win all games." The only possible explanation was that a "dwarf" hid in the machinery. Though his logic was wrong, Poe's deduction was verified several years later when a new owner published a full account of its fraud. Poe was largely correct except for one detail: William Schlumberger, the man who hid inside the machine, was over six feet tall.[18]

Despite their similarities, audiences responded differently to the chess player and Ajeeb. While Maelzel's automaton was a "Wonder of the World" that offered an "astonishing exhibition," Ajeeb only offered audiences an "amusing experience."[19] While viewers lauded Maelzel's automaton, they poked fun at Ajeeb's lack of authenticity. "There is a story about Ajeeb," a Philadelphia paper began, "which, like all other stories, must be true. A player went into the museum and beat the automaton several games and then remarked, 'I can easily beat Ajeeb.' The automaton became very angry, and all at once a voice was heard to issue from his insides, saying 'Oh you can, can you? Well, just wait till Pillsbury gets back.'" Harry Nelson Pillsbury, the most famous chess player in America, was correctly and widely suspected to be one of the automaton's operators.[20] Already knowledgeable of how an automaton could play chess, and incredulous owing to the era's sciences, commentators could see Ajeeb as a joke.

Race was far more central to the allure of turn-of-the-century chess-playing automatons than it had been before the Civil War. Though Maelzel's automaton would later be known as "The Turk," people at the time were not especially drawn to its race. Even if they noted its "Turkish" appearance, most accounts of the original referred to it as the "Automaton Chess Player." Fifty years later, however, exhibitors highlighted the vaguely non-European name "Ajeeb" in some cases by repeating it tens of times in a single advertisement. Though Ajeeb was always exotic, visitors were uncertain of its specific national origins. For some, Ajeeb was Turkish, for others, a Moor, Arab, or Egyptian; still others saw it as a "Hindoo." Another chess-playing automaton of the era, Chang, drew specifically from Chinese iconography. Per one 1894 report, Chang "seems to be as soulless and brainless as his famous Hindoo predecessor, Ajeeb, but he plays even a stronger game of chess and checkers than that well-known expert. There is a trapdoor in his breast through which any one may look apparently and convince himself that Chang is simply filled with wires and machinery, and Chang is moved around the room and wound up after every game to increase the observer's belief that the automaton plays without the aid of any human agency." Yet, the reporter concluded, "there is a strong suspicion prevalent that there must be and is a gentleman, colored or otherwise, in this automaton woodpile."[21]

Depicting automata in exotic forms was part of a larger transatlantic effort to distinguish modern "civilized" cultures from those "primitive" cultures increasingly controlled by Euro-American empires.[22] In an age of mechanical standardization, regimented labor, and scientific disenchantment, writers and performers romanticized less industrialized cultures as sources of magic, leisure, and spirituality. Inspired by works such as Robert Hitchen's extremely popular 1904 adventure novel, *The Garden of Allah*, merchants commercialized exoticism to convince consumers to pursue rather than deny their desires for pleasure and leisure.[23] Cloaking automatons in such aesthetics enabled exhibitors to tap into the same vein by suggesting a reconciliation between magic and machinery in the body of an exotic other that came alive at the command of the consumer. Ajeeb and Chang pretended to be machines but were not orderly or standardized; they varied their movements and responded to their opponents; they seemed to operate by a magic unleashed by the dropping of a coin in a slot—by the magic of capitalism.

The magic of capitalism that Ajeeb conjured, however, was for a far different audience than its predecessor. For almost all its existence, Maelzel exhibited his automaton on a stage for audiences of elite and middle-class men. The places he chose to exhibit it—the opulent National Hotel in New York, Julien Hall in Boston, Franklin Hall in Providence, and the Masonic Hall in Philadelphia—were lecture halls for the entertainment and education of gentlemen. The price he charged—fifty cents for an adult and twenty-five for a child–would have been exorbitant for ordinary laborers.[24] His was a device for master craftsmen, professionals, and successful businessmen, not the masses. When he performed in front of such men, Maelzel introduced the device, chose the player, wound the machine, and reported its opponents' movements; the entire performance and many of the theories of its operation stressed Maelzel's role as its controller.[25] Ajeeb, however, was typically exhibited in a museum that likely attracted a far wider clientele and did not depend on an external human controller.[26] Customers walked around the museum and then dropped a coin in a slot to start the device. As it played, the automaton seemed to respond to their moves on its own; its attendant was not a boss like Maelzel but a servant. While power over the "Turk" resided with Maelzel, power over Ajeeb seemed to reside with customers. In the early years of American industrialization, the automaton chess player was a tamed worker; in a commercial society, Ajeeb was a consumer good.

Ajeeb's role as a product coincided with the spread of toy automatons in American life. Initially imported from Europe, these devices first appeared in America early in the nineteenth century but grew more popular after the 1840s as part of a new ritual in American life: the giving of Christmas presents to children. As Christmas transformed from a public carnival into a domestic holiday when parents gave children presents through the mythical figure of Santa Claus, a variety of windup mechanical toys flooded American markets.[27] A 1860 advertisement for Piffet's in New Orleans publicized, "OLD SANTA CLAUS has established his Headquarters at this well-known establishment. He has brought for LITTLE GIRLS A Most Extraordinary Assortment OF DRESSED WAX AND KID, PAPIER MACHE, AND MECHANICAL DOLLS. For LITTLE BOYS, he has MECHANICAL TOYS such as Menageries, Farms. Railroads. Cities, Tea-Pins. Noah's Arke, Horses, Sheep, Goats, Cats, Dogs, Mice &c."[28] Reflecting the larger intellectual tradition in which the bodies of beasts were mechanical, most windup toys of the period came in the form of animals, but manufacturers increasingly produced moving dolls for girls and industrial-themed objects such as steam engines for boys.

Before the Civil War, such toys inspired nightmares. In 1859, the *Atlantic Monthly* published a short story by Irish American writer Fitz-James O'Brien entitled "The Wondersmith" in which "gypsies" plot to murder Christian children using magically empowered automatons. O'Brien set his story on "Golosh Street," a small alley with a "Hebraic taint" off of Chatham Street in New York, a foreign entertainment district between the Bowery and Broadway. While the "dirty" neighborhood contained numerous stores—including a fortune-teller, an artificial-eye manufacturer, a bird shop, and a used-book seller—the most intriguing was that of the "Wondersmith," a snakelike toy-automaton manufacturer named Herr Hippe. One night, Hippe invites the fortune-teller and the eye manufacturer over for a secret meeting. The fortune-teller brings a bottleful of "devilish" souls harvested from the gallows. The plan they hatch is simple: the fortune-teller will release the souls into the mannequins; the Wondersmith will sell the toys to Christian parents eager to please their children with Christmas presents. At night, these "brave automata" will awaken and slaughter the children with poisoned blades. Fortunately, the plan fails. As a hunchbacked manager of the used-book store named Solon and the girl he loves watch from a keyhole, Hippe drops the bottle of souls. As it shatters, the souls escape into the murderous automatons, which immediately come alive and stab their creators. Vengeful to the end, Hippe sets fire to his shop and burns the entire murderous group, "gypsy" and automaton alike.[29]

Combining racialized magic with consumer goods, O'Brien's tale was horrifying, especially given the unfolding transformation of Christmas and children's leisure in American culture. For much of the nineteenth century, Northern middle- and upper-class parents—the typical readers of the *Atlantic Monthly*—followed advice derived from John Locke to make play a learning experience. Preferring leisure that could prepare children for adulthood, they avoided toys, games, and fantastical fiction that stimulated the imagination rather than teaching social roles. Accordingly, the most common Christmas presents were books, but an increasing number of tin and mechanical banks—which could teach thrift—and similar toys were available for parents to purchase by the 1850s.[30] The Wondersmith did not sell such educational toys. O'Brien's introduction to the shop focuses on a "miniature theatre" replete with orchestra, conductor, and dolls set in a scene of a Templar saving a woman from a bandit. Centered on a violent but romanticized Catholicism, the scene itself was scandalous in age of nativist hostility, but such toys did not even have the purpose of teaching mechanics because their operation was inscrutable. Moreover, O'Brien named the blonde-haired, blue-eyed, and fair-skinned Solon after the Athenian statesman famous in the Anglo-American world for his efforts to preserve republican government. That Solon sold used books was key: he sold gifts, but they were not luxurious items like those sold by Hippe or those offered in the neighboring entertainment district. In O'Brien's tale, automaton toys were dangerous because they were objects of pleasure and amusement enchanted by even more dangerous immigrants than O'Brien's fellow Irish.[31]

Despite such wariness, the availability of toy automatons increased dramatically after the Civil War as toy manufacturers industrialized and middle- and upper-class families grew more open to amusement. Goodwin's Automaton Toys company advertised automatons in the shape of girls, boys, and horses as "perfect as living models; no winding up required."[32] In the late 1880s, Thomas Edison manufactured a doll that used a small windup phonograph to speak. Modeled after French dolls, the toy had white "skin" with brunette hair, "jet-back curls and sparkling brown eyes," and was "remarkably intelligent."[33] In 1886, a report on new toys spread nationwide advertised a dancing girl and a "lady in ball-room costume who fans herself languidly and raises to her eyes at intervals her lorgnette in a most affected manner"; another toy offered a "lady at her toilet" who "stands before a mirror applying powder to her face and neck" while still another offered "a lean and slippered pantaloon," trying to smash a mouse with a ladle."[34] Such toys were likely too fragile for the play of children but often took the form of satirical or horrifying images that could amuse adults while fascinating children with their movement.[35]

Like stage automatons, many of these toys used racial stereotypes to provoke laughter. In 1882, the Automatic Toy Works company advertised numerous automatons, most of which appeared in the forms of African or Chinese Americans. One depicted a black "Woman's Rights Advocate"—almost certainly a poor attempt to caricature Sojourner Truth—who the caption claimed believed in her equality with white men but would "not insist upon the last word." Two others depicted black ministers that stood behind pulpits while gesturing

—13—

The Woman's Rights Advocate.

In presenting this advocate to the public, and remembering with satisfaction the cordial reception our sterner suffragists and preachers have received, and believing in every respect she is their equal, we shall hope to receive as many calls for her. This woman will not insist upon the last word. Societies supplied with advocates on short notice.

Price, $2.50.

Fig. 2.3. A racist and misogynist caricature of a black suffragist—likely Sojourner Truth—from the Automatic Toy Works 1882 catalog. Library of Congress, LC-TS2301.T7 A8, accessible through Archive.org.

wildly with their arms. The accompanying description of one device promised that its "motions are so life-like and comical that one almost believes that he is actually speaking. The face and dress alone provoke irresistible laughter." The company's Chinese-themed automatons portrayed feminized laundrymen under the titles of "Fing Wing, A Melican Man" and "Ah-Sin, 'The Heathen Chinese.'" The former referenced a recent story in *Scribner's Monthly* about a Chinese servant who learned to steal from an Irish servant while the latter referenced Mark Twain's and Bret Harte's 1877 play *Ah-Sin* that the authors had meant to satirize stereotypes but had instead reinforced them.[36] Other companies throughout the period patented and sold numerous "Negro" dancing automatons to provide a home version of blackface minstrelsy.[37] Like Falconi's Indian, such grotesque toys offered a fantasy of winding a spring and laughing at a regimented body that reinforced stereotypes of peoples subjected to both violence and exclusion from American society.

Toy automatons were part of a much larger process of commercialization in which people purchased factory-made rather than handcrafted goods. As retailers sought to attract customers around the turn of the century, they constructed elaborate store and window displays using colored glass, light, mannequins, and automatons. Typical was a shoe-display automaton that appeared in the form of a white, Victorian woman. The company renting out the automaton would "supply the figure complete with motor and all parts, and necessary skirts," but allow retailers to choose the attire from their own wares.[38] Once turned on, an electric motor would drive the device to raise its dress "to a modest height," extend a leg to reveal its shoe, and then retract the leg, lower its dress, and change its shoes and hosiery. "It is next to an impossibility," one paper reported, "to make passage through the crowds which swarm in front of the window and watch the automatic marvel."[39]

Advertising automatons generally mimicked the form of the intended customer because they were models for consumer behavior.[40] The Bissell Carpet-Sweeper Company created a "pretty little lady" automaton sweeper that, per a trade magazine, "works incessantly, industriously," and "should prove a good advertisement . . . and an excellent example to all housekeepers who make her acquaintance."[41] Yet such modeling could imply that the use of store-bought products might destroy individuality and transform the customer into a mindless machine. The employment of human actors to perform as automatons magnified such a meaning. Periodicals across the country in the early twentieth century reported huge crowds in front of store windows as onlookers tried to decipher if the moving body in the window was human or machine. At a drugstore in Roanoke, Virginia, two policemen had to move a crowd that had gathered to watch a man with his flesh painted to look like wax make "mechanical" movements while he smoked.[42] At J. Goldwater and Bros. in Phoenix, a large crowd watched a man perform as a wax automaton modeling clothes with "short, quick, jerky movement" from side to side.[43] While stage automatons depended heavily on exotic stereotypes for their attraction, advertising for these automatons depended on the suggestion that, in an age of store-bought goods, both salesman and customer were practically machines. Perhaps, they joked, the magic of industrial capitalism turned everyone into automatons.

Transforming the Steam Man

White, Victorian men and women could also see themselves in Dederick's steam man and its descendants. Worried that the potentially monstrous device would frighten horses and likely customers, its inventor strove to "give it, as nearly as possible a likeness to the rest of humanity." This meant clothing the machine "in pants, coat, and vest, of the latest styles," and a top hat that doubled as a smokestack. It also meant molding its face "into a cheerful countenance of white enamel, which contrast[ed] well with the dark hair and moustache," and adding a smoking pipe.[44] Such a gentlemanly facade, newspapers dryly remarked, transformed the machine into "a person of commanding presence" and, ultimately, a man of fashion.

Whitening the machine created an opportunity to tell barbed jokes about romantic relationships. One told of a husband who "wishes to exchange his wife for one, provided it is speechless" and of five ladies who wish to purchase one each to serve as their husbands.[45] For the *Memphis Daily Avalanche*, this "New Man under the Sun" meant that "an old maid can buy her a man, ride him home and put him to bed without any supper"; the next morning, "a few coals and a bucket of water will answer for breakfast and run him in the interests of shopping sewing societies or the anti-cruelty-to-animals society until dinner time, when the meal can be repeated without dessert of finger-bowls."[46] Mocking women's participation in consumerism and reform movements, the *Daily Avalanche* insinuated that women wanted a man as easy to care for as a machine.

Other comments suggested that the steam man could replace workers. The *Daily Avalanche* identified it as "more wonderful than a Ku-Klux," apparently because it might ensure black men's subservience in a post-Emancipation world by providing a competing labor source.[47] A California paper fantasized that the steam man might replace workers and their legislative representatives. If the devices were given citizenship, it taunted, "the Democratic majority of the State would be of course proportionately increased, but if a few were sent as representative to the Legislature, there is little doubt that consistency and the State would be the gainer. The Steam man acts but does not talk." The steam man was preferable to human workers because "the cost of living would be inconsiderable," and he would always show the amount of pressure he was carrying, and could be relieved by a safety valve."[48] Employers would never have to fear worker rebellion because they could release a valve to cool off a steam man.

Each suggestion that mechanical beings could replace people was laughable, but the humor acknowledged a growing longing to replace unruly people with a machine that would provide deference and companionship. While industrialization seemed to encourage progress, it also seemed to make people less controllable and the community less harmonious than elite and middle-class men imagined it should be. In both the workplace and the home after the Civil War, patriarchal authority threatened to break down as former slaves, women, and workers demanded equality in relationships. At the same time that Drederick's steam man garnered headlines, African Americans were fighting for their rights, often with the same weapons they had used to secure emancipation in the Civil War. Though the women's suffrage movement splintered after the Civil War, activists continued to press for voting rights and challenge for political, economic, and social equality. The unruliest were workers who toiled in factories and mines. Workers in virtually every major industry rebelled against the authority of capital in the period, especially as ownership of industries moved from local hands to boards of directors in distant cities.

The steam man and its symbolic kin resonated with middle-class white male writers because of the way they connected the possibilities of machinery to the rebellion of other people who had long been seen as subordinate.

By the early 1870s, all references to Drederick had disappeared from newspapers; but later inventors sought the same enthusiastic reception with other powered men. Throughout the late nineteenth and early twentieth centuries, humanized machines enlivened by steam, oil, and electricity entertained Americans, though none earned as much attention as the original. In the 1890s, Phillip Louis Perew designed his electric man that pulled a carriage so that it came "dressed in a white duck suit and presents the appearance of a typical American with a walk as perfect as a human being. Close inspection is necessary to discover that it is not human."[49] In 1900, newspapers labeled an "Electrical Man" a "dude" because of its modern attire. Invented in 1911, John W. Belcher's "Automaton Woman" stood five foot eight, weighed 185 pounds, came "dressed in a red silk gown of the latest design," and, at least according to reports, could be mistaken for a person.[50] Few of these figures worked well, some were undoubtedly frauds, and all were impractical; nevertheless, their proliferation signaled a transformation in how at least men imagined the machinery shaping their lives. Increasingly, white Americans looked to machinery not as a slave or god but as a representation or caricature of themselves and their spouses. That was not the case, however, in literature aimed at working-class audiences.

Soon after Drederick's invention appeared in newspapers, educator Edward S. Ellis published a dime novel entitled *The Huge Hunter; or, The Steam Man of the Prairies*, the first story of a nonmagical mechanical man written by an American. Dime novels had been part of American culture for over two decades before Ellis wrote his story. Using cheap printing, standardized stories, and anonymous authors who had difficulty bargaining for higher wages, publishers dramatically reduced both the time required to write and the price of literature and thus made it available to larger audiences. Such low prices made these books especially attractive to working-class families. Middle-class children certainly read dime novels, but their parents as well as moral reformers frequently denounced the literature for encouraging vice. Though publishers marketed some dime novels to girls, the pri-

Fig. 2.4. Cover of the 1868 edition of Edward Ellis's *The Huge Hunter; or, The Steam Man of the Prairies*. Note that Ellis transformed the white enamel of Drederick's original into a much darker facade with a hooked nose. This steam man was a racialized caricature harnessed to a wagon carrying both "Yankees" and "Irishmen" into an expanded American empire. Rare Books and Special Collections, Northern Illinois University.

mary audience of Ellis's book and its descendants were likely working-class boys.[51]

Ellis treated the steam man as a triumph of American ingenuity capable of uniting "Yankee" and Irish Americans through the conquering of the frontier and the harnessing of black bodies. He opened his story with a scene of terror in which a Yankee and an Irishman observe a giant "divil" with black smoke rising from the machine approaching at an alarming speed. As it approaches, the two men notice a boy holding the device's reins. As the machine comes to a halt, it unleashes a screech like a locomotive that terrifies the Irishman. Though the calm Yankee reassures him that he had seen a similar device in Colt's armory before the war, the Irishman had good reason to fear: ten feet tall with a face of iron painted black, the creature appeared as a gigantic man with a hooked nose and a stove pipe that spewed exhaust. Once assured of the machine's safety, however, the two men board the wagon and marvel as Johnny, the young inventor, pilots "the monster with rare skill" across the prairie.[52] The steam man is not a "divil," as the poor Irishman has assumed; it is a machine in the form of a caricature of both African and Jewish Americans that, when harnessed by a Midwestern genius, could pull both Yankee and Irish Americans into a prosperous future.[53]

To emphazise Johnny's self-made nature, Ellis made the boy's origins humble. The son of a mechanic killed by one of the frequent steam boiler explosions that plagued industrializing American, Johnny is "humpbacked" and "dwarfed" but nevertheless possesses an "amiable disposition" and an unrivaled skill with mechanism.[54] "There was no limit," Ellis wrote, to Johnny's "inventive powers." After building his steam man, the young boy embarks on a fantastical journey of taming the West by killing buffalo and battling the "savages." Like in a conventional Western, Johnny's mission is an individualist pursuit; his steam man carries one man or a small band of friends anywhere they wish to travel, rather than a mass of men and women to pre-established destinations, as railroads did. Together, the deformed, but brilliant Johnny and his machine conquer and civilize the West.[55]

Ellis's greatest legacy was to connect the tropes of Western novels to fantastical inventions to Americanize a genre that would later be called science fiction. This fusion drew on the much more respected fiction of French author Jules Verne and British author H. G. Wells but added a much more explicit interest in the violent conquering of territory. Mimicking the characterization, plot, and themes of Ellis's story, dime science fiction tales concentrated on a young male inventor who creates an amazing device—a robot, a flying ship, a submarine—and then relies on that device to enjoy amazing adventures in an exotic locale and frequently subjugate local populations.[56] Women rarely appeared in the stories and, when they did, typically served as damsels-in-distress. Written while the United States conquered the West and began to expand overseas, the stories identified the inventor as the driving force of American empire.[57] Built by hard working, rugged white men and powered by steam engine or electric dynamo hearts, mechanical men were not just machines; they were America itself.

Eight years after *The Steam Man of the Prairies*, publisher Frank Tousey printed Harry Enton's *Frank Reade and the Steam Man of the Plains*, a retelling of Ellis's novel with a new hero who would become the patriarch of a long-running series. Between 1876 and 1881, three more Frank Reade stories used the steam engine to give life to machines: *Frank Reade and His Steam Horse*, *Frank Reade and His Steam Team*, and *Frank Reade and His Steam Tally-Ho*. At the conclusion of the original series, Reade, like a fictional version of Thomas Edison, became what Drederick had hoped to become: a business tycoon. After a dispute with Enton, however, Tousey turned authorship of the series over to Luis P. Senarens, who, like Enton, published as "Noname." Later nicknamed the "American Jules Verne" because of his prolific career, Senarens wrote over a hundred books and introduced new elements to the series, including a new hero, Frank Jr.[58]

The plot of Senarens's stories followed the standardized model but updated it to reflect the pursuit of an overseas American empire, consolidation of industrial capitalism, and the transformation of invention into a business. While Verne emphasized traveling to fantastical places—the center of the earth, under the sea, the moon—Senarens emphasized travels to places subjugated by European or American troops.[59] Frank Jr. explored Australia with an electric man, Africa with several different inventions, and Mexico, Central America, and the American West with his father's steam man. Yet, in imagining Frank Jr., Senarens constructed a protagonist who succeeded because of the economic legacy bequeathed to him by his father, not just his own prowess. Unlike Frank Sr. or Johnny of *The Huge Hunter*, Frank Jr. is not a self-made man. Instead, he is the son of a wealthy and famous inventor and, unlike Johnny, a "handsome . . . young man."[60] Frank Jr. does not invent the steam man alone in a run-down shack; he worked in his father's laboratory and streamlined his father's design. Such changes reflected real transformations in American capitalism. During the late nineteenth century, business owners erected corporations to consolidate power over the new machines and hired professional engineers to design and improve their factories while overseeing a labor force of less-formally educated mechanics and laborers. As engineers became hired workers, they lost their ability to tinker independently; immersed within a larger system in which machines had to serve the interests of the company, they could not devote their time to building trifles.[61] Released amid such changes, the Frank Reade Jr. series perpetuated myths of the independent and individualistic inventor even as he became part of a larger system.

Senarens's story did maintain one element of Ellis's and Enton's original tales that separated the literary symbol from Drederick's original: illustrations depicted the steam man as black.[62] During Reconstruction and later legal and violent efforts by white Americans to ensure that emancipation did not mean the loss of control over black labor, the shift in the automaton's apparent race created a fantasy by which both white inventors and consumers could re-enslave black men. Like minstrel shows and toy automatons, black steam men mocked the idea that African American men could become gentlemen—but also offered white working-class readers a vision of a group of people they often viewed as a threat to their jobs reduced to the level of a horse.[63] The racial difference between the consumer steam man and the literary one was critical to their meaning. Consumer steam men offered an illusion of companionship and equality; literary steam men were slaves.

Gender and the Humanized Machine

While children read about the blackface mechanical men helping to spread an American empire, adults read stories about humanized machines that inventors intended as replacements for people. Late nineteenth-century literary magazines printed several stories of Victorian families introducing mechanical men and women into their homes, either as replacement workers or wives. Such stories merged satire with horror to both parody the era's enthusiasm for mechanical improvements and fantasize about the possibilities and consequences of mechanizing a space that was supposed to remain a respite from the chaos and competition of the economic world. Written by both men and women, these stories established a key dichotomy centered on gender. Stories about mechanical men, including those found in dime novels, praised the machines for ensuring social progress, while stories about mechanical women always ended in disaster. That difference in the gender of the machines paralleled differences in the gender of the author. Stories by men imagined the devices as slaves; stories by women critiqued men's desire for slaves.

Several late nineteenth-century stories lampooned the era's enthusiasm for mechanical improvements by imagining mechanical women supplanting household servants. In Howard Fielding's 1889 short story "Automatic Bridget," a "New York capitalist" conman forms a corporation to mass-produce steam-powered artificial women invented by a naive mechanic. To create a more efficient way to knead bread, its inventor connected an upright cylindrical steam engine to a board shaped like a face, wooden arms, and three wheels on legs. The form did not necessitate a female identity, but, presumably to better reflect its function, "some modest hand" draped "the legs of the machine in a calico skirt, below which appeared three old-fashioned pantalets." Placing the machine in women's clothing made it "side-splitting" and helped the conman convince naive businessmen to invest money, even though the machine did not yet work. When the investors demand to see the device functional, the newly hired engineer miraculously fixes it. Once turned on, however, it proceeds to violently swing its arms and beat the stockholders, several of them to death. The conman, seeing the destruction and that the surviving investors intend to lynch him, flees town. Much like responses to Drederick's steam man, the story satirized men's fascination with mechanical improvements but, appearing twenty years later, Fielding added a critique of the corruption in capitalist investment culture. The story's most interesting innovation, however, was its shift from a steam man to a steam woman—one given the Irish name Bridget.

This was not a female clockwork automaton that would be wound by an observer to give it the power of movement, but a machine granted the appearance of life by an internal engine. Bestowing agency to a machine made in the form of an Irish woman transformed the tamed automaton into an agent of destruction.[64]

Ten years later, the Boston literary magazine *The Black Cat* printed Elizabeth Bellamy's "Ely's Automatic Housemaid," which made a similar point about the danger empowered mechanical women posed, but her story critiqued men rather than industrial capitalism. In the story, a father, angry at the "incompetence" of his two servants, tests two mechanical replacements in his house as a favor for his friend, the inventor Harrison Ely. Much more obviously female than in Fielding's vision, the devices have molded breasts as well as the requisitely gendered clothing. Though horrified by the creatures' lifelike appearances, the family replaces their servants with these mechanical counterparts, even to the extent of naming the machines after the dismissed. The new servants initially perform their tasks well but soon threaten the lives of the children, fight with each other, and destroy the home. The father returns the machines and consults Ely, who proudly tells him that the problem is "really a merit—merely a surplus of mental energy. They've had too big a dose of oil. Few housekeepers have any idea about proper lubrication." Ely promises to rectify the problem of mechanical women who are too smart and energetic; but, four months later, the devices remain at the factory, and his dream of forming a "syndicate" remains unfulfilled.[65]

Where Fielding mocked women to critique industrial capitalism, Bellamy satirized men's efforts to mechanize and control one of the few spaces that women in Victorian America could command, the household. Unlike her husband and eldest son, the wife is consistently and rightfully skeptical of the devices. She vastly prefers human servants whom she can directly tell what to do and, when necessary, fire. Her husband programs the machine without consulting her because he believes that "women do not understand machinery" and decides when to get rid of the new servants, not her. She tries to feed the machines oil once and is so horrified by their appearance that she refuses to do so ever again. For that effort, she earns the rebuke of Ely who assumes, like her husband, that her effort to feed the machines was the problem, not the machinery itself. Rather than to mock women,

Bellamy used her story to criticize assumptions that they could not understand machinery and defend their power over the household.[66]

A year later, *The Black Cat* published W. M. Stannard's "Mr. Corndropper's Hired Man," a response to Bellamy's story that imagined that Ely perfected his device in the form of an "Automatic Farmer" which is purchased by Josiah Corndropper and named "Tom." According to a note attached to the machine, it comes "guaranteed to do twenty-four hours' work a day—seven days a week, if necessary—without strain. He can perform any ordinary task that an intelligent man can do." Such fantastic capabilities are made possible by feeding it an oil "embodying all the essential nutritive elements which, acting upon our improved substitute for cerebral tissue, contained in the farmer's cranial cavity, results in a faculty which cannot be distinguished from ordinary common sense." Apparently, switching the form of the empowered automaton from female to male removed the problem of surplus "mental energy" that had plagued Ely's earlier efforts.[67]

When Corndropper takes the man home, he appraises it as "tall, broad-shouldered and robust looking, with a wonderfully intelligent and life-like countenance" but his wife lays out its essential benefit: "He don't talk back, like some hired men." The rest of the story confirms that initial evaluation. The mechanical man performs almost all the farm's labor and even protects Mrs. Corndropper when two outlaws threaten her. The family pays off the mortgage and Josiah is free to engage in politics in town. The story's final line captures its central message: Mr. Corndropper "never tired of boasting of a hired man who could do the work of three, on six cents a day, and earn his employer a five-hundred-dollar premium the first year."[68] Writing in a period when laborers in both the North and South demanded control over their wages and workplaces, Stannard imagined mechanical men replacing workers much like newspapers did with Drederick's invention. While stories such as *The Steam Man of the Prairies* reassured white workers that their jobs were safe, "Mr. Corndropper's Hired Man" endorsed the demands of owners and managers for more efficient and controllable production. It also, however, spoke directly to contemporary concerns of farmers that their independence and lifestyles were at risk owing to the development of industrial capitalism. While social and political movements such as the Patrons of Husbandry and the Populist Party advocated cooperation and mutualism to ameliorate the plight of farmers, Stannard suggested a far more individualistic solution: better machinery.[69]

Two other stories offered more scandalous visions of mechanical women: replacement mothers and wives. In George Haven Putnam's 1894 story "The Artificial Mother: A Marital Fantasy," a man laments that his nine children deprive him of time with his wife. As Putnam uncritically has the husband exclaim, "I might as well have no wife at all as to have one who, instead of belonging to me, her lawful lord and master, is at the beck and call of sundry small specimens of humanity." He asks his wife what the children want from her. She responds with "*me*" to which he dismissively replies: "Bosh. . . . They want merely something soft to touch, a swinging motion to addle their brains, and a monotonous din in their ears, and these would be just as well satisfied by a steam-engine as a mother." This inspires him to imagine a novel solution: a mechanical mother that can rock his young children, which would allow his wife to serve him. He borrows a mannequin from a local artist and then endeavors to empower it. Unlike in other stories of mechanical women, he does not use steam or electricity but clockwork to ensure a regular and uniform rocking motion. To give it an ability to croon to the children, he combines clockwork with a turkey gizzard and drumheads. He hires the artist to paint his wife's visage on the mannequin, places it in her clothes and notes that it "wore an air of material affection and solicitude almost surpassing that of the original." The machine convinces the babies of its authenticity but fails horrifically, owing to his wife's actions. When his real wife sees her visage on a machine, she names it a "demon." As she tries to take one of the children, the machine fights back, but, in the ensuing melee, breaks and collapses, sending both children across the room. Lacking true motherly sentiment and devotion, the machine does not care about children, but it still performed its central purpose well; the only problem was his wife's resistance, her horror at seeing a doppelgänger duplicating what she saw as her fundamental purpose.[70]

A year later, the Boston literary magazine *The Arena* published another story about a mechanical spouse, this time written by a woman, Alice W. Fuller. In "A Woman Manufactured to Order," a forty-year-old bachelor is intrigued by a sign that promises "Wives made to order! Satisfaction guaranteed or money refunded." The man loves a real

Fig. 2.5. Illustration from George Haven Putnam's *The Artificial Mother: A Marital Fantasy*, in which the husband watches his wife, on the left, encounter the mechanical doppelgänger he made of her while it rocks their children. In the ensuing battle over the children, the automaton collapses, sending one baby against a wall and the other to the floor. University of California Libraries, accessible through Archive.org.

woman, Florence, but is put off by her "strongminded-ways, and inclinations to be investigating woman's rights, theosophy, and all that sort of thing." He enters the shop, but must wait because the inventor is helping "a great politician who had to separate from his wife," because she "wanted to hold the reins herself." To solve the problem of a man whose "soul was not his own," the inventor is manufacturing "a quieter sort of wife, some one to do the honors of the home without feelin' neglected if he happens to be a little courteous to some of his young lady friends." Of course, the wife is also beautiful enough "make your mouth water." Intrigued by the vision of a beautiful woman "who would never reproach" or "find fault" with him, the bachelor purchases one and marries it. But, he soon grows tired of its intellectual and sentimental limitations. When he faces financial difficulties, she can only repeat a few stock phrases of love that do not alleviate his stress or provide him with sympathy. His mechanical wife having failed him, he runs to the still unmarried Florence and tells her of his plight. "I see now," he summarizes the lesson he has learned and the lesson that Fuller wished to impart, "that it is only a petty and narrow type of man who would wish to live only with his own personal echo. I want a woman, one who retains her individuality, a thinking woman." Convinced of his sincerity, Florence agrees to marry him after he has been divorced for a year.[71]

These stories of mechanical spouses connected contemporary tensions over gender and the family with the era's technological improvements. To many late nineteenth-century Victorians, the family seemed to be unraveling as relationships between men and women broke down. The rate of divorce more than doubled in the final three decades of the nineteenth century. Young couples delayed marriages and, once married, having children. Middle-class women in particular felt discontented with the strictures of domesticity, especially as servants, machinery, and purchased goods lessened the amount of their labor required to maintain a household.[72] As women challenged domesticity by working outside the home, fighting for the right to vote, and joining radical and reform organizations, many men fretted that their society was becoming feminized and looked for ways to reinvigorate their power and authority.[73] Both stories addressed these concerns. With no sense of irony, Putnam dedicated his tale about a man who longed for a wife he could control to "the oppressed husbands and fathers of the land and to the unknowing young men who may be contemplating matrimony."[74] Fuller's tale directly challenged such rhetoric with its derision of men who want nothing more than a beautiful, unintelligent object.

When combined with the era's dime novels, these four stories suggest a significant gender difference in late nineteenth-century visions of mechanical men and women. When employed in areas outside the home by replacing horses or male workers, mechanical men promoted both material and moral progress. But, whenever writers imagined mechanical creatures replacing women, the result was unhappiness and destruction. Part of this difference came from fears of empowering women, but it also stemmed from efforts to maintain the home as a refuge from the chaos, danger, and machinery of the industrial world. The home was not such a refuge; since the early nineteenth century, the introduction of sewing machines, heating stoves, central furnaces, hand tools, and more formalized relationships between families and servants had industrialized the home in lesser but similar ways to that of the factory.[75] Despite such transformations, turn-of-the-century Victorians sought to maintain the home's antimechanical character. When famed author of the utopian novel *Looking Backward* Edward Bellamy—no relation to Elizabeth—and early home economist Charlotte Perkins Gilman considered a future of women liberated from housework, they did not conceive of a home filled with machines; instead, they imagined families turning to professionalized services outside the home.[76] But, as Elizabeth Bellamy's tale indicates, opposition to mechanizing the home could stem from efforts to preserve women's power. Because the devices primarily worked outside the home, mechanical men were beneficial. Because they worked inside the home, mechanical women were horrifying.

This spatial difference, however, had a larger significance for human identity. Machinery, in these stories, could easily duplicate men's work; they could never duplicate women's. Even Putnam's artificial mother failed to do anything beyond rocking the children and cooing to them. At the beginning of the nineteenth century, clockwork automatons had frequently appeared in the form of women as a joke about their supposed lack of rationality. But, by the end of the nineteenth century, these visions of empowered automatons suggested an emerging reversal: in the industrial world, men could be mechanized but women could not.

This turn-of-the-century exchange of ideas about the gendered nature of humanized machines existed because it involved both men and women. Unfortunately, that relative gender parity did not last in the next century. American discussions about automatons, mechanical men and women, and, eventually, robots exploded in the twentieth century, but those that participated were overwhelming men. This does not mean that women ceased to encounter, imagine, or write about such devices. But mounting efforts to exclude women from the worlds of science and technology meant that their ideas and critiques were largely missing. Conscious of the way that men subjugated and objectified women, both Elizabeth Bellamy and Fuller used their stories to critique male desires to control both women and the household; rather than endorsing, as each of these male authors did, the slave fantasy, these women ridiculed it. As the voices of those left out of power were erased from the discussion of robots, the slave fantasy offered by the humanized machine mostly went uncontested through the twentieth century.

3

Mechanizing Men

Few writers found humor in the horror of industrial life better than the satirist and Civil War veteran Ambrose Bierce. Bierce witnessed his first combat at the Battle of Shiloh and his final at Kennesaw Mountain, where he was shot in the head. Afterward Bierce moved to San Francisco, where he wrote essays and short stories that drew on his war experiences, in part to critique the effects of mechanization on human heroism. As he wrote about a Confederate bayonet advance during the Battle of Shiloh: “Lead had scored its old-time victory over steel; the heroic had broken its great heart against the commonplace.”[1] In the ensuing decades, Bierce applied mechanical imagery more regularly. In his first collection of fiction, he included an anecdote about a lawn mower chopping off the head of the man controlling it.[2] In his most famous story, “An Occurrence at Owl Creek Bridge,” he recounted the hallucinations of a southern man condemned to hang from a railroad bridge for daring to commit the ultimate crime in industrial America: destroying railroad tracks.[3]

In 1894, Bierce turned his pen to automatons with the short story “Moxon’s Master.” Closer to *Frankenstein* than *The Steam Man of the Prairies*, the story imagines the danger of competing with machinery in a materialistic age. An unnamed narrator begins the story with a question to both the old scientist Moxon and readers: “Are you serious? Do you really believe that a machine thinks?" After defining a machine as "any instrument or organization by which power is applied and made effective, or a desired effect produced," Moxon asks his friend, "Well then, is not a man a machine?" From that basic premise, Moxon speculates that all matter possesses consciousness. "When soldiers form lines or hollow squares," he lectures his friend, "you call it reason. When wild geese in flight take the form of a letter V, you call it instinct. When the homogeneous atoms of a mineral, moving freely in solution, arrange themselves into shapes, mathematically perfect . . . you have nothing to say." From this, he concludes, "All matter is sentient, that every atom is a living, feeling conscious being." To prove his point, he quotes biologist and philosopher Herbert Spencer's definition of life as "a definite combination of heterogeneous changes, both simultaneous and successive, in correspondence with external coexistences and sequences." Logically, this means that "if a man during his period of activity is alive, so is a machine when in operation."[4] According to the era's scientists, Bierce was suggesting, man is no different from any other form of matter because—at least at the atomic level— everything moves, everything changes, and everything applies power purposefully.

Throughout this opening dialogue, the narrator occasionally remarks on a pounding sound in the adjoining machine shop. When asked about it, Moxon replies that it is "nobody: the incident you have in mind was caused by my folly in leaving a machine in action with nothing to act upon. . . . Do you happen to know that Consciousness is the creature of Rhythm?" Perplexed by this response, the narrator leaves but continues to think on the comment because it means that "all things are conscious, for all have motion, and all motion is rhythmic." That night, he returns to Moxon's home and ventures into the machine shop, where he sees the scientist playing chess with a mysterious opponent. Soon he realizes that the opponent is not a man but an automaton. After Moxon excitedly yells "Checkmate," the machine sits, seethes, and smashes its hand upon the scientist's head—remaining the physical master even in mental defeat.[5]

Though the image of an Ajeeb-like automaton bludgeoning its creator to death was horrifying, Bierce's tale primarily satirized late nineteenth-century conversations about the material nature of humanity. Simultaneously inspired and haunted by scientific discoveries that indicated that even mental processes had material causes, the duplication of human motions by empowered machines, and social transformations that undermined notions of autonomy, late nineteenth-century Victorians debated the nature of free will and human purpose. In a culture that celebrated the autonomous individual's abilities to exercise self-control and pursue self-improvement, such transformations were deeply threatening because they implied that the notion of an independent self was illusory. If all people were merely "conscious automata," as Charles Darwin's disciple Thomas Huxley called them, then no one was independent, no one had free will, no one had choice, and no one had purpose beyond fulfilling their biological urges.[6]

In focusing on Spencer, Bierce adopted another critique of materialism. A pre-Darwinian evolutionist and materialist, Spencer coined a phrase that would reverberate throughout the era and beyond: "survival of the fittest."[7] Spencer meant it to refer to the biological process of evolution, but it quickly acquired a much wider meaning as educated Americans reflected on the growth of competition between individuals, groups, and nations. In the decades after the Civil War, the economic unification of the nation, mass immigration, rising tensions between and among social groups, and the competition for foreign territories by Euro-American powers seemed to herald a new society in which everyone fought for what was best for themselves rather than the good of all. In that context, reformers and academics used "survival of the fittest" to characterize the sentiments of selfish individualists who advocated laissez-faire policies that would allow only the most competitive to survive. Though few Americans accepted that "survival of the fittest" should characterize human life, the phrase captured a pervasive feeling that competition, rather than cooperation, characterized industrial existence.[8]

Yet, as "Moxon's Master" suggests, no man was as competitive or successful as the machine. As it gained energy and vitality with the harnessing of steam and electricity, machinery began to exceed men's capabilities. On the battlefield, new industrialized weaponry proved far more effective at killing than human soldiers. As workers in practically every industry learned, machines could perform physical labor for more efficiency and regularly than they could. Even in historically mental jobs such as clerking, the first punch-card tabulating machines and cash registers enabled the calculating of information far faster than the human mind.[9] By the twentieth century, machinery seemed to set the pace and standard of success in the two realms most associated with manhood in the nineteenth century: warfare and work. Such a development might have meant the liberation of men from such endeavors, but it threatened their sense of identity and purpose, and justifications for power over others. Throughout the late nineteenth century, middle-class and elite American men worried that they had grown "overcivilized," too lacking in energy, resolve, and purpose. Confronted with the awesome power of machinery and a growing sense of their own inadequacy, men at the turn of the century rethought the nature of individualism and idealized the empowered machine as a model to emulate. While Bierce's tale allowed for Moxon's victory in the realm of reason, his quick destruction by the superior physicality of the clockwork automaton indicated that the day of the autonomous individual was over; in an industrial and commercial age, men either must succumb to the machine or turn themselves into machines.

The Autonomous Individual versus "The Machine"

Bierce's mechanical horror drew from a much longer tradition that contrasted the independent, typically male, and vitalistic individual who possessed the power to choose with the mechanical automaton bound to a set path. The divergence between men and automatons that emerged in the eighteenth century grew larger as social and religious transformations appeared to give more people control over their lives. While science and mechanics offered greater control over nature, market expansion and urbanization increased the range of both employment and consumer choice.[10] This enlargement of the choices available to American, especially white middle- and upper-class men, found support in the growing popularity of evangelical Christianity, which stressed the moral importance of both the choice to accept God's grace and the requirement to then embark on a process of self- and social improvement.[11] No longer feeling bound to a fixed future as Puritans and deists had once suggested, both men and women appeared less a part of the material world. As Calvinist turned Methodist philosopher Thomas Cogswell Upham wrote in his textbook *Mental Philosophy*, "Matter yields a blind and unconscious obedience; but the mind is able to exercise a foresight; to place itself in new situations; to subject itself to new influences; to surround itself with new motives, and thus control, in a measure, its own laws. In a word, mind is free. . . . But matter . . . may justly be characterized as a slave."[12] In an era that venerated choice and the possibility of improvement, autonomous individuals could conquer fate; automatons could not.

The mind's ability to exercise self-control by setting its own laws was central to notions of the autonomous individual and a harmonious society.[13] The breaking of a clockwork world and celebration of white male independence threatened to destroy all of society if each person pursued selfish individualism. Since the end of the eighteenth century, American writers typically had rejected such Hobbesian fears in favor of the optimistic liberal individualism of John Locke and Adam Smith that called on sentiment to constrain self-interest and violent passions. Antebellum ministers, reformers, advice-book writers, novelists, and politicians additionally preached an ethic of self-control through hard work, frugality, self-discipline, and family life. Focused on taming impulses to violence and unhealthy forms of sexuality, the Victorian creed of self-control required the management of emotional expression—especially negative feelings of anger, fear, and jealousy that might disrupt society and harm relationships.[14] To be sure, few among even their middle- and upper-class advocates fully conformed to such strictures, but the ethic of self-control created an ideal that those in other groups also sought to emulate as a path to success.[15]

Distinctions between individuals and automatons drew from an emerging literary use of "the machine" as the antithesis of the individual soul. Popularized by British conservative Thomas Carlyle's 1829 essay "Sign of the Times," "the machine" symbolized how the era's material transformations echoed through the internal worlds of male identity and purpose. In this "Age of Machinery," Carlyle claimed, "men are grown mechanical in head and in heart, as well as in hand. They have lost faith in individual endeavor, and in natural force, of any kind. Not for internal perfection, but for external combinations and arrangements, for institutions, constitutions, for Mechanism of one sort or other, do they hope and struggle. Their whole efforts, attachments, opinions, turn on mechanism, and are of a mechanical character." In Carlyle's distinction, men lost their individualism and became machines when they defined themselves by only what they could see and forsook the unquantifiable and unverifiable spiritual components of existence. Yet he primarily directed his ire not at machinery but Locke's individualism. Locke's "whole doctrine" was "mechanical," Carlyle wrote, because it ignored "the grand secrets of Necessity and Freewill, of the Mind's vital or non-vital dependence on Matter, of our mysterious relations to Time and Space, to God, to the Universe." In the Lockean age of machinery, men had ceased to care about the "Beautiful and Good" in favor of the "Profitable" and made "Mechanism" their "true Deity." Including a reference to a literary "wood and leather man," Carlyle's metaphor of "the machine" critiqued the secularization of a liberal, capitalist culture that saw profit as men's purpose.[16]

Few antebellum Americans welcomed Carlyle's skepticism, but they shared his vitalistic concern with inward elements of human identity. Carlyle's most famous American opponent, Massachusetts lawyer Timothy Walker, praised machinery for its ability to enable all people to develop their spiritual identity and purpose. When "we can compel inert matter to do for us," he fantasized that "there would be nothing to hinder all mankind from becoming philosophers, poets, and votaries of art. The whole time and thought of the whole human race could be given to inward culture, to spiritual advancement."[17] The political difference between the two was vast. Carlyle—who soon would publish his famous *On Heroes, Hero-Worship, and the Heroic in History* that trumpeted the power of individual men to transform the world—worried about the moral cost of mechanization to men. Walker idealized mechanization for its democratic potential, for its ability to free both men and women, upper and lower classes, so that all could forsake the material for the spiritual. Though the men disagreed over the political meaning of mechanization, both accepted—as did Ralph Waldo Emerson and Nathaniel Hawthorne and much of Victorian society—a fundamental distinction between matter and spirit, between outer appearances and internal character. That distinction would remain fundamental to American visions of robots throughout the twenty-first century.

The distinction between internal and external, authentic and inauthentic was a key theme in Victorian culture, as critics feared that men and women could exhibit virtue without internally possessing it. In Edgar Allan Poe's 1839 short story "The Man That Was Used Up: A Tale of the Late Bugaboo and Kickapoo Campaign," a military general exudes perfection. Amid numerous satirical paeans to the wonders of "invention," other characters extol his virtues; women love him for his physicality, wit, and courage. At the end of the story, however, it is revealed that the general is almost entirely mechanical; when his body had become damaged beyond repair in the Indian wars, the army replaced each of his parts with mechanical, cork, and wood substitutes. While he appears a perfect specimen of Victorian manhood, the general could not even assemble himself; he depended on a slave to do so. A commentary on how the military "uses up" its soldiers, the story also suggested that the external markers of the autonomous individual were false.[18] As in Julian Hawthorne's "Mullenville Mystery," a beautiful and humanoid exterior could hide the void of a machine.

Such critiques of the inauthenticity of polite society also appeared in stage automaton performances catering to working-class families. While Maelzel's chess player entertained elite audiences in Boston, J. D. Dexter presented an "Omnibus of Fine Arts" in Dock Square with several automatons, including "rope dancers, waltzers, housewives, gentlemen of leisure, military companies, and professors of the mechanic arts."[19] While Maelzel exhibited his automatons in expensive indoor halls and charged fifty cents, Dexter appealed to a working-class audience by exhibiting his automatons outside and charging only twelve and a half cents. Including waltzers, housewives, and gentlemen of leisure mocked the regimented and inauthentic nature of the middle and upper classes that sought to control working-class men. Adding automaton professors of the mechanic arts to the performance only underlined the critique as it insinuated that unlike real mechanics, who could only afford to attend performances such as Dexter's, professors of the mechanical arts were just going through preestablished motions.

The most extensive criticisms of the way that the emerging world destroyed human bodies and souls, however, came from critics of industrial labor. After a trip to England in the 1850s, Emerson remarked that "machinery has been applied to all work, and carried to such perfection, that little is left for the men but to mind the engines and feed the furnaces." Work with machinery, he observed, gave "a mechanical regularity to all the habit and action of men. A terrible machine has possessed itself of the ground, the air, the men and women, and hardly even thought is free." For a man who preached self-reliance and imaginative freedom, such a development posed a direct threat to white manhood. As he concluded, "The machine unmans the user. What [the user] gains in making cloth, he loses in general power. . . . The robust rural Saxon degenerates in the mills to the Leicester stockinger, to the imbecile Manchester spinner. . . . The incessant repetition of the same hand-work dwarfs the man, robs him of his strength, wit, and versatility."[20] Where Emerson saw machinery destroying white manhood, Herman Melville worried about the destruction of white womanhood in the 1855 story "The Paradise of Bachelors and The Tartarus of Maids." "Machinery—that vaunted slave of humanity," his narrator notes of a New England paper factory, "here stood menially served by human beings, who served mutely and cringingly as the slave serves the Sultan. The girls did not so much seem accessory wheels to the general machinery as mere cogs to the wheels."[21] Rather than turning white men and women into virtuous citizens, such literati suggested, industrialization turned them into mindless and soulless automatons bound to the regimented and monotonous rhythms of the empowered machine.

Conscious Automata

Even as Victorian culture separated human from machine, the development of modern science bound them together.[22] In the eighteenth century, interest in automatons stemmed from the way that they fused ideas from physiology, the physical and natural sciences, moral and political philosophy, and theology into a piece of art. Automatons were symbols of the universal nature of knowledge. Over the nineteenth century, however, intellectuals subdivided knowledge into disciplines that each focused on a facet of the material or spiritual worlds; only rarely did writers cross the boundary between the two. As the pursuit of knowledge industrialized, scientists abandoned vitalistic notions of existence in favor of methods emphasizing that only observable phenomena could count as evidence. Though some scientists rejected this "positivism" for its deterministic tendencies, others embraced an automaton theory of human nature that ignored unobservable elements such as the soul.[23]

The first specialty to undermine the distinction between humans and automatons was thermodynamics, a subfield connected to the rise of the steam engine that studied the energy required to make both living and nonliving matter function. Scientists had investigated the relationships among heat, force, and work for centuries, but only in mid-nineteenth century Europe did they create a unique field. With the ascent of steam power after the Civil War, American interest in thermodynamics spread rapidly; though inaccessible to most, its language and insights slowly seeped into the country's culture. When Melville satirized a Civil War battle of ironclad steam ships in 1866, he included a brief reference to the "calculations of calorie" to show how science, industry, and philosophy had turned war from a heroic into a quantifiable and mechanical endeavor.[24]

Like their eighteenth-century predecessors, the scientists at the center of thermodynamics were fascinated by the similarities between people and automatons. One of the field's founders, Hermann von Helmholtz, trained as both a physician and physicist and wrote, in his first essay translated in America, about automatons. The essay opens with a discussion of a supposed eccentricity of the recent past: the willingness of serious scientists to devote themselves to the manufacturing of "trifles." Yet, Helmholtz proposed, such scientists had a far more serious goal: "Though these artists may not have hoped to breathe into the creature of their ingenuity a soul gifted with moral completeness . . . there were many who would be willing to dispense with the moral qualities of their servants, if at the same time their immoral qualities could also be got rid of; and accept, instead of the mutability of flesh and bones, services which should combine the regularity of a machine with the durability of brass and steel." Earlier automaton manufacturers, he claimed, had "boldly chosen" not just to amuse the elite but to provide them with machines to replace their immoral and irregular workers. Helmholtz, however, acknowledged that such efforts were misguided. "We no longer seek to build machines which shall fulfill the thousand services required of one man," he wrote, "but desire, on the contrary, that a machine shall perform one service, yet shall occupy in doing it in the place of a thousand men."[25] What separated people from machines, Helmholtz argued, was degree of specialization. While modern machinery achieved efficiency through specialization, people could perform a multitude of tasks.

Lack of specialization did not mean that human beings were not machines. "To the builders of automata of the last century," Helmholtz concluded, "men and animals appeared as clockwork which was never wound up, and created the force which they exerted out of nothing." However, the steam engine's burning of fuel indicated that a similar process occurred in people. He theorized that food was made of "combustible substances, which, after digestion and being passed into the blood, actually undergo a slow combustion, and finally enter into almost the same combinations with the oxygen of the atmosphere that are produced in an open fire." The animal body, he concluded, "does not differ from the steam-engine, as regards the manner in which it obtains heat and force."[26] The body, to Helmholtz, was a food-powered, general-purpose automaton.

Because almost all early scientists interested in thermodynamics were European, Americans depended on periodicals to learn of the new field until universities began establishing physics departments in the 1870s. In popular thought, thermodynamics combined with the steam engine to reinforce a broader equivalency between food and fuel. In 1871, the *Youth's Companion* taught children that food was "only fuel to warm the body," explaining that "we require food frequently for just the same reason that, a fire requires coals frequently, and a lamp oil,—because we are burning away."[27] An 1897 *Los Angeles Times* piece made a similar point: "Remarkable experiments prove that mechanically we are the same as a steam engine. . . . The one thing that it [the body] never ceases doing for an instant, from the moment the individual comes into existence until he dies, is to burn fuel."[28] A year later, the *Grand Rapids Herald* proclaimed, "The Body a Machine, It Requires Food as an Engine Does Wood or Coal."[29] When the Wesleyan University professor E. B. Rosa described in 1900 the human body as a "living engine . . . guided by the brain as its engineer" with food as its fuel, he was laying out ideas that had been preached since the 1870s.[30]

The similarities between food and fuel were only one of the many insights spread by thermodynamics. Just as important was a phrase that became central to both the new science and conceptions of people: "motive power." The phrase originally referred to the source of work in engines; in the mid-nineteenth century, however, people began to apply it to spiritual elements. In 1855, a writer for a California technical journal identified hope as "the great motive power of man; it is courage, energy and perseverance."[31] The phrase recurred particularly in the lexicon of Protestant ministers. For the Baptist Reverend Edward Judson, the "love of Christ" was the "sustaining and motive power of the Christian" soul.[32] The "will of God," the *Christian Advocate* professed in 1889, was the source of the "strength and motive power" of people's "moral power."[33] The spiritualist poet Ella Wheeler Wilcox similarly noted, "Love is the law of the universe. It is the motive power underlying all existence."[34] The fusion of scientific language to spiritual and emotional elements of human identity showed the vitalistic potential of thermodynamics. Initially, the field had linked vitalism to materialism by arguing that energy was the source of movement, agency, and work in all organisms.[35] But energy was also an essential component of nonliving matter. By removing distinctions between living and nonliving matter and suggesting that all work came from energy rather than willpower, the field raised profound questions about the differences between people and the material world.

No idea had as important a role in expanding the automaton theory as evolution, particularly in the version espoused by Thomas Henry Huxley. Well known for his controversial claim in 1874 that people were "conscious automata," Huxley was willing to propose what few were willing to accept: that conscious thought was mechanical. For him, the consciousness of brutes appeared "to be related to the mechanism of their body simply as a collateral product of its working, and to be as completely without any power of modifying that working, as the steam–whistle which accompanies the work of a locomotive engine is without influence upon its machinery." Applying the same logic to people, he concluded that "all states of consciousness in us . . . are immediately caused by molecular changes of the brain-substance. . . . There is no proof that any state of consciousness is the cause of change in the motion of matter of the organism. . . . We are conscious automata."[36]

In 1876, Huxley engaged skeptical audiences in America.[37] For many, his claim was ludicrous and socially destructive. Much of the opposition came from ministers, but some came from scientists. *Popular Science Monthly* printed numerous attacks on his theory, including one by the industrialist Rowland G. Hazard that rejected it on the grounds of liberal individualism. "Every being is an independent, self-active power in the universe," Hazard claimed, "freely doing its part and cooperating with all other active intelligences in creating the future."[38] Even if Huxley had evidence, Hazard argued, the idea that people were simply machines would deprive them of "the dignity of consciousness, and with it of all the cheering and elevating influences of the performance of duty; for that which has no power can have no duties."[39] If people believed that they were mere automatons, he feared, no one would perform their social roles.

The most vigorous assault on the automaton theory came from the philosopher and psychologist William James. In an 1879 article, James rejected Huxley's contention for several reasons, but mostly because Huxley failed to consider that consciousness was necessary for individuals to control the unpredictable nature of life. Using Darwinian theory, James argued that if consciousness had utility in the evolutionary struggle, the conscious automaton theory was incorrect. The mind, James stated, was not a machine, because a "machine itself knows nothing of wrong or right: matter has no ideals to pursue."[40] However, human beings could decide between right and wrong. If people were simply conscious automatons, they would lack the purpose and power of choice necessary for the construction of "complexion of the character." As James wrote, "The problem with the man is less what act he shall now choose to do, than what kind of a being he shall now resolve to become."[41] For James, the difference between people and machines was clear: people could choose to improve; automatons could not.

Despite such opposition, scientists and intellectuals continued to explain thought and behavior without invoking the soul or free will. In the late nineteenth century, the growing interdependence of an urban, industrial society combined with social science to undermine Victorian notions of a single, coherent self. In cities, men and women seemed to play a multitude of roles and demonstrate different characteristics in each. Though James rejected the automaton theory, he observed this phenomenon and concluded that "a man has as many social selves as there are individuals who recognize him and carry an image of him in their mind."[42] At the same time that notions of the coherent self became fragmented, thinkers became more aware of the limitations on self-control posed by hereditary and unconscious thought. Writers of the late nineteenth century questioned whether people could be held responsible for behaviors they seemed unable to control. The conclusion that human behavior was determined by a combination of biological and social factors rather than conscious choice grew increasingly obvious to social reformers.

Few men condensed the new findings of human identity and behavior as succinctly as the inventor of the alternating current, Nikola Tesla. In 1898, Tesla demonstrated his "Telautomaton," a boat remotely controlled by radio waves. While not humanoid, the telautomaton originated in Tesla's determination to manufacture "an automaton which would mechanically represent me, and which would respond, as I do myself, but, of course, in a much more primitive manner, to external influences." The automaton "had to have motive power, organs for locomotion, directive organs, and one of more sensitive organs so adapted as to be excited by external stimuli." To perform its functions properly, Tesla reasoned, the automaton had "to have an element corresponding to the mind, which would effect the control of all its movements and operations and cause it to act in any unforeseen case that might present itself, with knowledge, reason, judgment, and experience." Though an external controller provided the machine with its mind, the device suggested that, in general, behavior might be shaped by "external influences." Early in his life, Tesla noticed that his mind was growing automatic in its ability to respond to external stimuli; from that, he concluded, "I am an automaton endowed with power of movement, which merely responds to external stimuli beating upon my sense organs, and think and acts and moves accordingly."[43]

Such logic undermined the distinction between the individual and the automaton established by the Revolution and reinforced in antebellum culture. If people were merely matter affected by external stimuli or driven by unconscious impulses, then they had no choice and could not demonstrate self-control; without choice, they could not morally improve themselves or the nation and therefore had no purpose outside of satisfying their desires. Advocating that people demonstrate self-control meant little if there was no inherent self. Yet, for many thinkers, the most pressing issue was not whether people were automatons but whether men like them were. When James started to discuss the ethical implications of Huxley's automaton theory, he turned to explicitly masculine language. The issue of choice, he wrote, "is of the utmost pregnancy, for it decides a man's entire career. When he debates, Shall I commit this crime? choose that profession? accept that office, or marry this fortune? — his choice really lies between one of several equally possible future *Selves*. What his entire empirical *Ego* shall become, is fixed by the conduct of this moment. . . . The problem with the man is less what act he shall now choose to do, than what kind of a being he shall now resolve to become."[44] Since the early nineteenth century, the central promise of the industrial automaton had been its ability to ensure that the every man could be his own master, that every man could choose his own fate. If Huxley and Tesla were correct, then no man could be a master because every man was a slave to matter.

Automatons of War and Work

The growth and consolidation of American institutions during and after the Civil War similarly threatened the autonomous individual. In business, companies merged to form huge trusts in such industries as steel, oil, meatpacking, tobacco, and even circuses to avoid the intense competition of the market. Factories grew larger and centralized as owners transferred control to managers and professional engineers who sought to ensure efficiency through stricter divisions of labor. Working in those factories were some of the millions of immigrants from Eastern and Southern Europe as well as China, who settled in cities and diversified the country's culture. Government, too, expanded after the Civil War as federal, state, and local authorities began to build both a regulatory and social welfare state to cope with the influx of people and the harsh effects of industrial capitalism. Before the Civil War, many white men could believe that they might become independent individuals—whether as farmers, craftsmen, or frontiersmen—who had to deploy a variety of skills and fulfill multiple purposes. As post–Civil War institutions attempted to place everyone in a specialized role where they repeated the same motions, this hope grew far less plausible.

Such transformations encouraged middle- and upper-class men to embrace the empowered machine and even the wind-up automaton as an ideal. An early impulse to turn men into automatons came with the Civil War. Traditionally, American culture had revered the "citizen solider"—the man who, in time of war, willingly sacrificed his freedom to protect the nation. The sheer size and speed of the war required both the Northern and Southern armies to demonstrate unprecedented levels of discipline and coordination.[45] To mediate between the citizen-soldier ideal and the demands of modern warfare, publishers printed numerous training guides to teach men how to coordinate. Among these were three by Union Colonel G. Douglas Brewerton: *The Automaton Regiment*, *The Automaton Company*, and *The Automaton Battery*. Containing manuals and colorful "blocks and counters" to represent units of the army, Brewerton's guides offered the "studious soldier" practical lessons on training and deploying troops. Ordinary soldiers, Brewerton claimed, could read the guides and become cogs of the giant war-making automaton. In choosing the automaton for a model, Brewerton expressed no willingness to deprive soldiers of soul or rationality. Presumably, he wanted soldiers to have some form of motive power, whether it was to save the union or demolish slavery. The premise of his manuals was to teach men maneuvers so that they might understand their place in the larger automaton of war and to so ingrain the drill that their battlefield movements would become automatic.[46]

Brewerton's suggestion that soldiers transform into cogs in an automaton was rare in the Civil War but in its aftermath soldiers on both sides remembered that they felt transformed into mindless machines.[47] One soldier caustically remarked that he was "but a machine by which fame and glory is manufactured for some great Gen"; another lamented that soldiers "have no right to think. Others have been appointed to think for us and we like the automaton must kick (or work) when the wire is pulled."[48] Other soldiers, however, found the discipline empowering and liberating. "Napoleon the first said 'a man to be a good Soldier must first be converted into a machine,' & I am inclined after some experience to concur with him," one Confederate observed.[49] Yet, imaging oneself as an automaton could also help soldiers deny responsibility for atrocities.[50] The average soldier, still another observed, "was not required to do much, if any thinking for himself, as his officers were expected to do that for him. He was simply expected to act a good deal as an automaton; so that his brain was never burdened about 'maneuvers,' 'advances,' retreats,' and 'strategy'; but he was simply expected to act when required and to obey orders."[51]

The US pursuit of an overseas empire in the waning decades of the nineteenth century encouraged a debate over the degree to which the military turned men into machines. For some, the uniformed soldier was an icon of strength, self-sacrifice, and order in a chaotic world that seemed more concerned with individual gain than national unity.[52] Explicit invocations of the automaton as a model were rare, but as the nation looked to expand to the Caribbean and Pacific in the 1890s, the term appeared in newspaper accounts of the military. In 1890, the Cleveland *Plain Dealer* praised a boy at a Grand Army of the Republic meeting for his automatonlike behavior: "Straight ahead he [the boy] looked, and when he reached the president's stand, up came the musket to carry and then to present, as if he were an automaton. Everybody got up and cheered the little fellow."[53] In 1891, the *Chicago Herald* similarly praised adults: "The Fifteenth Infantry . . . will henceforth occupy an enviable place in the memory of the half a million of more Chicagoans who saw the Fort Sheridan garrison parade—eight companies with 300 men, each one like an automaton, and the whole a superb body of soldiers whose perfect alignment, precise marching tempo and unerring manual of arms quickened the pulse, stimulated the nerves and delighted the eye of the spectator."[54]

Opponents of American militarization criticized mechanization and regimentation for turning citizen-soldiers into automatons with no moral responsibility. The most detailed critique came in the 1902 novel *Captain Jinks, Hero*, by Ernest Howard Crosby, the son of a Presbyterian minister and the country's foremost promoter of Leo Tolstoy's philosophy of pacifism and manual labor. A direct satire of the militaristic spirit surrounding the Spanish–American and Philippine–American Wars, the novel recounted the life of Sam Jinks, who, after given a set of tin soldiers, devotes his life to the military. Immersed within a hierarchical and undemocratic culture, Jinks learns to love violence while denying his own culpability.

While Jinks recovers from injuries sustained during the war against the "savage" "Cubapinos," a military inventor visits him to discuss the mechanization of war. The perfect soldier, the inventor postulates, is a machine without "conscience or reason." Why, he asks the young hero, "shouldn't a machine be made to take the place of a soldier? A torpedo is simply an iron soldier that swims under water and needs no breath, and does as he is told." Sam suggests that men are essential because they possess courage, an element no machine could duplicate. The inventor, however, scoffs at such a claim: "Courage! Why, what is more courageous than a piece of steel? It wouldn't be easy to frighten it. And it is just so with all soldierly qualities. Do you want obedience? What is more obedient than a machine?" A solider, the inventor argues, "must be obedient, and he must be without fear, conscience, or a mind of his own. In all these respects a machine can surpass a man." Clockwork, he concludes, is "the highest military praise possible."[55]

While few publicly wished to turn soldiers into automatons, some worried that competition with foreign powers might necessitate it. Americans had long associated automaton soldiers with European—particularly German—militaries. Reporting from Europe for his hometown paper in 1893, the future Republican senator Frank L. Greene told readers that "it is worth the time of any one interested in military matters, to watch a drill of German soldiers" so that they could see different ways of constructing a military. "There is apparent, even in the fine discipline," he detected, "the spirit of crushing the man into an automaton that would never be tolerated in our country. American soldiers were never distinguished for drilling, but it is because they will never submit to the utter and complete annihilation of their own individuality, as must be the case in Germany." Such drilling was impossible in America, because there was "a little too much of that sense of the dignity of 'Roman citizenship' for a common soldier to quite surrender his body and mind to his superior." Consequently, he concluded, "we do not drill like machines, and are rather, more or less conspicuous for lack of discipline. But, for all that, can we fight? . . . Can we?" Simultaneously impressed and repulsed by the German army, Greene wondered whether individiualistic Americans could compete in an orderly but violent world.[56]

Worse still, in 1894 newspapers across the country reported that the Spanish government had produced a remote-controlled version of the steam man to expand its empire. Made of iron and capable of shooting forty rounds a minute, these "Automaton Soldiers" could be "blown to pieces by means of an electric discharge acting upon a quantity of dynamite stored in his head."[57] If American soldiers might have to contend with regiments of mechanized men and humanized machines, no wonder military enthusiasts wished to harden the bodies of the American man. Worried that men had grown too weak from the closing of the frontier in the 1890s and the rise of luxury and wealth, President Theodore Roosevelt called for the fusion of man and matter: "We need the iron qualities that must go with true manhood. We need the positive virtues of resolution, of courage, of indomitable will, of power to do without shirking the rough work that must always be done, and to persevere through the long days of slow progress."[58] Two years later, the militarist praised pioneers for possessing both an "iron body" and "iron soul" because they had the willpower to overcome hardship.[59] What America required, Roosevelt argued, was not men made of malleable tin but men made of strong iron, men who could remain individuals despite the pressures of industrial life.

This belief that to succeed men had to fuse with matter and the machine extended into the workplace. The inventor and entrepreneur George Westinghouse earned praise as a "Human Dynamo" because of his efficiency and determination.[60] The Boston Congregational minister Edward Everett Hale earned the same nickname, as did Roosevelt.[61] In 1903, John G. Shedd, vice president of Marshall Field, told a group of young men that success in business required them to "be a human steam engine. . . . Be wound up tight, ready to hurl the whole energy of your soul into the thing that you have at hand." In the modern age, he contended, "mere strength of character will not make you a success. You must be able to wield this strength at a moment's notice, with concentrated power, and you must be able to do it continuously."[62] Such imagery was indicative of a new form of manhood available only to the wealthy and powerful in which the engine or dynamo became an ideal metaphor for the heart and "motive power" of a successful individual.

While managers compared themselves to mechanical sources of power, they identified workers with machinery in general. No one better articulated this than the efficiency engineer Frederick Winslow Taylor. A wealthy Victorian saddened by a perceived loss of discipline in the world, Taylor was one of the world's most famous and infamous engineers of the period. Influenced by the systematic management movement and the writings of Charles Babbage, Taylor proposed to increase efficiency by creating a stricter division of labor in which middle-class supervisors would, "do the thinking for the men." Placing the mental aspects of a task under the control of the manager, Taylor claimed in his 1911 treatise *The Principles of Scientific Management*, could raise productivity and living standards much like the introduction of automatic machinery. Greater efficiency would then lower the cost of all consumer goods. For Taylor, the replacement of individualized "rule of thumb methods" with standardized methods would enable laborers to "work together like a smoothly running machine."[63] To win their competition for jobs with machines, Taylor proposed, workers had to learn to act like machines.

The idealization of the machine extended even into the mental realm. In 1905 the young fiction writer Jacques Futrelle introduced Americans to Professor Augustus S. F. X. Van Dusen, the "Thinking Machine." Though modeled after Sherlock Holmes, Van Dusen was even more of a machine than the British detective. Futrelle made this clear by implying that Van Dusen is a human version of the automaton chess player who could defeat a grandmaster despite never having played the game before. As Van Dusen tells the master, "Chess is a shameless perversion of the functions of the brain. . . . It is a sheer waste of effort." The game was simply logic. "Of course logic will solve it," he continues. "Logic will solve any problem—not most of them but any problem. A thorough understanding of its rules would enable anyone to defeat your greatest chess player. It would be inevitable, just as inevitable as that two and two make four, not some times but all the time."[64] While the larger culture prized the "iron body," Van Dusen's body is feeble, "child-like" and excessively pallid.[65] Refusing to follow social graces, he is aggressive, impolite, and egotistical; he is, ultimately, a misanthrope who does not care for the feelings of others. Yet he is also, like a machine, always right and, in a society that valued success and "survival of the fittest," always a winner. For a "Thinking Machine" what matters is not an often cruel and dismissive demeanor but the result: victory. Much as Carlyle had warned in the 1820s, external rewards seemed to matter more than internal character.

The divergent application of mechanical metaphors to people reinforced criticisms of industrial labor that came from radicals seeking a complete upheaval of the economic system. Drawing on the work of Karl Marx and the English social critics John Ruskin and William Morris, American radicals complained that the problem with industrial labor, as the anarchist Emma Goldman stated, was "machine subserviency." In modern industry, she lamented, "man has been degraded into a mere part of a machine and all that makes for spontaneity, for originality, for the power of initiative, has been either dulled or completely killed in him until he is but a living corpse dragging out an aimless, spiritless and idealess existence." She also remarked, "The division of labor has never before reached such height, nor has man ever before been so much degraded to a mere machine. . . . His activities are mechanical; his work, instead of liberating him, is riveting his chains still deeper into his flesh. The iron necessity of eking out a living imposes such occupations which are remunerative, though in no way related to his nature or inclinations. The question is not as to what could give the greatest satisfaction or joy; rather is it what gain, what material results can accrue therefrom."[66] Or as Eugene Debs, the labor radical and perennial socialist candidate for president, told listeners in Kansas, "The machine you work with has to be oiled; you have to be fed; the wage is your lubricant, it keeps you in working order."[67] The price of the material gains experienced by the rich, these activists argued, was the destruction of the bodies, minds, and souls of workers.

By the twentieth century, the worlds of war and work were almost completely mechanized. To be sure, they depended on people to perform essential functions; but regimentation and discipline compelled men and women to accommodate the machine. The problem for American men was acute. To be successful and powerful in the industrial age, men had to become machines.[68] However, as the late nineteenth-century stories of mechanical women suggested, the same requirement did not constrain women whose bodies, spiritualism, and emotions seemed far less easily mechanizable than the behaviors of men. At the beginning of the nineteenth century, women's supposed lack of independence and reason appeared to make them more mechanical than men; by the beginning of the twentieth century, though, their ideological separation from the competitive world of industrial capitalism made them seem more authentically human.

Humanized Machines versus Mechanized Men

In the final few decades of the nineteenth century, the emergence of modern science, the rise of affluence and leisure, the development of powered machinery, and the growth of large institutions challenged what it meant to be an individual and a citizen. Enchanted and repulsed by the increasing materialism of their culture, Americans debated how to preserve individualism. In the process, they turned to two symbols, the mechanized man and the humanized machine. Men turned into machines, mechanized men stressed the regimentation, specialization, and standardization of modern life. Devices made into the shape of men or women, humanized machines stressed the growing urge to replace people with more efficient machines. By imagining such symbols, writers tried to both comprehend and construct a new understanding of the relationship between the individual and the material world. It was fitting, then, that the most important turn-of-the-century evocations of each symbol appeared in a series dedicated to celebrating the wonders of material abundance: L. Frank Baum's Oz books.

From the publication of *The Wonderful Wizard of Oz* in 1900 to his posthumous *Glinda of Oz* in 1920, Baum was one of the country's most popular authors. A onetime window dresser who adored using mechanical mannequins in displays, Baum was attuned to the imaginative longings of a commercializing America in ways, that, say, Senarens was not. The result was a multimedia Oz empire that included novels, musicals, films, and merchandise. Set in an often-amusing fantasy world, the Oz stories offered readers the chance to escape the drab realities of contemporary life for the wonders of a land of abundance. The stories, however, also offered readers a set of modern fairy tales for an industrial and commercial republic that worshipped—like the New Thought movement, of which Baum was a part—the power of the immaterial mind to transform self and society.[69] The Land of Oz lacked steam, oil, or electric-powered machines, but it did have magic—which, in a culture where Thomas Edison earned the nickname of "Wizard," often served as a metaphor for the new sources of energy.[70] Within the Land of Oz, readers found a visual ode to the wonders of industrial civilization, the unique powers of the human mind, and an argument for the compatibility of abundance, desire, and individualism.

Baum's most famous metal man was the Tin Woodman. Baum endeavored to leave the "nightmares" out of his fairy tales, but the Tin Woodman's origins were horrific. Born fully human, the Tin Woodman fell in love with a munchkin girl who worked for the Wicked Witch of the East. Afraid of losing her servant, the Witch enchanted his ax to slip at each swing. Each chop of the ax severed another part of the Woodman's body: first limbs, then torso, and finally head. Fortunately, a nearby tinsmith replaced the missing parts with ones made of metal. The new body initially gave the Woodman confidence because he could now withstand the Witch's power. But a sudden rainstorm exposed him to another vulnerability: rain. As he stood rusted and still for a year, he decided that because he lacked a heart he could not love; the Wicked Witch had destroyed his soul by turning him into a machine.[71] In observing his quest to restore his capacity to love, though, the reader learns that the Woodman is the most sentimental of the characters. He cries out of joy as well as at the death of a beetle; his tears are so frequent and profuse that they threaten to rust his hinges. At the quest's end, he regains his feelings with only a silk and sawdust heart and words of reassurance from a fake wizard. The Woodman never loses his capacity for love; he just thinks too hard, illogically and materialistically. In believing that his capacity for love was bound to his physical body rather than the power of his immaterial mind, the mechanized man fails to understand the nature of his soul.[72]

In a later adventure, the Woodman meets another man made of tin, a soldier named Captain Fyter, who has exactly the same origin story, except with a sword.[73] Like the Woodman, the soldier retains his capacity for sentiment and even continues to pursue the same munchkin girl with whom the Woodman originally had fallen in love. Though their bodies have been hacked to pieces by an empowered tool, neither man has become an automaton. In a country debating the effects of military and industrial regimentation, Baum showed that the physical price people paid for abundance was not too great because of the overwhelming power of the human mind. Though workers and soldiers might be butchered by empowered tools, they could never lose their essential humanity. The problem of individuality caused by

Fig. 3.1. Illustration from the original 1918 edition of *The Tin Woodman of Oz*, by L. Frank Baum, in which the Tin Woodman meets his soldier equivalent, the Tin Soldier. Reflecting the way that workers and soldiers lost autonomy to empowered machinery controlled by engineers during that period, these two characters are the product of accidents in which their tools attacked them.

industrial abundance, Baum suggested, was illusory because people were more than machinery. Vitalism justified the physical suffering of modern life.

Baum's novels also include a humanized machine: an automaton named Tik-Tok, introduced in *Ozma of Oz*, which begins on a steamship as Dorothy and Uncle Henry travel to Australia, where Henry can enjoy a rest cure. Henry, the story explains, is "not very well, because he had been working so hard on his Kansas farm that his health had given way and left him weak and nervous." During a storm, Dorothy ventures on deck and is washed overboard. She washes up in the Land of Ev, a neighbor of Oz. Hungry, she and the talking chicken that accompanies her find a tree that grows lunch boxes filled with a carefully wrapped "ham sandwich, a piece of sponge cake, a pickle, a slice of new cheese and an apple." The contrast with the life of Uncle Henry could not have been clearer. In Kansas, Henry destroyed himself growing food; in the Land of Ev, an entire meal grows conveniently on trees.[74]

After lunch Dorothy and the hen find a door in a mountain; inside stands a mechanical man labeled Tik-Tok. Powered by a windup spring rather than steam or electricity, Tik-Tok is a nostalgic automaton more appropriate for an agrarian America. Yet, through magic, it can perform actions that no automaton could. The card left with Tik-Tok by its creators, Smith and Tinker, identifies it as "Patent Double-Action, Extra Responsive, Thought-Creating, Perfect Talking MECHANICAL MAN . . . Thinks, Speaks, Acts, and Does Everything but Live." Like other writers of mechanical men stories, Baum never claimed that Tik-Tok is human. Upon seeing the device, Dorothy compares it to her tin friend and exclaims, "This copper man . . . is not alive at all." As Tik-Tok later tells the Woodman, "I am only a ma-chine, and can-not feel sor-row or joy, no mat-ter what hap-pens." Tik-Tok, Baum indicated, is not a mechanical monster. Smith and Tinker did not intend to create life; they intended to create a machine to assist human life. As Tik-Tok tells Dorothy, "From this time forth I am your o-be-di-ent servant. What-ev-er you com-mand, that I will do will-ing-ly—if you keep me wound up."[75] In Baum's fairy-tale America, Tik-Tok is the mechanical slave who would free people from the burdens of daily life. It would not threaten a child, as stories of mechanical women suggested; instead it would serve and protect Dorothy so she could enjoy the abundance around her.

Collectively, Baum's metal men signified a long-term transformation in what distinguished people from machines. Before the Civil War, American writers emphasized that white, middle-to-upper-class men, unlike machines, possessed independent, conscious, reasoning minds. With Baum's metal men, that distinction goes missing. Tik-Tok can think by way of the simple winding of clockwork. The Tin Woodman longs not for a brain—which he, like the Scarecrow, lacks—but for a heart. Certainly turn-of-the-century writers still denied that a machine could think; people did not have to assert that Ajeeb was a fake because everyone assumed it was. However, in most of the iterations of the robot in the late nineteenth and early twentieth centuries, people seem less certain that reason was the unique province of people. Brewerton's military automatons could still think. Bierce's scientist struggled to defeat his chess-playing automaton. With its fluid brain, "Mr. Corndropper's Hired Man" was as intelligent as any workman, and men believed that "Ely's Automatic Housemaid" was apparently too intelligent to duplicate the roles of women. The "Thinking Machine" was so precise and logical in his reasoning that he was practically a machine. The contrast between Baum's men of tin and his man of bronze was emblematic of a far larger cultural shift: what increasingly separated human from machine, spirit from matter, and even—for at least some elites—poor from rich was not brain but heart, not quantifiable reason but unquantifiable love, not the world of work but the world of fun. That transformation would have profound implications as twentieth-century Americans confronted both the emergence of a mass society dedicated to leisure and consumption and the rising cultural authority of people long denied access to the full rights of citizenship.

PART 2

Masters and Slaves, 1910–1945

One day, the science fiction writer and psychologist David H. Keller envisioned, robots will play football while "small, studious-looking" men control them. Spectators will watch on television in their homes rather than in a stadium with thousands of cheering fans. Companies will plot to use robots to replace inefficient workers while police robots beat the rebellion out of the displaced humans. One man, however, will stand against such machines: Ed Ball, a multimillionaire ex-football player and "misogynist" who spent his early life in "places where the white man was almost unknown." Ball, Keller imagined, will return and, horrified at the decline of individualism and community, restore power to powerful men. He will invest millions in companies building robots and then use the media to threaten virtually all workers with technological unemployment. As the ire of those workers grows, Ball will pay an inventor to develop a machine that can remotely shut off power to all robots while he trains men to play football. When the device is ready, he will stage a match between the robots and his men. After his well-trained team proves that men can triumph over machines, he will announce an ultimatum to the country's businessmen: stop replacing workers with robots or else a board of humanitarian men—employed by Ball—will push the button to destroy their robots.[1]

Published in June 1929, Keller's "The Threat of the Robot" captured a conservative man's anxieties about the growing prominence of machinery and selfish consumerism in American life after World War I, the Russian Revolution, and the ascendance of a mass consumer economy built on the worship of automatic machinery. Appearing five months before the stock market crash that formally began the Great Depression, the story presciently addressed one of the most talked-about causes of unemployment: the replacement of workers with machines.[2] Yet the story's most pointed critiques emphasized more intimate anxieties. "Gentlemen . . . ," Ball excoriates his fellow businessmen, "I little realized the changes in human life that were the direct result of your commercializing the inventions of science. . . . You had made possible a vast extension of the use of the radio, television and the wireless control of machinery, which you dressed in the form of men and women and called robots." By building such devices, he monologues, "you threatened the very life of American labor. You destroyed the best there was in sport, took away the pleasure of attending amusement in mass, and, by placing all entertainments in the home, you turned mankind into a selfish, introverted, anti-social animal who cared for little save his own entertainment. You did this to make money—there was little of the altruistic, the love of humanity, in your efforts to popularize these scientific discoveries."[3] A love of money and an indifference to the effects of technology, he claims, have made people narcissistic and too concerned with selfish desires to feel sentiment, connect to others, and build a loving community.

This was a radical critique of the unrestrained pursuit of wealth rooted in a conservatism that drew from early American ideologies of work, individualism, and community. Born in 1880 in Philadelphia, Keller studied medicine at the University of Pennsylvania before entering private practice and enlisting in the medical corps during the Great War. Remaining in the United States, he specialized in neuropsychiatry and dealt with the psychological trauma that afflicted soldiers long after the armistice. Keller later accepted a position as the chief medical officer at a state hospital where he studied the effects of shell shock on veterans. Haunted by the war, Keller retreated into the past. In 1922, he published his family's history of improvement through hard work, self-denial, and other supposed traits of nineteenth-century individuals. Though predominantly set in antebellum America, the book could not escape World War I. Keller opened the book with a photograph of himself in his uniform and concluded it with a list of family who had joined the military, including fifteen who had also served in the recent conflict.[4]

By 1928, Keller was exploring a career as a writer that would allow him to escape into the future rather than the past, via science fiction. The genre dated at least to the nineteenth-century writings of Mary Shelley and the later adventure stories of Jules Verne and H. G. Wells but remained undefined until the 1920s, when the publisher Hugo Gernsback introduced the first pulp magazine devoted to what he named "scientifiction:" *Amazing Stories*. Reaching a circulation of over 100,000, *Amazing Stories* was an almost immediate success. Because he intended to use the magazine to educate a young audience, Gernsback insisted on the scientific plausibility of all stories and preferred that they show the amazing, rather than the horrifying, potential of new technologies. Gernsback initially reprinted older stories by such writers as Verne and Edgar Allan Poe; however, he soon began to print original material submitted to him by readers such as Keller, whose medical training fit the publication's mission.[5]

Yet Keller's fiction was deeply pessimistic. In his first story, he worried that automobiles might physically and morally weaken American men and women. In another, he fretted about how technologies—especially a blackface "mammy" robot—liberated women from housework and motherhood and destroyed human relationships; in still another he worried about the possibility of using hormones (a recent scientific discovery) to create an artificial woman.[6] As in "The Threat of the Robot," Keller's deepest fears focused on how science and technology were disrupting manhood and womanhood and turning everyone into self-absorbed, overcivilized, pleasure-seeking machines. The problem he identified was not predominantly one of technology but of culture, of values that seemed to privilege individual gain over family and community. Yet, the solution Keller proposed in "The Threat of the Robot" was revealing: Ball effectively uses his radio signal to establish a dictatorship, a solution that many Americans in the 1930s dreamt might resolve the problems of the Great Depression.[7]

In his extreme and blatant misogyny, hostility to modernity, and endorsement of dictatorship, Keller was atypical; but his response to the machine age was only a more hyperbolic version of common anxieties that resulted from a confluence of fears about gender, race, and class relations in a culture that increasingly located leisure, not work, at the center of human identity and purpose. Between World War I and World War II, three revolutionary developments coalesced. First, by the 1920s, American culture appeared overwhelming dedicated to leisure and pleasure. Certainly, large swaths of the population continued to work hard; but as compensation for the decline of creativity and independence in and control over work, people looked to leisure to give them a sense of identity and purpose. Second, culture seemed more democratic, more welcoming to the disenfranchised and skeptical of the power of elite, white, and old men. With white women free to vote and participate in popular and consumer culture and African Americans, immigrants, and young people creating the bulk of the most popular trends, the cultural authority of older white men appeared to wane. Finally, the rapid advance of machinery appeared to make technology—a recently popularized word—the driving source of change.[8] In such a world, the people most accustomed to power and authority used stories of out-of-control robots to address their fears that they lacked the power they once wielded in society and the purpose that came with it. Once masters, they now imagined that they had become slaves. As Ball argued, "Mankind must never again be threatened by the crushing weight of machinery. The robots must be the servant and not the master."[9]

The question of who would be the slaves of tomorrow and who would be the masters reverberated throughout the period as writers, filmmakers, musicians, artists, and engineers pondered how to adjust male identity and purpose to a machine age that seemed to make war unheroic, labor either degrading or nonexistent, amusement the center of human life, and other people ignore them. To the businessmen, engineers, and commentators who celebrated the automatic production of goods, the answer to that question was obvious: machines would be the slaves and men the masters. But to workers whose jobs were threatened by machinery and to men such as Keller who worried about cultural and moral decay and the potential decline of the white race, the answer was more troubling. If they did not, as the future Senator Ralph E. Flanders suggested at the beginning of the Great

Depression, learn to "tame their machines," neither the individual nor the nation would survive.[10] As the creators of American intellectual and popular culture debated the causes and solutions to the decline of white men, they turned to a new symbol potentially far more potent than either the automaton or the mechanical man: the robot.

4

Symbolizing the Machine Age

On October 9, 1922, the New York Theater Guild performed a play unlike any previously seen on American stages. The play's author, a Czechoslovakian named Karel Čapek, was virtually unknown in the United States. The play's title, *R.U.R.*, was indecipherable; the explanation of the acronym, *Rossum's Universal Robots*, did nothing to illuminate the subject. Unlike in Eastern Europe, where the term was a Slavic variant of a word that referred to serfs, most Americans had no idea what a "robot" was prior to that night; by the end of the play, audience members had at least a vague understanding of the term and, within a decade, so did most of their compatriots. While only a small number of people saw *R.U.R.*, the word *robot* percolated through US culture during the decade. Spread by fiction writers, reporters, businesses, filmmakers, and intellectuals, the robot had fully entered the lexicon by the start of the Great Depression as the quintessential symbol of the machine age—and it never left.[1]

R.U.R. and its robots began as the idea of a cosmopolitan Czechoslovakian attempting to comprehend the Great War and the Russian Revolution. Born in a small Bohemian town in 1890, Čapek trained as a philosopher and critic but soon began writing fiction, often with the help of his brother Josef, a cubist painter. Before the war, the brothers lived in the modernist meccas of Paris and Vienna, where they encountered the ideas of men and women like themselves attempting to adjust art to the machine. During the war, Čapek avoided military service due to a spinal injury and instead attended lectures, including some by disciples of John Dewey. For Čapek, the crisis started by World War I seemed like the final stage of the same disease: the decay of the soul in the name of mechanical efficiency. The combination of the war and the Russian Revolution seemed to herald the fate of all societies if the disease were not cured. Eschewing both capitalism and socialism in favor of pragmatism and the ascetic spirituality of Leo Tolstoy, Čapek was skeptical of utopian goals, whether promoted by religious figures, business leaders, scientists, or radicals. He looked to human tolerance and labor to improve the world, not technology, luxury, or revolution.[2]

In *R.U.R.*, the titular company mass-produces and sells artificial, biological laborers that their inventors engineered for maximum efficiency. Echoing modern optimism for science and invention, Rossum's scientists, managers, and engineers proclaim that the robot heralds a new post-scarcity world in which, as the company manager, Harry Domin, argues, "there'll be no more poverty. . . . Everything will be done by living machines. People will do only what they enjoy. They will only live to perfect themselves."[3] Though predictions of abundance and self-improvement come true, they yield dangerous consequences. After robots take over the economy, the birth rate drops so precipitously that virtually no children are born. Governments replace soldiers with robots; wars, now with minimal human costs, grow common. Yet, as robots become essential for production and armed with weapons, they develop consciousness once Helena Glory, a social reformer determined to give them rights, has a scientist manufacture a soul for them. This proves disastrous when the robots—now conscious of their enslavement—rebel. The ensuing war unfolds offstage but concludes on Rossum's island headquarters with its human inhabitants surrounded by a mob of faceless robots. The rebellious slaves kill all but the one resident with whom they identify, a manual laborer named Alquist who preaches that the toil of labor forms the basis of the soul. But before the robots' victory, Helena, feeling guilty for the consequences of her antislavery rhetoric, burns the only copy of their formula. As the epilogue begins, the robots, lacking sexual

Fig. 4.1. Photograph of the original Theater Guild production of *R.U.R.* included in *Labor Age*'s positive review. Note both the appearance of the robots in the typical worker uniform and the way that the chosen caption—a quote from Domin—frames the story as about a manager's conception of robots: "A working machine must not play the piano, must not feel happy, must not do a whole lot of things."

organs, appear doomed to extinction. While Alquist vainly attempts to recreate the formula, Providence steps in. Two robots miraculously learn to love and overcome their biological deficiency to become the new Adam and Eve. Though humanity is extinct, the new children of God inherit the Earth.

Part satire, part monster, part revolutionary, and part slave, Čapek's robot was an industrial horror appropriate for the aftermath of World War I and the Russian Revolution. But not all Americans accepted the play's pessimism. Far more distant than their European counterparts from the previous decade's slaughter, Americans of the 1920s appropriated the robot to fight their own battles and tell their own stories. Drawing on ambiguities from the play, some Americans saw the robot as a symbol for the inherently degraded nature of the working class; others more sympathetic to workers saw the robot as a satire of the way that managers and engineers stereotyped and objectified unskilled laborers; still others rejected the robot as a symbol of workers and adopted it as a symbol of technology. As commentators debated the robot's meaning, they transformed it into the central symbol of the dangers and possibilities of an era self-conscious about the growing might of automatic machinery, and both frightened and enchanted by the emergence of a society dedicated to mass production and leisure.[4]

Americans already had symbols that could have filled such a role. Numerous writers invoked Frankenstein's monster to address the seemingly suicidal nature of modern life, especially when describing the Great War and its aftermath. Indeed, the organic nature of Čapek's robots seemed to suggest that the playwright had merely updated Shelley's tale for an age of mass production and consumption. Yet, the robot was not solely a Frankenstein monster. American adaptors, critics, commentators, fiction writers, and even corporations preferred to imagine and depict it as a metal machine closer to the mechanical men of late nineteenth- and early twentieth-century culture than an organic being. But even that antecedent seemed inadequate. For many, the robot was not a mechanical man but a mechanized one, a person turned into a mindless, emotionless machine by the degrading effects of assembly line labor. Both meanings—the robot as machine and the robot as worker—tapped into long-standing debates previously synthesized by the automaton. But though some used the terms interchangeably, the robot was not exactly an automaton either. Real automatons were orderly and predictable in ways that an emerging Jazz Age celebrating spontaneity and accepting of uncertainty was not. Powered by an internal source and occasionally displaying intelligence, robots could move seemingly at will. As halting and jerking as their movements could be, their random rhythms conveyed autonomy much better than the regular loops of automatons.[5]

The robot's debut during the waning of Victorianism and the waxing of modernism enabled critics from multiple perspectives to embrace it as the central symbol of the problems and possibilities of a new era.[6] Though the play is dystopian, the robot itself could also be utopian; it could stereotype workers or satirize those stereotypes; it could critique mass production for destroying workers' humanity or support it for bringing them leisure. Nearly human, the robot could address class, race, or both; it could be gendered and sexualized in ways that no other machine could. It could be dumb or intelligent; emotionless or driven by emotion. It could be evoked in popular culture or by intellectuals writing in scientific journals and speaking at conferences. Though people disagreed on its precise meanings, the robot quickly became an icon that most Americans could understand. Amid great anxiety about destruction and uncertainty and hope for leisure and the cultural revival it might inspire, the robot synthesized the divergent strands of a machine age culture that people desperately longed to comprehend.[7]

The Uncanny Machine Age

If any man inspired consciousness of the machine age, it was Thomas Edison, the celebrated inventor of the electric light bulb and power generation, recorded sound, and motion pictures.[8] As he sat with a reporter in 1910 to tell of his plans for an "automatic clerkless shop," Edison articulated the ethos that had driven his adult life and that would soon echo through the obsession with robots: "This is the machine age. . . . Wherever man's power or horsepower can be eliminated, speed, accuracy, and economy are the result. . . . Eventually, nearly everything in this world will be got down to a mechanical basis. That will mean that we can live easier and cheaper." His new shop would fulfill that vision, for it would contain "no shopkeepers, no clerks, no boy to wrap-up packages." Without human workers, Edison prophecized, "there will be no wasting of time in talk, no pricing of articles nor any sampling. Shopping will be an exact, speedy, and business-like proposition." Such a shop could reduce prices for every good and enable the poor to enjoy a purchasing power equivalent to his, a hugely successful businessman. The problem of class divisions would be solved by removing merchants and making the efficient machine the only element between producer and consumer. This "Samaritan Market," he concluded, was a "sure thing" to help the poor because it would "help a man to help himself. He puts his nickel in the slot and has no one to thank." Shorn of their dependence upon unnecessary people, Edison argued, people would finally be free to help themselves.[9]

Stated baldly, Edison's ethos of efficiency sounded ridiculous, but it echoed real and imaginative efforts to rationalize consumption. In the late nineteenth century, farmers' movements sought to remove the middleman between producer and consumer who charged producers more for factory-made goods and gave them less than they could receive for the products they sold.[10] In Edward Bellamy's contemporaneous *Looking Backward*, future Americans shopped at stores that replaced the shopkeeper with machinery.[11] Real life mimicked such visions. Early twentieth-century Philadelphians and New Yorkers could purchase food at automats, a German invention that relied on machinery rather than cooks and waiters to distribute food and drink. By the 1920s, an emerging vending machine industry preached of its ability to offer "automatic retailing" in which automatons would offer pleasure at the drop of a coin.[12] Amid mass advertising and opulent department stores, a simple device that could distribute goods without "waste" was appealing to those who continued to believe in the simplicity and frugality of republican life.

Though dreams of automatic retailing persisted, such stores never dominated the landscape. Automats remained confined to the largest of cities, while the vending machine industry concentrated on small items and mechanical arcade games.[13] Despite numerous attempts to create automatic stores, the luxury and opulence of conventional retailing remained attractive. Indeed, America in the machine age seemed more dominated than ever by advertising, luxury, excess, and the wasting of time in frivolous pursuits.[14] Edison's specific vision failed, but the tension between his ethos of mechanical efficiency and the excesses of mass consumption and leisure would inspire much of the debate over robots and the cultural transformations they might cause.

In the machine age, practically every realm of human life seemed mechanized in the pursuit of efficiency. Since the late nineteenth century, work in mines, on farms, in factories, and in offices had centralized and accelerated. Despite labor union efforts to provide workers with power on the shop floor, the pursuit of lower costs and larger markets encouraged the electrification of machinery and new organization techniques—including the assembly line—that rapidly increased productivity.[15] The home, once seen as a refuge from the mechanical world of work, was transformed as electrification enabled middle-class families to add washing machines, refrigerators, vacuum cleaners, electric irons, toasters, radios, and alarm clocks. The number of Americans who owned such machines was limited—in 1930, only an estimated 25 percent owned a washing machine, 30 percent a vacuum cleaner, and 40 percent a radio—but an advertising industry that grew from an estimated $682 million in sales in 1914 to nearly $3 billion in 1929 extended technology's influence with messages of efficiency and the liberation from labor.[16]

Efforts to increase productivity combined with workers' movements to reduce the number of hours of work per week while raising wages. Average work hours dropped from 48.3 hours per week in 1900 to 44.8 in 1915 in the unionized building trades, and from 59 hours to 55 in manufacturing. Across the entire economy, work hours fell from about 60 hours per week in 1900 to below 50 by 1920. By the end of the decade the eight-hour workday—long desired by workers and their unions—had become standard, and many salaried workers gained annual vacation time. Despite the lower hours, manufacturing employees saw their wages rise from an average of $435 a year to $568 across the same period. Much as advocates had predicted, machinery had ended the age-old choice between increased leisure or increased wages. In the machine age, men and women could have both.[17]

The man most associated with the new ways of organizing work was Henry Ford. In 1913, Ford opened the first full-fledged assembly line at his Highland Park, Michigan, plant to produce the Model T. Synthesizing elements of production found in other industries, Ford and his engineers connected a highly subdivided labor force, interchangeable parts, electrical drive shafts to ensure a continuous supply of power, the sequential ordering of tasks, and the ultimate symbol of the line: the belts and slides that moved tasks to the worker. The line could produce an automobile in an astonishing 93 minutes (down from twelve hours when the cars were assembled by traditional methods in 1909) that could then be sold for as low as $300 in the 1920s—cheap enough that Ford's workers could purchase one. Recognizing the potential of mass consumption far better than earlier businessmen, Ford famously paid his workers $5 a day, double their prevailing wage before the advent of the assembly line. But efficiency also meant product standardization; until 1925, consumers could purchase one model in only one color, dark blue, or, later, black. By standardizing both process and product while paying workers more, Ford offered a simple solution to the problem of degrading industrial labor: compensate workers through leisure and consumption.[18]

Leisure became commercialized and mechanized as well. In cities, the spread of dance halls, movie theaters, professional sports teams, and amusement parks accompanied the growing centralization, standardization, and regimentation of labor as workers welcomed the liberating potential of amusement. In riding the first roller coasters at Coney Island or dancing the cakewalk, grizzly bear, or fox-trot to the syncopated rhythms of ragtime and improvisation of jazz, young Americans' bodies mimicked the acceleration they experienced in the workplace. The mechanization and individualization of transport with first bicycles and later automobiles gave men and women a sense of empowerment, speed, and pleasure. In attending films that intercut time and space and shifted perspectives, they saw a reality of motion projected on a screen in ways that seemed vaguely unreal.[19]

The unreal nature of much of the modern amusements encouraged critics to draw stronger lines between human and machine that centered on authenticity. Responding to the emergence of Edison's phonograph, the "March King" John Philip Sousa wrote in 1906 that "talking and playing machines . . . reduce the expression of music to a mathematical system of megaphones, wheels, cogs, disks, cylinders, and all manner of revolving things, which are as like real art as the marble statue of Eve is like her beautiful, living, breathing daughters." Contrasting "the mathematical and mechanical" with "the emotional and the soulful," the band leader suggested that no machine could play the nationalistic and romantic music that he favored with the depth found in the soul. Eighteenth- and early nineteenth-century listeners had frequently praised the precision of automaton music performances, but precision was not enough for Sousa; music needed soul.[20]

The need for soul in an age of efficiency further encouraged the growth of mass amusements and a celebration of leisure as the defining feature of identity.[21] In the early twentieth century, reformers placed play at the center of human life and identity, especially for children. Far more than the subdivided and deskilled workplace, play allowed children the freedom to develop themselves while socializing them for an urban society. Yet, the kinds of play that reformers recommended differed substantially from commercialized amusements. Advocating playgrounds, sports such as basketball that could teach teamwork, hobbies, nature trips, and traditional dances, reformers sought to maintain Victorian links between leisure and self-improvement. By the 1920s, interest in children's leisure had extended into the adult realm as corporations such as the National Cash Register Company provided workers with recreational outlets as compensation for the monotony of work, and as a means of preventing unionization through the building of a community in the seemingly more equitable pursuit of fun.[22]

Yet, to critics, the growth of commercialized amusements threatened the possibility of using leisure to improve self and society. Throughout the era, intellectuals and social critics spoke of the "leisure problem," the sense that people devoted too much time to mindless, individualistic pursuits rather than more productive forms of leisure that could reinvigorate the self and community.[23] This was part of the anxiety that shaped David Keller's "'The Threat of the Robot," as he was horrified that people might watch football alone in their homes rather than in a stadium with their fellow citizens. Magnifying this anxiety was that many of the most popular trends—including dances, jazz, and film—originated in the cultural expressions of African Americans and immigrants and were frequently most enjoyed by the young, regardless of race, class, or gender. To be sure, older, white, and male cultural authorities exercised general control and censorship, but in the dance halls, cinemas, and amusement parks, they could not contain everything.[24]

The growing cultural authority of disadvantaged groups was part of a much larger blurring of the lines and categories that had dominated Victorian life. By the twentieth century, science had shattered predictable laws of the universe and suggested a world that was less deterministic and purposeful, less Sousa-esque and more improvisational like jazz. In biology, Darwinism merged with the science of hereditary to transform evolution from a study of the development of higher forms of life into a study of inheritance, a transformation that gave scientific credence to arguments for biological and racial determinism. In physics, quantum mechanics and Einstein's "special theory of relativity" meant the world was both more complex, random, and indecipherable than previous generations had assumed. In mathematics, philosopher and scientist Charles Sanders Peirce advocated replacing the search for absolute proof with statistical probabilities that acknowledged a degree of uncertainty, a transformation that echoed through the philosophy of pragmatism he shared with William James and John Dewey that emphasized adjusting idealisms, preconceptions, and beliefs to real-world results. By the time the German physicist Werner Heisenberg codified his uncertainty principle in 1927, uncertainty had been a dominant element of intellectual life for three decades.[25]

The growth of uncertainty and sharper distinctions between the real and the artificial inspired feelings of the uncanny, especially from automatons. Coined in the nineteenth century, the term *uncanny* was theorized in the early twentieth by psychologists Ernst Jentsch and Sigmund Freud who each associated it with German author E. T. A. Hoffmann's 1816 short story "The Sand-man," in which a man falls in love with a harpsichord-playing, singing-and-dancing female automaton only to be driven mad when he realizes her mechanical nature. Locating its origins in the tension between the human need for "intellectual mastery" over the environment, Jentsch described the uncanny as "a dark feeling of uncertainty" that arises in moments of "doubt as to whether an apparently living being is animate and, conversely, doubt as to whether a lifeless object may not in face be animate."[26] Focusing on the automaton, he observed that such a "feeling of unease" grew worse as the mechanism improved and the device appeared "truer to nature."[27] Freud extended this analysis by arguing that the "uncanny effect is often and easily produced by the effacing the distinction between imagination and reality, such as when something that we have hitherto regarded as imaginary appears before us in reality, or when a symbol takes over the full functions and significance of the thing it symbolized."[28] Despite their differences, both Freud and Jentsch associated the uncanny with a tension between the scientific ethos to master nature and the persistence of the unexplainable—particularly, the animation of material objects that should not, according to reason, move. Unifying anxieties about the uncertain and the unreal, the uncanny captured a sense that modernity was turning the fake real and the real fake.[29]

Americans rarely associated uncanny feelings with automatons prior to the late nineteenth century. In Julian Hawthorne's 1868 story of a dancing automaton, the audience responded enthusiastically, not horrifically, to its lifelike nature. Yet, early twentieth-century Americans frequently invoked the uncanny to describe the feelings inspired by both the automaton and people who acted like automatons. One reporter noted that a shoe-salesman automaton's "eyes are remarkably strong, and looking intently at his expressionless orbs gives one an uncanny feeling."[30] When storefront dress models threatened to strike, newspapers reported on French automatons that might take their jobs. "Naturally it is uncanny to see these figures move or even to see them lounging in such extraordinary life-like positions," one article stated. "Everybody who sees them has the same sensation at first—and it is a sensation with a twinge or something startling and awkward in it." Yet, they continued, the uncanny feelings created by the appearance of life in matter were not permanent: "One gets the joke, and especially one gets the final comfortableness of having wax, rubber and machinery do the work."[31] Uncanny feelings could arise around people who seemed inert. In the syndicated "Confessions of a Bride" column from World War I, a wife described the situation with her husband who had recently returned from the war in France. "He is no nearer recognizing me as his wife, or any of the family associates of the months before the war," she wrote, "than he was the first day he returned from France. It is as uncanny to his physicians as to the rest of us. Like a rational, flesh-and-blood automaton he moves among us. . . . Sentiment is dead within him."[32] In America at least, the uncanny was a feeling tied directly to the way that modern life—including both commercialism and war—seemed to empower the machine and destroy the individual soul.

The fear that the machine age had destroyed the soul led American writers to call for a reinvigoration of the nation's culture. Drawing on nineteenth-century intellectual traditions, critics such as Randolph Bourne, Van Wyck Brooks, Waldo Frank, and, later, Lewis Mumford worried that industrial life destroyed both the individual and community. As Bourne noted in 1914, "The clerk dulled and depressed by the long day, and the factory worker—his brain a-whirl with the roar of the machines—must seek elation and the climax which the work should have given them, in the crude and exciting pleasures of the street and the dance and the show."[33] Frank made the connection even clearer. The machine, he argued, was "the god of the American world," while industrialization had destroyed "the American soul and made it poor"; this "sagging, uncreative world," he concluded, "bore witness to the fate of human spirit in a civilization which could persist alone by the denial of experience, by the mechanization of Desire."[34] Only cultural reinvigoration through a synthesis of the spiritual and material, such intellectuals argued, could reinvigorate America's "beloved community."

When *R.U.R.* premiered, the tensions between empowered machinery and the seemingly disempowered individual had been growing in importance for decades, but it was World War I and the Russian Revolution that underlined their potential danger. As commentators searched for meaning, they frequently turned to Mary Shelley's *Frankenstein* as a symbol of the out-of-control nature of modernity.[35] But the preindustrial *Frankenstein* was not quite the right symbol. It was materialistic but not consumerist; ugly and monstrous, it was not quite uncanny like the dolls and automatons Jentsch and Freud discussed; an individual rather than part of a mass crowd, it threatened its master, not all of humanity. In *R.U.R.*, audiences found their modern *Frankenstein*. As John Corbin caustically wrote in his negative *New York Times* review, "Having been worn threadbare in the country of its origin, the Frankenstein metaphor has reached Czechoslovakia and has there achieved a social consciousness—has become, in fact, class conscious."[36] Shelley's tale had been an intimate affair between father and son, but Čapek offered a grand melodrama with a conclusion perfect for a world confronting not just war and revolution but an entire culture transformed by mass production and consumption: mass extinction.

Class, Race, and the American Robots

The year 1922 was an inauspicious one for a radical Czechoslovakian play to open in America. Suspicion of foreigners had only increased after the Great War arising from a series of bombings that suggested immigrants were intent on transporting the Russian Revolution to the United States. In the South and the Midwest, the revived Ku Klux Klan had won supporters by proclaiming "one hundred percent Americanism" and attacking African Americans, Catholics, Jews, and other "un-American" groups. In the 1924 Johnson-Reed Act, the federal government implemented a system of quotas to limit immigration of Eastern Europeans, among other groups. Though the nativist heat would dissipate, its energy empowered a broader conservative ascension that repudiated earlier reforms and decimated labor unions and other groups that threatened the interests of capital.[37]

Despite, or perhaps because of, the political environment, *R.U.R.* was a hit that spread from New York to other cities, including Los Angeles, Chicago, and Berkeley. The *New York Sun* called it a "magnificent melodrama, superbly portrayed and directed." The *Oakland Tribune* dubbed it "the best melodrama to be presented on any stage this decade" and "the strongest piece of writing that has graced the American stage in ages." Favorable reviews appeared in the literary magazines *Colliers Weekly* and *Harper's Bazaar*. Worker organizations such as the Industrial Workers of the World (IWW) praised the play. The League for Industrial Democracy's *Labor Age* claimed that "for anyone interested in the questions of the present day, the play is thoroughly worth while." American author and socialist Carl Sandburg called it "significant, important, teasing, quizzical, funny, terrible, paradoxical," and compared it to "the strongest plays of Henrik Ibsen."[38] In a syndicated review originally written for *Vanity Fair*, radical journalist Heywood Broun praised the play's combination of melodrama and ideas, even though he found the ending "self-consciously sweet."[39] Religious magazines—particularly those outside of evangelical Protestantism, such as the *Christian Century*, *Jewish Forum*, *Catholic World*, and *Universalist Leader*—echoed this enthusiasm. Just a year later, Doubleday published the play to similarly rave reviews. Within a year, the Goldwyn Company acquired the movie rights, though the play would not find its way to the silver screen.[40] Revived throughout the 1920s and 1930s and periodically thereafter, *R.U.R.* was one of the most successful radical plays of the era.

But what its robot symbolized was unclear. Physically and thematically, Čapek's robot resembles artificial humans such as Frankenstein's monster more than the mechanical men of American culture. Even so, *R.U.R.* merged the two types. The original inventor, "old Rossum," used chemistry to create an artificial human; but his engineer son, young Rossum, streamlined the design to make the robot more fit for work. "A man," Domin later quotes young Rossum, "is something that feels happy, plays the piano, likes going for a walk, and in fact, wants to do a whole lot of things that are really unnecessary. . . . When he wants . . . to weave or count." Removing such capabilities, according to the factory managers, turned the robot from an artificial person into a machine. "Robots are not people," Domin insists. "Mechanically they are more perfect than we are, they have an enormously developed intelligence, but they have no soul."[41] Unlike the inventor, his son and the later executives believe in a soul, but associating that soul with culture and leisure rather than work enables them to dismiss concerns about worker well-being. In the satire of the play, they, like Frederick Winslow Taylor, see workers as material objects to control rather than autonomous individuals capable of their own creativity. In its original production, the Theater Guild reinforced that viewpoint by costuming the actors as uncannily uniform men with identical haircuts, black pants, and gray shirts with a triangle-shaped logo.[42] While Domin and his coworkers might think the robot is a machine, the performance stressed the devices' similarities to human workers.

Most writers initially understood the robot a symbol of potentially revolutionary workers, but they disagreed on the nature of the problem and its solution. *Colliers* framed Charles W. Wood's review under the headline "Workmen Made to Order" because he spent much of it lamenting how "practical education" merely fits workers "scientifically to their jobs" rather than teaching the whole person.[43] In such a reading, the problem of rebellious workers was caused by their lack of culture, a deficiency that education, not workplace reforms, could cure. Others, however, saw a need for a more significant restructuring of society. In 1926, a manager at the Philadelphia Rapid Transit Company, which was experimenting with worker ownership of the company, pointed to a copy of *R.U.R.* and said, "This is what we are coming to—Robotism—if we continue to let men produce like machines, with no more incentive than to increase their speed and to get higher pay. . . . Make this vast army of workers the owners of the machinery of production and they will become more sturdy of character. Then we can rest assured that they will never do as Russians did."[44] America, the manager suggested, did not need to merely improve the education of workers; it needed to empower them by giving them control over the workplace.

The organization to see the greatest revolutionary potential in the play and its central symbol was the IWW, the era's most radical and inclusive of unions.[45] Writing in the organization's *Industrial Pioneer* magazine, Rosa Knuuti described the play as a "lurid and . . . ironical melodrama around the theme of the class struggle—the degradation of the present-day industrial worker into a veritable mechanical working machine."[46] Čapek, she judged, had "rendered a great service to mankind by so dramatically calling the attention of all who have eyes to see and brains to understand, to this aspect of the class struggle."[47] Later that year, another writer for the magazine identified the workers in Ford's plants as "robots," not because they were dehumanized, but because Ford wanted labor that was "Robotized," made more efficient through the removal of components not necessary for production.[48] The next year, the publication used the term to describe the "contempt" the press held "workers in" by printing ridiculous beauty and health advice as well as literary and philosophical "bunk."[49] In a later issue dominated by stories about technological unemployment, the publication reprinted a speech by its editor in front of a branch of the IWW. "Labor," he argued, "is regarded as a Robot, a mechanically adjusted creature, devoid of understanding and emotion, whose only mission in life is to produce wealth and die in warfare when necessary for those whose profit it is waged, namely, the capitalist class."[50] The robot's ideological power, IWW writers argued, came from its ability to satirize how employers and their supporters dehumanized and objectified workers.

But middle-class radical publications rejected the play for engaging in the same type of stereotyping that IWW writers thought it ridiculed. In the *Nation*, Ludwig Lewisohn praised the Theater Guild's production but found the play's logic to be a "complete absurdity" because "on the one hand we have the assumption that men can be reduced to the level of mere machines which, in the nature of things, would not revolt at all; on the other hand we are told that these helots will revolt against slavery, oppression, their own soulless estate, which at once reinvests them with all the passions, powers, and thoughts from which the triumphs of civilization . . . draw their origin."[51] The *Single Tax Review* similarly found the play "sophomoric" and asked, "Are we to understand that Mr. Čapek would have the R.U.R. represent the workingmen of the world rising in rebellion against their masters? If so why should he suggest that their victory will be speedily followed by their own extinction? That doesn't seem plausible." The only explanation the reviewer could imagine was that "even the progressive Theater Guild would not have stood" for a complete victory by the workers.[52] Unlike Knuuti, such critics found nothing humorous or ironic in the play. Taking it seriously, they saw the robot as a misguided symbol of the proletariat, not as an absurdist satire.

The problem with imagining the robot as a worker was that if viewers did not understand the joke, the play risked implying that nothing could fix the problems of industrial society. This was the crux of Corbin's critique of the play in the *New York Times*. Five days after his initial review, Corbin contrasted *R.U.R.* with the, in his view, superior anti-machine satire of New Zealand writer Samuel Butler's 1872 novel *Erewhon* that had become an increasing part of the conversation about machinery over the preceding decade. While Butler had critiqued the importance of the machine in human life, Corbin claimed that Čapek made "the great enemy of mankind . . . the human beings that civilization has doomed to an eternity of feeding it—the so-called proletariat." Citing the example of the Russian revolution as evidence, he argued that "the true enemy of civilization is not the machine, but the mechanized human being—dwarfed in intelligence, stunted in sympathy, swayed by the only idea one can ever derive from the seamy side of the industrial fabric, the idea of soulless mastery, sheer physical power." Though he assumed that Čapek was a socialist, Corbin called him "constructively a Nihilist" and questioned whether he really meant "to tell the Workers of the World that better men than they can be made out of salt, sand, and the white of an egg?"[53] Later, in response to reader criticisms that he dismissed the play too quickly, Corbin reiterated that the play's logic suggested that "all effort to improve the lives of factory workers" will lead to "the annihilation of civilization."[54] By constructing a monstrous icon that symbolized workers rather than machinery, Čapek implied that nothing except extinction could resolve the issues of industrialization—and Corbin was far too optimistic to accept that fate.

What Corbin grasped that few other reviewers appreciated was that in America, race was essential to understanding the robot. *R.U.R.* did not explicitly address race but as biologically distinct laborers, robots could easily encourage viewers to adopt a racial view of the creatures. This was especially true in a diverse country such as the US, where employers frequently segregated jobs by race and justified harsh treatments of workers with claims of their innate inferiority.[55] American industrial advocates had long associated labor they found degrading with those whom they deemed racially inferior, just as they had long applied machine metaphors to those same workers. One of the era's most frequent defenses of Fordism, even among liberals, was that only people biologically suited to mindless work would end up working on assembly lines. As one engineer wrote in *American Machinist*, "Most people do not want to 'express themselves,' and are much happier with somebody else taking the responsibility. The stolid look is a generic one. Such men are built stolid; they would be stolid on a farm."[56] After *R.U.R.*, people had a new term with which they could dehumanize entire groups of people based on a presumption of inferiority: robot. "If 'R.U.R.' is an indictment against the stupidity of our capitalistic class," a Broadway producer wrote, "it is equally an indictment against the crass ignorance of our working class."[57] Or, as both a critic and a supporter of industrialization later explained, the industrial laborer is a "natural robot."[58]

Corbin recognized the centrality of race to the robot in America by entitling his essay "The Revolt against Civilization," a reference to a recent book by the eugenicist, biological-determinist, and anti-immigration activist Lothrop Stoddard. Corbin, or his editor, chose that title because he thought that the robot "concretely symbolized and dramatically focused" Stoddard's arguments.[59] *The Revolt against Civilization* expanded claims Stoddard made in his more famous 1920 book *The Rising Tide of Color against White World-Supremacy*, which explained how Chinese and Japanese industrialization as well as nonwhite population growth would imperil the global supremacy of white nations. Stoddard also worried, however, that materialism had created a "profound *malaise*" that led, in a direct parallel to *R.U.R.*, to a declining birth rate "which affected nearly all white nations."[60] In *Revolt against Civilization*, he looked within industrial societies themselves, especially Russia, to explore how a biologically inferior underclass threatened the core of white civilization. Stoddard's arguments were deeply controversial, but they reflected a much larger racial anxiety felt by white Americans of the time, including, at least partially, Corbin.[61] In that context, the robot could easily merge stereotypes of race and class in a way that confirmed assumptions of the inherent biological inferiority of others. A potentially revolutionary icon sanctioned by some of the most radical writers in America, Čapek's robot could have deeply challenged the central ideology and class divisions of industrial capitalism. But it ultimately did not; instead, it foundered on the same shoal that had wrecked many American radical and labor movements: race.[62]

Culture and the Autonomous Machine

But the robot also faltered as a radical symbol because people quickly associated it with the amazing power of technology. Even before *R.U.R.*'s premier, American writers invoked the mechanical man to symbolize the autonomous power of machinery. In 1922, after observing Ford's assembly line, journalist Arthur Pound announced, "This is the century of the automatic machine. The social problem is to accommodate the use of automatic machinery to the well-being of the masses; our political problem is to avert class and state wars growing out of quarrels over the profits, powers, and privileges accruing through the production and marketing of goods." He then explained his chosen symbol and the title of his book: "Much of our modern heart-searching leads down to the Iron Man at the base of the industrial structure. He claims the twentieth century as his; the social and economic forces that he releases are those most likely to carry on into the future the reality of our day." Anthropomorphizing the automatic machine, Pound gave it agency, power, and manhood. It—not the people who tended it or even who owned or managed it—determined the course of material and moral development.[63]

The stakes Pound established were high. Because automatic machinery would make everyone efficient, it could level wages to give everyone a similar level of consumption. Echoing the optimism of Tench Coxe and later advocates of industrialization, he argued that America could achieve equality by letting the iron man work its mechanical magic. But, he warned, the iron man could also destroy democracy, especially if government did not direct its development. "America's high duty," he professed, "is to guide the continuing evolution of the Iron Man intelligently. . . . Governments now stake their existences upon controlling men: in the dawning age, the acid test of sovereignty may be control of machines. Through such control, the leveling tendency, inherent in automatic production and reinforced by popular education, may be directed toward the goal of true democracy; whereas, undirected, it may push the human race into a new slavery, or stampede it into a new anarchy."[64] The choice for Pound was clear: government had to exercise "moral control" over the iron man if democracy was to survive.[65]

Yet the specter haunting Pound's analysis was not machinery, but alienated workers. Observing men feeding materials into an automatic machine, he reiterated earlier warnings: "There is a stirring within him [the worker] which informs him, even before the voice of the agitator reinforces the conviction, that this is no life for a real man. He gets, literally, no fun out of his labors." By removing joy and creativity from work, Pound argued, repetitive movements "starved" men "in their souls," which led them to seek "sedatives" in the form of mass amusements. But because those sedatives did not teach thrift or self-restraint, people were growing selfish and destructive. Viewing modern work as unfulfilling and leisure as destructive, Pound joined numerous other intellectuals in embracing "education for leisure" as a way of promoting self improvement. "Unless our American gentlemen and gentlewomen appear in due time and in sufficient numbers," he concluded in words that would reverberate through *R.U.R.*, "civilization will be wrecked by machine-made barbarians, unable . . . to replace what they have destroyed."[66] For Pound, the only possible response to the danger of the alienated worker was cultural reinvigoration.

Published shortly before *R.U.R.* debuted in October, *The Iron Man in Industry* raised many of the issues that would dominate the play as well as later interpretations of the robot, including the way it drew attention to technology rather than capitalism as the problem of modern life. Pound did recognize the power of capitalists to shape the introduction of machinery, but his symbol and title transformed the machine into the central character. The contrast with the Progressive Era's most influential economist, Thorstein Veblen, was telling. The titles of the works that framed Veblen's life employed the terms *leisure class* and *price system* that directed attention to class and capitalism. In the 1920s and early 1930s, works such as Stuart Chase's *Men and Machines*, Floyd Dell's *Love in the Machine Age*, and Ralph E. Flanders's *Taming Our Machines*, joined Pound's *The Iron Man* in directing attention to technology.[67] Much like Pound's, such works critiqued other elements of American life—including capitalism—but embraced the machine as the dominant source of change and, frequently, cultural transformation as a solution.[68] In doing so, they shifted popular conversation away from capitalism to a topic that was far safer—for both businesses and the writers—in the aftermath of the Russian Revolution and the Red Scare: the relationship between technology and culture.

The first American film about an automaton was typical of this rhetorical shift. In 1919, Rolfe Photoplays released a fifteen-part serial entitled *The Master Mystery* that starred escape artist Harry Houdini as Quentin Locke, a Department of Justice agent investigating International Patents, Inc., for antitrust violations.[69] The company's executives made their vast fortunes by accumulating patents from independent inventors, promising to manufacture and market the devices, and then simply letting the inventions rot in a basement. In the first episode, the film introduces an automaton in which a human brain can be transplanted to give the machine consciousness and the human a more powerful body. Upon seeing a prototype, the company president tells the inventor that the device is "ridiculous, and even if possible, it would be of no use except as a terrible engine of destruction." Yet, after another executive and the automaton's inventor attempt to cooperate with Locke's investigation, viewers see a life-size version of the automaton trudging up the stairs to murder the two men. Locke spends the entire serial building his antitrust case while escaping from various traps set by henchmen and the automaton. When Locke finally defeats the device and unmasks it, he reveals that rather than a full automaton or even a human brain implanted into a machine body, it is merely a disguise: Houdini's most frightening enemy is not a machine but a wealthy monopolist so committed to destroying the potential of machines that he dressed up as one.

The plot and characterization of *The Master Mystery* indicted monopolistic corporations and capitalists, but its iconography indicted the machine. The very definition of the "idle rich" that Veblen critiqued in his *Theory of the Leisure Class*, the villainous executives of the series luxuriate in opulent mansions while depending on the hard labor of others and preventing the manufacture of technologies that could aid the public.[70] The celebration of Houdini's Locke as a bureaucrat nobly performing the public's work to break up a monopoly

(C) *B. A. Rolfe Productions* *Scene from the Photo-Play*

IN THE CLUTCHES OF THE IRON TERROR

Fig. 4.2. Promotional image for *The Master Mystery* that shows the fake automaton, or "Iron Terror," directing its henchmen as they capture Harry Houdini's investigator and his love interest (played by Marguerite Marsh) Image captured by Robert Zinck, Widener Library, Harvard University.

similarly drew on Progressive Era endorsements of neutral experts as protectors of the public interest. Yet, Rolfe's advertisements and most of the film identified the automaton as the agent of horror. Numerous posters and published stills showed Houdini and his love interest menaced by the monstrous "Iron Terror."[71] Though the final episode revealed the truth, the bulk of the film's materials focused on the machine.

While *The Master Mystery* continued to indict the leisure class, others suggested that the primary solution to the problem of the machine age had to be cultural.[72] In his 1922 book *Social Change with Respect to Culture and Original Nature*, Columbia University sociologist William F. Ogburn coined the phrase *cultural lag* to refer to the tendency of cultural values and institutions to adjust slowly to changes in material conditions. For Ogburn, "the rapidity of change in modern times" caused by the "increase in inventions" created "maladjustments" between cultural values and social institutions, and perhaps even between human nature and modern life. "Where one part of a culture changes first, through some discovery or invention, and occasions changes in some part of culture dependent upon it," he noted, "there frequently is a delay in the changes occasioned in the dependent part of a culture."[73]

Ogburn relied on an emerging anthropological definition of culture that was far more inclusive than Pound's definition. Whereas Pound wanted individuals to acquire culture, Ogburn quoted the anthropologist Edward B. Tylor's definition of culture as "that complex whole which includes knowledge, belief, art, morals, law, custom and any other capabilities and habits acquired by man as a member of society." Ogburn included "material culture" in this definition, but cultural lag depended on distinguishing it from other forms. Changes in material culture, he argued, "force changes in other parts of culture such as social organization and customs, but these latter parts of culture do not change as quickly. They lag behind the material-culture changes."[74] In making such a distinction, Ogburn adopted a social form of mind/body dualism in which material culture provided a society's body and immaterial culture provided its mind and soul. What was needed to solve the problems of modernity was not revolution but a gradual reconciliation between the two.

German modernist Fritz Lang's 1927 film *Metropolis* similarly called for a cultural solution to the problem of the machine. Inspired by Lang's view of the New York skyline, *Metropolis* depicts a mechanized society divided into a white ruling class that plays in the skies above the earth and a darker working class that toils underground.[75] Fully dehumanized, workers shuffle as a faceless army with their heads bowed in servitude to the clock until their bodies are sacrificed to a fiery mechanical god the film names after the Old Testament's Moloch. The workers are known by numbers rather than names and lack all individuality and independence. Even this mechanization is not efficient enough for the wealthy. When the workers, influenced by the rhetoric of a saintly woman named Mary, begin to plot a revolution, the Henry Ford–like leader John Masterman hires the scientist Rotwang to mass-produce a "machine man" to replace his workers.[76] To thwart the revolution and rid the world of workers, the two kidnap

Mary, transfer her form and visage onto the machine man, and then use the doppelgänger to inspire the workers to attack the machines and unleash a flood that will destroy the entire underground city and its inhabitants. His workforce exterminated, Masterman can replace it with Rotwang's machine men and the ruling class can enjoy even greater levels of pleasure.

While Masterman and Rotwang hatch their plan, Masterman's son Eric ventures into the underground city in search of Mary, whom he has envisioned in a dream. Once there, he discovers that his pleasure depends on the harsh lives of the workers and begins to try to ameliorate their situation. Before he can convince his father, however, Rotwang kidnaps Mary, transforms the machine man into a highly sexualized duplicate of her, and unleashes it.[77] It then incites the workers to attack the machines and unleash floodwaters upon their homes and children.[78] Eric, however, saves the real Mary, the workers, and the children before killing Rotwang. Upon discovering the reality of machine Mary, the crazed workers burn it at the stake and remove any threat to their jobs. As Masterman, Eric, and a nameless worker clasp hands to make peace, Lang's concluding message appears on the screen: "There can be no understanding between the hands and the brains unless the heart is the mediator."[79] Only a religiously inspired sentimentality, Lang proposed, could solve the problems of the industrial age.

When framed in the socialism/capitalism debate, *Metropolis*'s message was muddled, especially after American censors cut lurid scenes of opulence. Reviewers found the film's technical achievements noteworthy but consistently complained about its "unimaginative" and "perplexing" story.[80] The film's depiction of the lives of the rich and poor pointed to a class analysis, while its focus on machinery—from the opening images of pistons, to clocks, the machine Moloch, and its mechanical woman—seemed to indict technology. Its sentimental ending endorsed stereotypes of managers as "heads" and workers as "hands" while the failure to include Mary alongside Eric as the "heart" suggested that only men's sentiment mattered. Yet, the film's repeated emphasis on spirituality and feeling drew from the broad critique of materialism and the call for new values that could bring a divided community together. Once again, culture became the solution to the problems of industrial capitalism.

But reviewers also thought that the machine man Lang indicted could be the solution to the plight of workers. The American version of *Metropolis* never uses the term *robot*; however, many of its reviewers did—to describe the mechanical Mary. In his *New York Times* review, Mordaunt Hall identified the fake Mary as a "Robot woman."[81] *Popular Mechanics*, which had never used the term previously, printed photographs of Mary in both human and robot form next to a blurb that read: "A new German Film, Based on the Old Robot Story, in Which a Mechanical Person, Created by an Inventor, Becomes a Frankenstein Monster, Has Been Produced in Germany."[82] In his review that newspapers reprinted across the country, H. G. Wells sardonically remarked that "Rotwang the inventor is making a Robot, apparently without any license from Čapek."[83] By identifying the machine Mary as a "robot," such publications shifted the meaning of the term from a worker to a technology. "Mechanical civilization has no use for mere drudges," Wells wrote. "The more efficient the machinery the less need there is for the quasi-mechanical minder. . . . The whole aim of mechanical civilization is to eliminate the drudge and the drudge soul." The film, he observed, failed to acknowledge that technology had brought wealth to the masses: "Unless the masses of the population have spending power, there is no possibility of wealth in a mechanical civilization. A vast, penniless, slave population may be necessary for wealth where there are no mass production machines, but it is preposterous with mass production machines."[84] In Wells's review, the robot is not a dehumanized worker but the technological savior of the worker. Domin in *R.U.R.* could not have said it much better.

Men and Machines

Few men captured the emerging hopes and anxieties for the humanized machine as well as the radical journalist Stuart Chase. Born in New Hampshire in 1888, Chase attended MIT and Harvard, where he acquired both technical expertise and a liberal arts education.[85] After graduation, he worked as an accountant in Boston but soon joined the Federal Trade Commission and later the Labor Bureau. Service in such institutions regulating the American economy fit well with Chase's political leanings. Before the war, he dabbled in Henry George's single-tax plan and socialism, and afterward embraced Veblen's suggestion that engineers, as experts on production, should have political power in the machine age. In 1925 he began a series on waste for the *New Republic* in which he found that America wasted 50 percent of its manpower in advertising, sales, and idleness. Concentrating on efficient production rather than wasteful advertising and selling, Chase argued, would ensure a higher standing of living for all Americans.[86] Praised by liberal writers across the country, the series transformed Chase from an obscure bureaucrat into a chief advocate of centralized planning.

In 1929 Chase published *Men and Machines*, a collection of essays that explored the question of whether the machine liberated or enslaved men.[87] Chase could not hide his obvious esteem for machinery, but he identified a litany of accusations that he considered valid, including the mechanization of warfare; worship of money; and increases in technological unemployment, advertising, mental diseases, accidents, and class divisions. His most vigorous critique, however, focused on the transformation of workers into "robots." Chase devoted an entire chapter to "the mechanism of flesh and blood first heard of in a Czechoslovak play, toward which, it is alleged, all men are moving." Connecting the robot to Ford's assembly line, he agreed that such work was degrading, unhealthy, and dangerous, but used statistics to show that only a small number of American workers encountered such toil. Of those small number, he continued, many were biologically suited for the toil because they lacked the physical capabilities to perform other jobs. Paraphrasing a French investigator of American factories, he argued that "many [workers] have the gorilla appearance, but on tabulating their nationalities, he [the investigator] finds them to be stolid peasants from Russia, Poland, Roumania. Many of the immigrants who are tending our machines were born with a dumb look."[88] American factories, Chase argued, transformed few men into robots, and most of those were so racially or physically inferior that they appreciated it.

Though Chase believed that industrialization had improved American life, he still depicted machines as "a billion wild horses" that threatened to stampede and destroy humanity. "Man is not the slave of his machines," he wrote, "but he has allowed them to run unbridled, and his next great task is, by one method or another, to break them to his service." To explain how to bridle machines, Chase posed a hypothetical: suppose that "you, dear reader," were an American dictator. Such a dictator should suppress machines where they harm human civilization; threaten the life, limb, and psychology of people; overly tax natural resources by overproduction; waste labor by selling products rather than making them; encourage passive rather than active entertainment; or produce shoddy goods. Furthermore, he concluded, the dictator should establish a central location where engineers, economists, and mechanics could oversee the national economy because "machines, like horses, can be tamed only by men who understand them." After establishing a government by experts, the dictator should seize resources and redirect scientific inquiry to researching "serious" devices that would free workers from drudgery. Chase admitted such centralization might hamper progress, but he was fine with that possibility. As he glibly remarked, "I am willing to put off that promised trip to Mars for a few years in exchange for a city planned for comfortable and civilized living."[89]

Chase was no fascist like Keller's fictional Ed Ball or the real dictators of Europe, but it is revealing that centralization was the only solution he could even imagine. Chase was fundamentally uncertain about what direction to pursue. In the machine age, it seemed, every person operated, as on the Western frontier, for his or her own self; but in the machine age it also seemed as though individualism and thus male identity had been destroyed. Something was needed to restore both the individual and the community and to reconcile machinery and soul, but Chase could not imagine how this could be done without greater centralization. Unlike cultural critics, he saw no possibility of creating a new culture that could fuse the material and the spiritual and give men the purpose and meaning they had supposedly experienced prior to industrialization.

Chase did, however, offer a solution that men such as Edison, Ford, and the fictional Domin would have appreciated—a device that people, without any sense of danger, had already started to call a robot. As the end of his history of the machine, Chase introduced "Mr. Televox," a white mechanical man produced by the Westinghouse Electric & Manufacturing Company that could flip electrical switches on or off by hearing orders over the telephone. "The ghost of James Watt must have shuddered down the years—for he was a kindly man—as the bodies of women, men, and little children were broken in the dismal caverns where his engines pounded," Chase stated, "but for Mr. Televox his back must straighten and his hand go gladly to the salute."[90] What America needed, Chase suggested, was not revolution or even cultural reinvigoration; all it needed was robots. And Westinghouse would provide them.

5

Building the Slaves of Tomorrow

In late 1930, a year into the Great Depression, Dr. Phillips Thomas, an engineer for Westinghouse Electric Company, demonstrated to an audience at Chicago's Armour Institute of Technology how machinery could aid humankind. As part of a campaign to sell Westinghouse products and recruit engineers, Thomas introduced new uses for a thirty-year-old technology, the vacuum tube. He carefully explained how vacuum tubes could extinguish fires, regulate room temperature, and build safer airplanes—all automatically. Such applications promised great improvements in safety and lifestyle, but in an era of engineering wizardry, they lacked the expected spectacle. Thomas, however, came prepared. Already an experienced showman, he had brought with him a device that could interest even the most skeptical observer: a black-skinned mechanical man he named Rastus.[1]

While Thomas and Rastus appeared on stage, other mechanical men were rampaging across pulp magazines, films, stages, and storefronts. Sometimes these gigantic, lumbering, metallic men were called "robots," but few symbolized workers. Continuing the transformations that had begun in the second half of the 1920s, Americans in the Great Depression increasingly used the term to refer to humanized machines that replaced workers rather than workers turned

Fig. 5.1. Two photographs of "Rastus Robot" from Hugo Gernsback's *Radio-Craft* magazine, in which S. M. Kintner uses his "photoelectric emitter" to reenact the legend of William Tell and to order Rastus to stand and sit.

into machines by industrial labor. Yet, the uncannily human Rastus looked nothing like these robots. Where earlier engineers and costume designers had relied on wood, wallboard cutouts, or metal to give their robots a vaguely human facade, Thomas and S. M. Kintner, Westinghouse's assistant vice president, molded rubber into a racist caricature that resembled the minstrel character after whom Rastus was named.[2] His black body covered by overalls, a white shirt, and a pail hat, Rastus embodied stereotypes of docile black men; making the racial dimensions of Čapek's robot explicit, he was, effectively, a "boy" slave for his engineer master.

During the performance, Rastus appeared on stage seated with an apple on his head while Thomas held a bow and arrow. In this reenactment of the story of William Tell, however, the arrow employed a "photoelectric emitter" that shot a beam of light into a cell embedded in the robot's eyes. Now activated, the photoelectric cell flipped a switch that lit a fuse just below the apple. As Rastus sat motionless, the apple magically fell to the floor—victim, it appeared, of a press of a button and a flash of light. The trick completed, Thomas pushed another button and Rastus bowed and mumbled a few words. The inventor then usually had the robot perform more routine tasks: sweeping the floors; switching lights on and off; sitting down and rising again. The message of this blackface performance could hardly have been clearer: the vacuum tube of yesterday had turned the rampaging robot of today into the slave of tomorrow.[3]

The conservative *Chicago Daily Tribune* certainly understood the message. "'Let Electrons Do It,' Motto for Moderns" proclaimed a headline in the next day's edition of the paper; "Dr. Phillips Thomas Tells of New Slaves." "For a nickel," the article stated, "you can buy 130,000 million million electrons and put them to work. An ounce of these slaves of the lamp of understanding represent 100,000 kilowatt hours of energy. You press a button and 160 million a second pass through the toaster wires on your breakfast table." To clarify the importance of such power during the discontent of the Great Depression, the newspaper compared electrons to workers: electrons, "unlike all other creations, are all exactly alike and can be depended on for their actions."[4] More than a new tool, electrons were the submissive laborers that would transform the way American society produced its goods; they were, like Rastus, humanlike in their ability to serve but without the individuality that enabled people to resist domination.

Thomas and Rastus performed their mechanical minstrel show for middle- and upper-class audiences across the United States and Canada in the early years of the Great Depression. After premiering at the National Electric Light Association convention in San Francisco in early 1930, the two performed in New York in November and then at the Armour Institute. From there, Thomas and Rastus traveled to Alberta for a meeting of the Canadian Electrical Association. Rastus reappeared in 1935 at the *Syracuse Herald* Progress Exposition for what seems to have been its final performance. The main archivist of Westinghouse's robots claims that Rastus's rubber skin caused it to overheat and melt, a horrifying end to the only black robot the company built.[5] After Rastus, Westinghouse never made another mechanical man that so uncannily crossed the boundaries between human and machine. Instead it maintained a strict division by building gigantic metallic-colored men which only marginally shared a shape with their controllers.

Rastus was one of six mechanical men and women built by Westinghouse between 1927 and 1939 to spread the company's message of robotic slavery to middle- and upper-class white families. During that time, Rastus and its "family"—Mr. Televox, Katrina van Televox, Telelux, Willie Vocalite, and Elektro—performed at management and professional clubs, fraternal lodges, department stores, technical colleges, and local and international fairs across the United States and Canada. Everywhere they appeared, the machines attracted large crowds and sensationalized newspaper coverage. Even people in small towns learned of the robots as local newspapers reprinted stories of performances, often with astonishing, attention-grabbing headlines and photographs.[6] These articles likely familiarized most Americans with the devices and, more important, taught them that these machines, rather than Čapek's creatures, were robots. Though Westinghouse mostly avoided the term *robot*, the press seized upon it to stimulate its readers' imaginations even further. When it included two photographs of Rastus and Kintner on a spread celebrating amazing advancements in electrical equipment, for instance, Hugo Gernsback's *Radio-Craft* magazine dubbed the device, "Mr. Rastus Robot, the most life-like of mechanical men," before excitedly explaining its control mechanism.[7] By the mid-1930s, practically any machine that seemed to duplicate human features or behaviors, regardless of its appearance, could be called a robot. More than any other device, Westinghouse's robots popularized Čapek's term, though that did not mean that audiences accepted the ideological message the company wished to convey.[8]

Though Rastus was neither the first of its kind (Mr. Televox was) nor the most popular (that was Elektro), as the only black-faced robot Westinghouse produced, Rastus highlighted the company's message better than any other. Deeply troubled by growing skepticism of the machine age, Westinghouse designed its devices to pacify the symbol of the rampaging robot. The company's white and metallic robots achieved this through humanization: they had unique names, smoked cigarettes, and, once given artificial voice boxes, had the ability to tell jokes—even about sex. But Rastus, with its minstrel show name and docile appearance, was itself the joke. While its fellow robots hid their enslavement beneath a veneer of friendship and equality, Rastus was an eager to please, crude caricature; explicitly a slave, it was far from the celebration of blackness found in the era's popular culture and places such as Chicago and Harlem. Since the nineteenth century, American advocates for industrialization had talked about machines as the new slaves that would enable the country to fulfill its democratic and moral promise. But the synthetic blackness of Rastus did not offer those benefits; instead, it offered an unambiguous fantasy of power over machinery and the bodies of black men.

Whitening the Machine

The televox that so enchanted Stuart Chase was built by another man who yearned to tame the machine, Westinghouse engineer Roy James Wensley. Born in 1888 in Indianapolis, Wensley exemplified the myth of the self-made engineer. After eighth grade, he left school to support his mother with two jobs, one as a railroad lineman and the other as an electrician's assistant. Not content with his positions but interested in electricity, he spent his evenings studying electrical engineering through a correspondence course. Upon completion, he joined Westinghouse in Pittsburgh and began researching automatic labor-saving devices in the switchboard engineering department. While there, Wensley grew interested in "remote control" operations and automatic substations and began to investigate a means of using phone lines to flip electrical switches.[9]

In a 1926 editorial for Westinghouse's *Electric Journal*, Wensley identified "unrest" in the labor market as his inspiration. Immigration restrictions, he assumed, had "placed a premium on the unskilled worker" while "higher standards of education" encouraged men to develop a "distaste . . . for operating jobs." Workers' demands for higher wages and more skilled positions, he believed, necessitated the invention of devices to replace human labor if civilization was to continue to progress. "Each time that inventive effort has released a certain amount of human labor from an essential industry," he wrote, "there has been a corresponding benefit to the cause of our modern civilization." By cheaply extending electricity into regions where it was not previously economical, automatic switches would allow power companies to accommodate the suburban market by solving one of the problems that had hindered progress since "ancient days": the unreliability of people.[10]

Wensley soon invented a machine that fulfilled his goals: the televox. Capable of controlling electrical devices remotely via sound, the televox consisted of two stacked rectangular metal boxes into which the engineer placed a machine capable of receiving instructions over an attached phone line, interpreting those instructions, and performing the requested action. Because of the difficulties associated with programming a machine to understand speech, communication with the device relied on a language of musical notes. When the manager telephoned the machine and blew a note on a whistle, the televox would turn the attached piece of equipment—whether an industrial substation or a vacuum cleaner—on or off. Such a machine, Wensley believed, would enable managers to control the flow of electricity remotely without the aid of on-site workers.[11]

Westinghouse executives saw advertising magic in the device. Since it had lit the 1893 Columbian Exposition, the company had relied on amazing displays of technology to sell its products.[12] As intellectuals, popular culture, and even engineers expressed unease with out-of-control machinery, with some even suggesting that government control innovation, the televox offered an alternative vision: consumer control over technology. To preach this solution, the company sent Wensley to middle-class clubs, where he sat the device flat on a table and demonstrated how to use it. Journalists quickly recognized the possibilities Westinghouse wanted them to see. The *Los Angeles Times* proclaimed, "Wife Cooks by 'Televox'" before explaining how the machine would revolutionize labor in the home and free women for other tasks.[13] Such efforts, however, failed to attract broad public interest. A performance for journalists in New York City, however, inspired the imagination of Waldemar Kaempffert, the science and engineering editor for the *New York Times* and former writer for *Scientific American* and *Popular Science Monthly*. Together, Wensley's machine and Kaempffert's imagination introduced America to Mr. Televox, the "Nearest to a Robot" ever produced.[14]

According to Kaempffert, the televox was more than a machine; it was an "electrical man" and a "mechanical slave." An accompanying strip of drawings made the point clearer. No longer just two conjoined boxes on a table, the televox stood upright, possessed jointed arms and legs, and performed tasks ranging from reading a gauge to turning on a light. "When R. J. Wensley, the engineer who designed this electrical substitute for humanity," Kaempffert began the story, "demonstrated its capabilities recently by ordering it to light and extinguish lamps, start and stop a fan and vacuum cleaner, and control a motor, his audience of sober business men imagined itself for a brief hour in that fantastic world of the future beloved of romancers—

a world in which men and women do little more than think and bid automatons to fetch and carry, manufacture the countless things a machine civilization requires, sweep streets, cook, wash and dig ditches." The televox, he continued, "can be called up on the telephone, asked questions, and given orders which it obeys without the usual human arguing, impudence or procrastination." With these words, Kaempffert trumpeted Westinghouse's message: the device's superiority to human labor, the domination of supervisors and engineers, and the possibility of a world of leisure and more meaningful work.[15]

Kaempffert's enthusiasm originated in the vitalist critique of the machine age. Sounding like Chase and Arthur Pound but especially the manager Domin from *R.U.R.*, he professed that "man is a highly complex organism adapted to live in a highly complex environment. In a factory, in an artificial environment, he is 90 per cent useless. So long as he sits at a machine he might as well be legless. So long as he does nothing but feed bars of steel into a machine . . . he might as well be earless and noseless. For that reason engineers like Wensley are not concerned with mere imitations of men but solely with a few functions that men are called upon to perform under special circumstances. An automatic shoe-pegging machine is not expected to play the grand piano."[16] Like Chase, Kaempffert loved the televox because it offered simple technological solutions to the problem of alienating labor. America did not need reform or government action to restore its preindustrial state and resolve the guilt people felt for the degrading conditions of labor; all it needed was technology.

With his next show scheduled for George Washington's birthday, Wensley cut a piece of wallboard into the shape of a person, painted it white with rivets along the sides, gave it a face, and cut a hole in the chest large enough to encase the bottom box of the machine. Much like Drederick, Wensley's first impulse was to make the machine a caricature of himself. The day before the performance, he assembled the device—standing upright this time—with its new attachments on the stage so it could perform for reporters. Next to the machine, he placed the customary vacuum cleaner and other appliances, along with a flag and painting of George Washington. He and the newly christened "Mr. Televox" then performed the usual act. Afterward, Wensley answered reporters' questions while photographers and filmmakers captured images of the new machine.[17] Soon stories of the wondrous robot appeared on periodical pages and cinema screens. For a company seeking to build enthusiasm for technology, the debut was an unqualified success.

The official unveiling offered greater fanfare. Wensley added an orchestra and ended the show with a choreographed sequence in which he whistled for the machine to turn a spotlight on the portrait of George Washington. After a few more notes, he turned the whistle over to a judge, who sounded a signal to unfurl an American flag. The orchestra then played "The Star-Spangled Banner" as the audience stood and clapped. The robot was, like Washington, unfurling a new America where people could be free from tyranny, though this time the tyranny came from other machines rather than the British.[18]

After its debut, Westinghouse sent the original Mr. Televox as well as copies across the country with Wensley and other engineers. Performing in professional organization meetings, men's clubs, department stores, universities, and other middle-class environments, the devices garnered newspaper coverage around the world. The company's engineers further humanized the machine. Wensley started to refer to it affectionally as "Herbert," after the world's most famous engineer, Herbert Hoover. Soon, Westinghouse attached a phonograph that enabled Herbert to utter prerecorded phrases that gave it a voice and personality while reinforcing its ideological power. When "Mr. Televox" appeared at a national meeting of booksellers alongside the historian and machine age advocate Charles Beard, the company had it read Beard's speech extolling the virtues of technology.[19]

The press responded with the fervor that Westinghouse desired. "Meet Mr. Televox, the Mechanical Man!" one headline exclaimed. Accompanied by photographs of the original machine—a close-up of its "larynx" and a shot of it flanked by a regiment of soldiers—the article hailed the wonders of the machine and joked that its new voice box allowed it to "talk back" to its masters."[20] Another headline announced "Televox, Automatic Servant Works at Master's Bidding," while another story noted that "he obeys his master . . . more faithfully than many modern servants."[21] Other newspapers reported that "some day there probably will be hundreds of thousands of him at humanity's beck and call."[22] The *San Antonio Light* assured readers that it would be no "more than a decade or so before a person can go to a store and pick out from the above case most any kind of automatic man or woman he or she might fancy—an ideal servant or workman who would ask no food or wages but a little current and an occasional drop of oil; or even a flattering admirer could be purchased who would whisper in a neglected wife's ear all the nice things that a busy husband forgets to say."[23] More than a new machine, such comments joked, the televox was a new man able to compete with real men in all realms of life.

The *Light*'s joke about neglected wives drew from the company's focus on female consumers rather than male workers. One of Wensley's first examples imagined that a woman playing bridge at a friend's house could order her televox to shut the windows in preparation for a storm.[24] One newspaper predicted that "those who sponsor him indicate that Mr. Televox soon may be expected to do about nine-tenths of America's housework, thus leaving all the time they want to pursue careers, destinies and what not. He promises to be particularly good at keeping the kiddies out of mischief."[25] The upper middle-class and elite women to whom Westinghouse marketed the televox, however, often performed only a small amount of housework themselves. For the first hundred years of American industrialization, elite and middle-class families could rely on a steady stream of low-wage workers to perform housework, although some families frequently complained about the unreliability of human servants, especially if the servants were nonwhite.[26] During the 1920s, the combination of more home appliances and the curtailing of immigration accelerated a decline in household servants; still, they remained a status symbol.[27] Mr. Televox offered the illusion and status of a human servant without the cost or anxiety.

Mr. Televox also had a more salacious appeal. "He has many of the qualifications of the 'ideal husband,'" one newspaper claimed of Herbert. The only problem: "He is plain to the point of ugliness."[28] Another envisioned a world in which even the problem of appearance had been solved: "The Love-Starved Woman Will Soon Be Able to Buy a Mechanical Man of Powerful Muscles, Resounding Voice and Any Color of Hair or Complexion She Desires—And She Will Bestow Caresses with Tireless Fidelity and Repeat Over and Over Again Any Tended Phrases His Owner Desires to Have Poured Into Her Enraptured Ears." The article itself joked, "In this happy future, no old maid need look under the bed for a man in vain. He would always be there and such a nice man, a perfect imitation of her favorite matinee idol or film star, with blond or dark hair, moustache or clean shaven, anything her heart desired."[29] Such commentary, however meant in jest, resonated with audiences enough for Westinghouse to endorse it. A publicity photo of Herbert showed it with a young woman caressing its shoulder and gazing longingly into its eyes while it did nothing. A joke Mr. Televox told captured the humor implicit in such photographs. When asked at a convention to name its favorite book, Mr. Televox quipped, "*Is Sex Necessary?*"—a recent book mocking sexology researchers.[30] Showing women's interest in Mr. Televox and its lack of interest in them ridiculously revealed the technology's superiority to workers: lacking lust, Mr. Televox was a man that women and husbands could trust.

Such visions exceeded even the televox-inspired fantasies in Hugo Gernsback's magazines. The headline for a 1928 article on the device in *Science and Invention* asked readers "Has the Automaton Arrived?" Based on a lecture broadcast on Gernsback's radio station WRNY, the article explained how the "robot" or "electric handy-man . . . will do any kind of a job and signal you that it has been done."[31] *Amazing Stories* printed an account of the invention at the conclusion of a story about the adventures of a mechanical man sent to explore the moon as a "proxy" for a human being.[32] "If you believe that the accompanying tale is too fantastic," Gernsback wrote, "may we call your attention to the fact that a great deal of scientific work has already been accomplished along similar lines."[33] That same year, *Amazing Stories* published David Keller's "The Psychophonic Nurse," about a black-faced mechanical mammy controlled by sound and built by the "Eastinghouse Company."[34] A year later, Gernsback's *Air Wonder Stories* published a short story entitled "Flight in 1999," which briefly included a reliable mechanical servant named "Televox."[35] The televox, in Gernsback's magazines, was a machine that could end household labor and extend the reach of humanity to the stars.

Initially, most writers hesitated to use *robot*, but within a few months of the device's debut, the *New York Times* began to connect the terms. "Robot is the proper enough name for Televox, the Mechanical Man," the paper claimed. "But the trouble with robot is that the word has been diverted from its original meaning. The first glimpse of Televox with his thorax full of electromagnetic coils and his incandescent eyes is a bit grisly until we recall that, after all, he is

Fig. 5.2. Mr. "Herbert" Televox with the wife of a Westinghouse engineer fawning over it. Women's interest in Mr. Televox was a particular theme emphasized by Westinghouse and commentators as a way to satirize fears of technological unemployment. The original caption dubbed the device "Romeo Televox, the ideal lover." Photo by George Rinhart / Corbis via Getty Images.

a machine being made into a man and not a man being turned into a machine. It is the latter thing which people mean today when they say robot; and it is our earnest recommendation that they stop doing so." The Čapek brothers, the *Times* contended, "were thinking fifty years behind the times" when they suggested that modern work turned workers into soulless machines. "Televox is not the modern worker," it concluded, "but the liberator of the modern worker. He is the gentleman who has turned the twelve-hour day into an eight-hour day. He is the newcomer who now has a great many sociologists wondering what shall be done to make the best use of the increasing leisure of the workers."[36]

Popular Science Monthly echoed such sentiments in an article that announced, "Mechanical Men Walk and Talk: Amazing Automatons Invented to Operate Mighty Machinery, Speak at Meetings, Make Lightning Calculations and Rid the World of Drudgery." While the headline referenced automatons, the text claimed that Mr. Televox, along with the "Product Integraph" (an early analog computer) and an automatic power distribution center heralded "the age of the robot," in which "legions of docile eminently useful robots" serve at the command of men.[37] Rejecting work as a defining feature of human identity, the article ended by paraphrasing of the words of a New York Edison Company official: "Men thus freed from unpleasant chores . . . never need fear unemployment in a well-organized society but . . . may look forward to a better opportunity for the development of their inherent talents and intellectual powers. They will receive the gift of leisure, which will enable them to apply their released energies to the achievement of a finer, fuller life than they can enjoy at present." Domin, in *R.U.R.* would have enthusiastically agreed.[38]

Labor Age was not so certain about that possibility. Below an image of Kaempffert's original article, the socialist magazine proclaimed that "American workers have a new rival. Karel Čapek's dream of robotry is quietly coming to pass." Taking seriously engineers' pronouncements of the robot's superiority to human workers, the article worried that "brains in a workingman have long ago become a dangerous asset. His think-box is supposed to have been taken from him, just after the lamented war for democracy. The employers can never be sure, however, that he may not have a serious relapse into thinking. The American robot brings to many a William H. Barr and Charles Schwab and John D. Rockefeller, Jr., a sigh of relief. The great day of brainminus workers is at hand!"[39] Companies, it feared, might even use one of the devices to control a battery of machine guns to put down strikes. Mr. Televox did not just destroy jobs but redistributed power to management and threatened the lives of workers.

Despite such warnings, Westinghouse embraced these connections with *R.U.R.* by building a new, three-dimensional mechanical man whose body of molded rubber could be mistaken for human. Named Telelux after the company's photoelectric "brain" that allowed for remote control via beams of light, Herbert's "baby brother" debuted at a Pittsburgh electrical exhibition in 1929, where it turned lights on and off at a signal from its "master" standing up to seventy-five feet away. Though the Telelux relied on the same technology as Rastus, its body was white and the company placed it in scandalous poses with women. As a caption of a photograph of the Telelux with a model summarized, "An Accomplished and Tireless Mechanical Man, but Singularly Unresponsive to the Charms of Miss M. P. Carr, an Attractive Model Who Stirs No Sentimental Feelings in His Metal Breast."[40] Much like Mr. Televox, the Telelux offered a dispassionate, asexual alternative to servants. With the Telelux, Westinghouse seemed to have perfected the robot as a replacement for the nearly human-looking automaton and utopian symbol of a consumerist America where people used remote-controlled machinery to free them for leisure and pleasure.

But that was before the start of the Great Depression. After October 1929, the warnings of Keller and *Labor Age* proved prescient. In 1927, Herbert Televox had promised managers a cheaper and more reliable workforce, and everyone else greater fulfillment through lives dedicated to meaningful work and leisure. Much like its presidential namesake, however, in the Great Depression all it seemed to have brought the masses was misery.

The Threat of Technological Unemployment

The Great Depression forced Westinghouse to respond to the consequences of its promise to liberate workers from their jobs. As unemployment peaked near 25 percent, the idea that people would find more meaningful work seemed dangerously outdated, while expanded leisure time seemed feasible as the federal government debated legislation mandating a thirty-hour workweek. Throughout the Depression, academics, politicians, business leaders, and labor unions debated how to resolve a crisis that many believed was caused by the persistence of "technological unemployment"—the permanent replacement of human workers with machinery. While critics imagined rampaging robots threatening to destroy American jobs, machine age supporters, including Westinghouse, deployed robots to suggest how

Americans could "tame their machines" and restore human power to a world that seemed to have grown too dependent on technology.

Over the 1920s and 1930s, workers in extraction, manufacturing, sales, and creative industries lost their jobs to new technologies. On farms, engine-powered tractors and combines displaced thousands of workers as banks consolidated small farms into larger parcels. Miners, too, lost their jobs to machines as the number of loading machines used in the coal industry tripled. Unskilled and semiskilled laborers in factories suffered as well. With new conveyor belts in iron and steel mills, one worker could perform tasks formerly done by twenty-five. William Green, the president of the American Federation of Labor, claimed in 1930 that machinery could make 73,000 bulbs in twenty-four hours while a worker in 1918 could make forty-eight in a day's labor and that over the 1920s, industrial production increased 42 percent, but industrial employment dropped 7 percent.[41] Though service jobs increased, these succumbed to machinery as well. Interest in the vending machine industry and other "automatic salesmen" spread throughout the period; automatic telephone switchboards displaced thousands of operators. Pilots—a key symbol of the ability of men and machines to work together in the 1920s—seemed destined for unemployment when companies such as Hughes Aircraft publicly demonstrated automatic pilots.[42] Even creative workers faced the threat of technological unemployment: the advent of motion pictures with sound cost thousands of musicians their jobs and inspired a large protest movement against the "talkies."[43] Finally, discussions of the first "mechanical brains," analog computers, raised the possibility that perhaps no job would be safe from mechanization.[44]

Workers, unions, and their supports frequently anthropomorphized and ridiculed the machines that threatened jobs as rampaging but humorous mechanical men. The American Federation of Musicians promoted a boycott of the "talkies" with images of a "robot" musician playing two fiddles and a cello while also mocking, like John Philip Sousa, the lack of soul in "canned music." The *Locomotive Engineers' Journal* included a picture of a giant mechanical man labeled a "Labor-Displacing Machine" sweeping workers off the payroll.[45] Despite such robotic imagery, workers and their unions largely refrained from condemning technological progress; instead, they called for government intervention to shape the introduction of machinery. As

Green noted, "Labor is convinced there is nothing inherent in technical progress that is dangerous," but "lack of planning and control has brought serious consequences which . . . will lead to complete breakdown."[46] Instead of destroying the rampaging robot, unions called for taming it.

The threat of government control over innovation frightened more conservative engineers and business leaders, including Ralph E. Flanders, a former editor of *Machine* magazine and president of the National Machine-Tool Builders Association and the American Society of Mechanical Engineers. In his 1930 essay "The New Age and the New Man," Flanders blamed capitalism for the horror of early industrial life. Machines, he claimed, "have brought us ugliness" and "a narrower, more concentrated, more intense, and less well-rounded existence for the mass of mankind and particularly for the more able among them." Such problems, though, were largely a result of a "selfish and doctrinaire political economy" and a complicated "process of distribution" that prevented the spread of material plenty with moderate toil to everyone. Machinery, he argued, could provide "plenty for all" with perhaps as little as a three-day workweek—if business leaders and engineers would create a "great synthesis of usefulness and beauty" into "one organic structure" that made "Life and Art one and indivisible." What this structure might resemble he did not say, but it would "give secure, remunerative employment to the largest possible number of people, under working conditions which give full opportunity to ability and likewise satisfy the requirements of human dignity and self-respect." In 1931, Flanders expanded these points in *Taming Our Machines*. Rejecting centralized planning because of its inability to predict human desires and detrimental effects on the individual, he echoed Hoover's calls for "associationism" by proposing a nongovernmental committee of businessmen, specialists, and engineers—not workers. These groups, he claimed, would succeed because of "*enlightened self-interest*" and thus did not need to be accountable to government or other people.[47]

The most radical political scheme for controlling machinery came from the "technocrats." Led by Howard Scott, an engineer formerly aligned with the Industrial Workers of the World, and Dr. Walter Rautenstrauch, the chair of Columbia's Department of Industrial Engineering, the movement named itself the Committee on Technocracy and popularized the idea of technological unemployment. In 1932, Scott declared that "the fundamental cause of the depression is not political, it is technological."[48] As one of their critics summarized their beliefs, the movement "dwells upon the increasing perfection of the machine and it cautions us of the danger that the machine may get the upper hand of mankind; that instead of man running the machine, the machine may be running the man."[49] Yet technocrats did not blame the machine for modern enslavement. In his 1933 book, *Life in a Technocracy*, Harold Loeb, a wealthy artist who penned some of the movement's most important ideological statements, claimed that "man is the slave of the machine only by his own volition. Technology is not capitalism."[50] Americans could have machines, the technocrats argued, but only if engineers and scientists—men like those in the movement—tamed them.

After the election of Franklin D. Roosevelt in 1932, technological unemployment became a central concern of officials who proposed numerous solutions that relied on government power to help workers. Officials in the administration, including the president himself, consistently focused attention on the machine as a dangerous force in modern life. The administration along with large Democratic majorities in Congress attempted numerous solutions to the problem of technological unemployment. Massachusetts Representative William Connery and Alabama Senator Hugo Black introduced an AFL-backed bill for the thirty-hour workweek to encourage employers to "share the work"; the bill came close to passing before Roosevelt withdrew his support to focus on public works programs that would reemploy displaced workers instead.[51] Other members of congress proposed a "Technotax" on innovation to help compensate workers replaced by machinery, but it too failed.[52] Yet, with the passage of the National Industrial Recovery Act in 1933 and the National Labor Relations Act in 1935, the government legitimized unions as a voice for workers in their relationship with employers and the government. During the Depression, the newly formed United Electrical Workers successfully unionized Westinghouse's nonengineering workforce and even promoted several strikes.[53] For a company dedicated to replacing workers with robots, the changing political context of the Great Depression posed a significant challenge.

White Men versus Machines

While workers concentrated on the loss of jobs, much of the middle-class discussion of technological unemployment continued to fixate on the threat machinery posed to the strength and virility of white men. In his 1930 book *Perhaps Women*, Sherwood Anderson claimed that overindulgence in the power of machinery had made men "impotent in the face of the machine."[54] A man embracing the power of the machine, Anderson continued, "is like a man who has indulged for a long time in self-abuse. He can no longer stand erect." The opening vignette of Flanders's *Taming Our Machines* envisions a boy awakened by a steam whistle and listening to the sounds of deformed men and women plodding to work.[55] Flanders ended the vision by contrasting the boy with a heroic Anglo-Saxon heritage: "Children of knights and yeomen; of the victors of Crecy and Agincourt; of Shakespeare and the Elizabethan sea rovers; of Roundheads and Cavaliers; of the Muscovy, East Indian and Hudson's Bay Company; of the heroes of the Plains of Abraham, of Trafalgar and Waterloo!" How had such a noble race "bred this cheery, plucky, but mentally limited and physically handicapped army which marched and countermarched daily outside a window?" The answer, he argued, was industrialization. William Ogburn voiced the same fears in a 1934 pamphlet written for the Civilian Conservation Corps, *You and Machines*. Though it rejected the idea of long-term technological unemployment, the pamphlet's introductory sketch depicted two giant mechanical men towering above a confused man. The pamphlet then asked a basic question: "How does the White man differ from the Indian?" The answer, Ogburn revealed, is that "the white man . . . can't live without machines. . . . Man is becoming more and more dependent upon the machine. He leans upon the crutches of civilization." The machine had robbed white men of their manhood.[56]

Science fiction magnified such concerns by transforming the robot into a rampaging mechanical man that threatened to destroy all humanity if it was not tamed by the right men. Sometimes, these robots were, like Westinghouse's, remotely controlled, but frequently they had various types of mechanical brains that allowed them to develop consciousness of their situation and revolt. These stories fused imagery of the mechanical man to Čapek's tale, to offer a vision of out-of-control machinery that directly challenged Westinghouse's vision. Yet, because such stories tended to be written by middle-class white men whose jobs were not directly threatened by machinery, the anxieties they expressed and solutions they proposed were far different than those of workers or other radicals. "The Threat of the Robot" by Keller was typical. Though it briefly mentioned fears of unemployment, its main threat was to existential meaning, purpose, and manhood. Consequently, where workers looked to the democratic taming of machinery and even Flanders and the technocrats suggested control by experts, Keller imagined a heroic and individualistic businessman saving the day through force of willpower. Like Chase describing technology as "a billion wild horses," science fiction deployed imagery and themes typically associated with Westerns to transform Čapek's play into a fantasy in which white men restored their manhood through overpowering the machine.

The 1930 novella *Paradise and Iron*, by the physician Miles J. Breuer, shared the concerns about the decline of manliness and purpose in an age of "mathematical robots." In this story, a doctor and former Texas Ranger named Davy Breckenridge travels to a distant island where squidlike robots from the "City of Smoke" perform all the work while people live in a "City of Beauty" where they engage in sports, art, music, and literature. Despite the prominence of such high cultural activities rather than mass entertainment, the residents' lives remain meaningless while their bodies grow weak. As the visiting doctor comments, "I had been having a queer impression of frailty and helplessness in the actions and appearance of these people. Like a baby in arms, was the vague idea I had. . . . They never had to work. They were never driven by necessity, never haunted by the shadow of want. They only played. They had no conception of danger, privation, and pressure. They were petted and pampered children."[57] Even traditional forms of leisure, Breuer implied, would not be enough to compensate for the loss of meaning and purpose that comes from struggle and work.

Capable of thinking but not empathizing, the story's main robot plans to exterminate the island's humans. Once Breckenridge learns of the threat, he leads the island's men in a revolt. When Mildred Kaspar, the woman he loves and the daughter of the man who invented the machines, is captured by the machine "Dictator," Breckenridge ventures deep into the City of Smoke, where he learns the machines' plans. "Men are hard to understand," the Dictator says to explain why it has been kidnapping young women. "I can understand a 'feeling' for a supply station, or an 'emotion' for a repair-machine. But why such an intensity of 'feeling' for a girl? Why do your young men become so disturbed on her account, and exert their soft muscles so energetically, and give up everything for her?"[58] Yet, the machine will return Mildred if the doctor helps it conquer the world. "In a short time," the Dictator informs Breckenridge, "I can cover the world with wonderfully organized machines, infinitely better than the feeble, foolish, incompetent humans that occupy it now."[59] Breckenridge rejects the deal and, after a brutal fight, rescues Mildred. With the robot dictator destroyed, the residents of the island relearn how to work and reclaim their vigor.

But not all stories about diabolical robots presented fantasies about the restoration of white manhood. Perhaps the most threatening pulp magazine criticism of robots came from within Westinghouse itself: that of Harold Vincent Schoepflin, a mechanical engineer who wrote under the name Harl Vincent.[60] In his 1934 story "Rex," Vincent imagined a diabolical robot that enslaves the human species so it can steal emotions. Unlike other robot stories, "Rex" did not focus on the threat machinery posed to manhood, womanhood, meaning, or purpose. Instead, like *R.U.R.*, it criticized how the technocratic pursuit of order and efficiency deprived people of their essential human trait, emotion. Set in the twenty-third century, when almost one billion metal robots serve the needs of three hundred billion humans, "Rex" focuses on the titular robot surgeon with a "body like a Greek God's" that develops sentience, curiosity, and reason. With its newfound abilities, Rex analyzes human civilization only to discover a society destroyed by class divisions. In a clear parody of rationalization, Rex blames such problems on emotions. After dissecting an engineer, it uncovers the source of emotions: a collection of brain cells that robots lack. With this knowledge, it plans to create the perfect being, a human-robot synthesis in which emotion can be controlled.[61]

To do so, Rex starts a "general strike of the robots" that ends with itself as a dictator. In its first public address, it announces its intentions: "I am Rex . . . Master of robots and of men. I come to you in the name of pure logic as the protagonist of a new era in which man, who created the machines, will obtain real rather than fancied benefit from them. I come to evolve a new race of beings and to promote the growth of knowledge and the advancement of science in United North America."[62] Like a technocratic engineer, Rex promises to use its capacity for reason to bring order to industrial society. But to do so, it must first destroy the emotions of people. Symbolically replicating the dehumanization of factory workers, Rex removes people's emotion centers and replaces them with a machine that transforms them into "human robots" totally under his control. After studying people to determine their suitability for different jobs, Rex turns them into single-purpose machines with "minds . . . capable of thinking nothing but mathematics, riveting, welding, food synthesis, or childbearing."[63] With these transformations, society grows more orderly and class differences erode. But Rex remains discontented. Trying to become the perfect being, it implants human emotional cells into its mind. When the surgery seems to fail, it commits suicide and leaves humanity—now transformed into mindless robots—to remain the slaves of machines. In "Rex," not even a heroic man can save people from the robot dictator.

Offering the broadest range of social criticism in Great Depression stories of robots, "Rex" maintained Čapek's dual focus on the plight of workers and the dangers of mechanization but added a critique of radical efforts to plan the economy. Unlike most other robot stories of the era, this was not optimistic about the possibility of restoring human control over the economy, society, or even the self. That such a dark vision came from an engineer within Westinghouse indicates the limits of the company's robotic fantasy. The company had succeeded in making the robot a machine, but it could not fully jettison Čapek's nightmare. With the start of the Great Depression, it needed a new strategy, and it found that in Mr. Televox's successors.

The Comic Relief of the Machine Age

The popularity of the rampaging robot icon raised a marketing dilemma for Westinghouse. Its target audiences continued to covet machines to replace workers, but the larger culture called for taming the out-of-control robot, usually through government control of invention. To resolve that dilemma, Westinghouse initially experimented with new forms—Rastus and a female televox—that could reinforce its message of mechanical servitude while allaying the fears of white male workers. When these forms failed to resonate with audiences as well as the original did, the company fully embraced the iconography of the mechanical man by building two metallic robots, Willie Vocalite and Elektro, which mocked the idea of rampaging machinery and turned the robot from a monster into the cigarette-smoking and womanizing comic relief of the machine age.

Westinghouse's executives directly challenged these anxieties in the company's professional magazine, the *Electric Journal*. In a 1930 article, president F. A. Merrick attacked the "The Machine Myth," the idea that "the use of labor-saving machinery tends to decrease the worker's opportunities for employment and to reduce his wages." The idea of technological unemployment, Merrick admitted, contained "a grain of truth," but the current situation was different because modern education had created a more fluid labor market and employers had learned the necessity of enhancing consumer-buying power. Now, he argued, machines did not threaten men; instead, they guaranteed a widespread distribution of power and wealth while enabling the replacement of "hundreds" of working-class men with "two or three engineers."[64]

At the same time its engineers were building Rastus, Westinghouse's journal used the language of slavery to describe technology. Slavery, Merrick suggested, had brought "civilization" to the world by enabling "certain strong peoples" to use "the muscles of others to supplement their own feeble strength." Without slavery, he continued, Americans needed machinery or else there could "be no art, literature, science, leisure, or comfort for anyone."[65] But machines could become the new slaves. A 1927 article by the journal's editor argued that the country's current prosperity was due "to the greater individual production of the necessities and luxuries of life. Through greater development of mechanical slaves we have achieved the apparently contradictory results of doing more work per man and having more abundant leisure for individual enjoyment and self-development than any other nation." With machinery, the editor argued, Americans could perform more work while enjoying the culture and leisure associated with the plantations of the Old South. The key to achieving this ideal state was empowering the engineer, "the modern slave driver," because he was "a public servant" who was "most likely to employ them [machines] solely for the betterment of mankind."[66]

When Westinghouse built and exhibited a black robot in 1930, it connected the machine age to a preindustrial era of slavery while trying to relieve fears that machinery would destroy the jobs of white workers and degenerate the "race." Writing in the same journal the same year, M. S. Sloan, the president of the National Electric Light Association and the New York Edison system, made the sentiment much clearer: "The workmen of this country have become bosses and foremen, not of other human beings, but of mechanical slaves."[67] The machine had not weakened men or brought widespread unemployment; it had allowed all men to become as strong, powerful, and cultured as slavemasters had been in the past. Manhood, per Westinghouse's journal, did not lie in physical strength or even work; it lay in controlling others. With Mr. Televox and Rastus, men did not depend on machines; they controlled machines while ridding themselves of their reliance on other peoples.

Westinghouse employed African Americans in low-skill positions, but the company's engineering culture ridiculed them as unfit for machine age employment.[68] In one *Electrical Journal* cartoon, an Uncle Tom–like figure happily served as an elevator operator while a smiling "electric eye" stood on legs outside the elevator reaching its hand to the man. The accompanying text explained how photoelectric tubes could prevent elevator doors from closing on a person. A later image depicted a rotund, singing mammy in front of an open electric refrigerator while ironing. When the refrigerator repairman arrives to investigate the homeowner's high electric bills, the mammy exclaims, "Boy, dat 'lectric refrigerator sure am fine. It gives such a nice cool breeze on mah back."[69] In both cases, the jobs—bellhop and servant—performed by the cartoon black figures were ones that the company was developing technologies to replace. The only African American fit for work in the machine age, the company's journal implied, was a robot invented and controlled by white engineers.

Westinghouse also experimented with a female robot—one not uncannily human like Rastus but a caricature modeled on Mr. Televox. Named Katrina van Televox, it came clothed in a white dress, apron, and bonnet. "With no mere assistance than a command from her master," one article claimed, "she turns on lights, starts vacuum sweepers, electric fans, and many other electrical appliances."[70] Despite such stories, interest in Katrina remained low. While an estimated fifty thousand people witnessed a performance in Chicago, Katrina appeared in only a handful of cities and garnered significantly less attention than any of the other robots.[71] Nevertheless, the experiment with a female form of the televox showed the company attempting to adjust to the changing context by emphasizing both the machine's inherent subservience and that it would not threaten the jobs of white men.

After Katrina and Rastus failed to attract as much attention as Mr. Televox, Westinghouse switched to building metal men that could mock the rampaging robot of popular culture. In 1931, Westinghouse debuted "Willie Vocalite," a seven-foot-tall, 260-pound robot with an eighty-two-inch chest. The chief technology behind "The World's Wonder Robot," as one promotional advertisement dubbed him, was a device capable of sensing both light waves and, even more remarkably for the time, human voices.[72] The most popular Westinghouse robot for much of the Great Depression, Willie appeared at expositions, club meetings, and department stores between 1931 and 1939, including the 1933 "Century of Progress" Exposition in Chicago. Later, Westinghouse added a Voder—a device that poorly duplicated human speech—built by Bell Telephone Laboratories to give Willie a voice without the aid of a record. As a 1939 advertisement for the robot professed, "The modern Frankenstein . . . walks, talks, dances, smokes cigarettes, counts on his fingers, distinguishes colors, makes love, and does any number of things through remote electrical control."[73]

Westinghouse and the press sexualized and joked about Willie even more than they had the telelux and televox. The same promotional article that called Willie "the modern Frankenstein," included a photograph in which Willie held a woman in his lap above the caption, "Adept at making love, is 'Willie Vocalite.'"[74] One article spread through small-town newspapers dubbed him a "metal Adonis" in contrast to the "ugly Mr. Televox."[75] A report about "Willie the Incorrigible" opened with a unfortunate joke about how the robot "bit a blond."[76] Another screamed that "Willie HAS VOICE/ Upon command he can make a political speech, sing an excerpt from an aria, or get 'hotcha' on the latest modern hit song./ Being an electrical sort of person, Willie is adept to the alternating current."[77] Such jokes were common for a device being used to mock anxieties of the machine age. While critics of the machine age worried about moral decay, Willie Vocalite seemed to enjoy all the rhythms and pleasures modernity could provide.

The contrast between Wille and another robot at the Chicago exposition was revealing. Outside the Hall of Science stood a white bronze statue of a mechanical man entitled the *Fountain of Science* and sculpted by Louise Lentz Woodruff, the wife of a Chicago banker. According to a promotional flyer, "The theme of this fountain—Science Advancing Mankind—is represented by the great robot-like figure typifying the exactitude, force, and onward movement of science, with its powerful hands at the backs of the figures of a man and a woman, representing mankind." In a fair with the theme of "Science Finds, Industry Applies, Man Conforms," the statue perfectly represented how the robot, even if a savior, could also be the master of humankind.[78] Yet, Willie comically suggested the opposite: that humanity remained the master of even the most frightening machines.

Westinghouse's efforts to transform the robot into comic relief paid off in two satirical Walt Disney short films. In the 1933 animated short *Mickey's Mechanical Man*, the titular mouse uses music—an invocation of Mr. Televox's control mechanism—to train a human-shaped machine to box an ape named the "Kongo Killer." While Mickey plays a calm melody on the piano, the machine boxes like a Victorian gentleman, but when the flapper Minnie Mouse honks an automobile horn, the machine transforms into a brutal killing device. Billed as a fight between beast and machine, this "Battle of the Century" pitted primitivism against modernity while mimicking larger concerns about the power of black boxers such as Jack Johnson and Joe Louis. Initially, the Kongo Killer easily dominates, but Minnie fetches her horn, returns, and sounds it to unleash the machine's inner monster. Reinvigorated, the mechanical man heals and deploys all its gadgets to annihilate Kongo. After the match, however, the machine explodes and leaves Mickey and Minnie to kiss amid its dancing wreckage. Despite the machine's victory, modernity appears just as uncontrollable as primitivism, at least when placed in the hands of an empowered woman.[79]

Fig. 5.3. An image from the battle between Mickey's Televox-inspired mechanical man and the Kongo Killer that demonstrates the machine's physical superiority to the savage beast, itself a racist caricature of the era's black boxers.

In 1933, Disney avoided the term *robot*, but four years later it embraced the term when Donald Duck fought against a "robot Butler" in *Modern Inventions*. Much like Mr. Telelux and Rastus, this robot relies on a photoelectric cell—shown in the film as a gigantic single eye—to see the world around it, including that Donald insists on wearing his hat indoors. When the robot forcibly removes the hat, Donald quickly replaces it with another, which prompts the robot to again remove it. This cycle unfolds throughout the rest of the film as the machine, programmed to enforce conventional manners, proves incapable of accommodating Donald's wish to keep his cap. As he tours the facility and fends off the robot, the duck encounters other machines. A robotic hitchhiker's thumb pokes him in the eye. Another machine gift-wraps him in cellophane. An automatic baby cradle wraps him in a diaper and bonnet. A barber-and-shoeshine machine entraps him and then parts and pampers his rear feathers while another set of mechanical arms blackens his face before telling Donald he looks "like a new man." Unable to accurately process reality or understand individual desires, the machines in *Modern Inventions* do not liberate Donald; they entrap, infantilize, molest, and blacken him.[80]

Each of Disney's depictions of Westinghouse-like mechanical robots expressed anxieties but ignored the concerns of workers. Instead of technological unemployment, these films—as did pulp writers and intellectuals—centered on issues of race and gender. Mickey's mechanical man plays as a caricature of a white gentlemen who is too calm to fend off the blows of the racist caricature of a black boxer until the discordant sound of an automobile horn arouses him to fight to victory. The machines in the hall of invention visited by Donald Duck are far too impractical and poorly conceived to replace human workers. More of Rube Goldberg–style satires of the mechanical pursuit of efficiency than threats to jobs, they undermined Donald's whiteness and masculinity. Like Westinghouse's devices, both visions of the mechanical man contained a hint of menace but undercut it with humor. Even when the robot seemed out of control, it posed no substantial threat to well-behaved white workers. In the most widely spread response to the Westinghouse robots, the company's choice to turn the robot into a joke-telling machine seemed to succeed.

The World of Tomorrow

The most popular Westinghouse robot was also the last: Elektro, the star of the Westinghouse Pavilion at the 1939/40 World's Fair in New York City. Entitled "The World of Tomorrow," the fair was the most popular public exhibition of the early twentieth century. The title encouraged Americans to look forward to a world where the Depression and war did not exist, a world in which Americans had learned to make science and technology their slaves. Forty-five million people passed through the turnstiles. Given the relatively high ticket prices, most of the attendees were members of the middle and upper classes. Unlike previous fairs, which had been joint government-business ventures, the 1939/40 fair was planned and managed by the large corporations directly responsible for the expansion of machines in American society, including General Motors, General Electric, American Telephone and Telegraph, and Westinghouse.[81]

The fair's central message connected individual liberty to technology. Opening on the sesquicentennial of George Washington's election, "The World of Tomorrow" fixated on the revolutionary tradition. In the center of the main concourse, Constitution Mall, stood a sixty-eight-foot statue of Washington. Proceeding down the concourse, visitors encountered statues representing freedom of speech, assembly, religion, and the press. On side streets and in exhibits, they experienced the fair's second theme: a promise that scientists, engineers, and the corporations for which they worked were devising mechanisms to liberate people from the tyranny of the machine.

Few exhibits captured the link between individual freedom and technology better than Elektro. Performing six to ten times a day, Elektro was likely seen by millions of Americans both at the fair and, through newsreels and films, in theaters across the country. The company's engineers in Mansfield, Ohio, built Elektro around the same size as Willie Vocalite—seven-and-half-feet tall, 260 pounds, with an eighty-two-inch chest—and incorporated controls by voice rather than light or music. Engineers later provided Elektro with a mechanical dog named Sparko and made plans to build an entire robot family that were waylaid by the war.[82] To increase the robot's ideological power, Westinghouse placed it atop a twelve-foot stage beneath a gigantic mural depicting an amazing technological future and hired women to serve as its main operators. Westinghouse advertisements further reinforced the gendered points the performances hinted at: "Unlike the man-made Frankenstein monster of fiction, the Westinghouse robot is all kindness and geniality in spite of his towering size and formidable appearance. The slip of a girl who puts him through his paces orders him around at will and even chides him when he doesn't snap to it."[83] If a girl could control such a gigantic creature, Westinghouse was implying, then surely anyone could.

To spread its message, Westinghouse produced a film entitled *The Middleton Family at the Fair* that takes a fictional family from Indiana to New York for the fair.[84] Tying company advertising to technological propaganda, the film focuses on the efforts of two white men—a Midwestern father and a young engineer—to convince women, children, and an Eastern European immigrant to keep their faith in American science and technology. As the film opens in Indiana, the son, Bud, is so disillusioned by the lack of jobs that he cannot listen to the news; instead he turns the radio to a jazz station and tries to dance away his sorrow. The daughter Babs's situation is even worse: she is dating an immigrant, artist, and communist—a trifecta of anti-Americanism per the company's logic—rather than the white engineer whom she dated in high school and who currently works for Westinghouse. To restore his children's faith, the father has the family and the immigrant meet at the Westinghouse exhibit, where the former boyfriend turned engineer will show them the benefits of the machine age. When they meet, the engineer and the communist debate the relationship between labor and technology. The engineer dismisses any concerns by dispassionately summarizing the pre–Great Depression conventional economic wisdom: mechanization has created new jobs, lowered prices, and increased demand for products. To underline the point further, he lectures that "electric control" has made the steel mill safer for workers than their homes. Dumbfounded by the logic of American engineering, the immigrant artist offers no substantial reply. Bud grows more optimistic, while Babs loses interest in her communist boyfriend.[85]

When the film introduces Elektro, the family watches in amazement as it performs its tricks "under his own power." Throughout the performance, the male controller and Elektro banter back and forth about who controls the machine. In this banter, Westinghouse acknowledged people's fears of out-of-control science and technology common in the Great Depression, but only to dismiss and mock them. The film's producers used the family's verbal and physical responses to ensure that the audience views Elektro as an object of wonder and humor. After witnessing the first moments of the robot's act, the father exclaims, "That is the most remarkable thing I have ever seen!" His son, looking up with just as much wonder, provides the humor: "Boy, what a guard that would make on my football team!" The grandmother is equally amazed. "Why, he's almost human!" she shouts. As the family reacts, the engineer calmly describes the scientific principles behind the spectacle.[86]

Elektro was the centerpiece, but the most telling moments in the film came in two kitchen scenes. At Westinghouse's pavilion, the mother and grandmother attend a "Battle of the Centuries" between a "Mrs. Drudge" who frantically washes dishes by hand and a "Mrs. Modern" who occasionally glances up from her magazine as an electric dishwasher completes the task. After Mrs. Modern's victory, the grandmother notes with approval that the winner will "look young when she's a hundred." Such an appeal to vanity was typical of Westinghouse's strategy, but the grandmother's love of appliances does not stem from a concern about her own looks; she already had a black servant who, later in the film, is seen drying dishes. Much as the company had implied with Rastus, the film insinuates that using a machine would be far preferable to employing a nonwhite servant.[87]

A more communal and democratic alternative to Westinghouse's consumerist vision existed even at the fair: one articulated by the most prominent critic and historian of technology in midcentury America, Lewis Mumford. Born in New York City in 1895, Mumford initially shared his generation's faith in technological progress. As a child, he attended Stuyvesant High School, which specialized in training young men for careers in engineering and science, and even published a short review of radio equipment for Gernsback's *Modern Electrics* in 1910. Despite this early interest in mechanical journalism, Mumford joined in Randolph Bourne's and Van Wyck Brooks's efforts to reinvigorate American culture and community by focusing on literary and architectural criticism. By the Great Depression, Mumford was an accomplished author and critic interested in using the arts to reconcile human and mechanical values. In 1934, Mumford emerged as the country's premier thinker on technology when he published *Technics and Civilization*, a history of the relationship between machines and culture over the previous thousand years.[88]

In *Technics and Civilization*, Mumford critiqued capitalists for usurping the benefits of the machine for their own power and argued that government must intervene to shape the adaptation of technology. "Instead of being utilized as an instrument of life," he claimed, the machine "has tended to become an absolute. Power and social control . . . have gone since the seventeenth century to those who have organized and controlled and owned the machine." Citing *R.U.R.* and connecting it directly to "Mr. Televox, the modern automaton," he denounced Fordist production for transforming men into machines. Still, he retained faith in human agency. "Contrary to the assumption of those who worship its external power," the critic wrote, the machine "is not an absolute. All its mechanisms are dependent upon human aims and desires: many of them flourish in direct proportion to our failure to achieve rational social cooperation and integrated personalities." If people could use their "imagination and intelligence and social discipline," they could "renounce useless machinery and burdensome routine." Such endeavors required individuals to use language and literature as "protection against the automatic processes of machine civilization" and to coordinate political and economic efforts to achieve a more equitable distribution of goods. "If such a control cannot be instituted with the cooperation and intelligent aid of the existing administrators of industry," he warned, "it must be achieved by overthrowing them and displacing them."[89] *Technics and Civilization* did not advocate control of machinery through consumer technologies or the leadership of engineers but through revolution.

Despite Mumford's radicalism, the World's Fair Corporation invited him to address an early banquet. Speaking to the assembled benefactors, he called for a fair that emphasized "this planned environment, this planned industry, this planned civilization" as a means of escape from the horrors of the machine age.[90] Mumford's focus on planning found its most popular exposition in *The City*, a documentary for which he wrote the narration.[91] Shown several times daily at the Science and Education Pavilion, *The City* was one of the most widely viewed documentaries of all time. Structured in four acts, the film examines American urban life in preindustrial, industrial, metropolitan, and postindustrial settings and suggests ways the country could transition from its current dystopia to a postindustrial utopia.[92] Combining technological determinism with machine age anxieties, *The City* synthesizes arguments that had been building in American culture for decades in support of government intervention to bring order to the chaos brought by the machine. But, unlike *Technics and Civilization*, here Mumford blamed the machine for the ills of the age, not capitalism.

The film opens with an idyllic vision of an eighteenth-century New England town. People work hard to produce goods for local markets; the small town enjoys both unity and democracy. The narrator recounts how life in the community is balanced, and art "isn't something foreign we look at in a showcase—it is in the artifacts of everyday life, the homespun blankets and hand-sewn quilts, the work of blacksmiths and other artisans." As the scene shifts to a blast furnace, the music crescendos to the discordant sounds of industrial life. Over Aaron Copland's score, Mumford's words blame mechanization, even Pound's iron man, for the turmoil of industrial civilization:

"Machines! Invention! Power! Black out the past! Forget the quiet cities! Bring in the steam and steel, the iron men, the giants! Open the throttle! All aboard the promised land!" Intermixing images of the hustle and bustle of city life with images of a restaurant's automatic pancake makers, bread slicers, and toasters, as well as several shots of mechanical traffic officers, the film reveals the horrors of mechanized life and demonstrates how people follow the rhythms and the dictates of machines: "Men and women losing their jobs, losing their grip unless they imitate machines, live like machines." The segment of the film ends unsubtly, with an automobile driving off a cliff.

As triumphant music returns and images of Hoover Dam brighten the screen, the film shifts to its vision of utopia, the town of Greenbelt, Maryland, a city planned by the United States Resettlement Administration, a New Deal agency. In Greenbelt, science and government have tamed technology to ensure a more humane life without degrading work or mass culture. Echoing Westinghouse's message, the film claims that here, "science turns on new currents—who shall be master, things or men. At last men take command. Here science serves the worker and they do the work together, making machines more automatic and the men who govern them more human." In this government-planned city of the future, preindustrial values return: "This new age builds a better kind of city close to the soil once more, as molded to our human wants as planes are shaped to speed." The new city is not overcrowded, and men, machines, and nature work closely together. There are no class differences, and everyone enjoys similar lives. Automatic machines no longer appear in the world; people engage in simple, more individualized forms of leisure such as baseball games and bicycling. Even family relations improve as parents spend more time with their children. In Greenbelt, democracy is reborn. "Here, life comes first," Mumford wrote. "Machines at last serve men and set them free for other tasks and other pleasures besides their work." Through planning, *The City* argued, government could tame the machines and restore preindustrial harmony.

Many of the most popular exhibits endorsed this critique of the unplanned nature of the modern life because it blamed technology rather than capitalism and corporations. Inside the iconic Perisphere was the "Democracity," a vision of a future in which technology, planning, and democracy create a perfect urban community. As visitors viewed the city from a perch on a moving sidewalk, they saw a carefully planned model city radiating out from a centralized skyscraper that contained no pollution or disorder. Streamlined and efficient, the city and the surrounding countryside fit perfectly together.[93] While the Democracity retained Mumford's democratic commitment, GM's Futurama exhibit jettisoned it for consumer technologies. Soon, the manufacturer indicated, a centrally planned highway system would liberate Americans from the confines of geography and traffic congestion and new technologies would prevent cars from escaping their lanes. Together, government planning and technology would enable individuals to drive safely at speeds of up to seventy miles an hour.[94] In GM's future of an American dominated by government-backed corporations and corporate-backed government, the out-of-control automobile of *The City* would be relegated to the past.

Elektro and Westinghouse, however, did not look to solve the problems of the machine with democratic or technocratic planning but with additional consumer technologies. Rather than critiquing the age's lack of order, the company suggested that the real problem of the machine age was that white families lacked power over devices that should be their slaves. Each vision offered reassurance that taming machinery was possible, but only Westinghouse's offered the possibility of personally controlling technology in a way that allowed them to ignore the wishes of other people. Westinghouse and its media supporters offered a preview of the paradoxical theme that would dominate the postwar era: to transcend the horrors of the machine age, Americans had to invent better machines. To restore the harmony of the preindustrial past, they needed to build the slaves of tomorrow.

6

Conditioning the Robot's Brain

In 1931, *Wonder Stories* printed an insurance advertisement that responded to Westinghouse's devices while capturing the conventional wisdom of the differences between people and robots. Beneath a sketch of gigantic, metal, boxy, vaguely humanoid figures, the advertisement proclaimed that "ROBOTS dance, sing, talk, smoke, calculate complex figures, and perform work of intricate character. But who wants to be a robot? A ROBOT DOESN'T FEEL. It has no appreciation of music, no sense of artistic enjoyment, no profound knowledge of the love of man for woman, mother for children or father for the family." Reiterating assumptions long dominant in American culture, the advertisement identified an internal human trait that gave meaning to external behaviors. Robots could perform human activities, but, because they could not feel, they could never appreciate their actions. The advertisement's depiction of machines with only vaguely humanoid bodies supported that distinction as such bodies could never inspire uncertainty over the boundaries that divided the material from the vital. This, too, was typical of the era. As the robot performed additional human behaviors over the 1920s and 1930s, its body—in both science fiction and stage shows such as Westinghouse's—resembled humanity less; it became, as in the advertisement, less of an android and more obviously mechanical as a way to visually separate human from machine. Aside from deranged robots such as Harl Vincent's Rex, most fictional robots of the interwar period lacked the bodies, minds, and souls to do more than duplicate people's physical behaviors.[1]

This vision of the obviously artificial robot helped reinforce lines between human and machine that science and engineering continued to erode. Throughout the first three decades of the twentieth century, the public learned of a series of discoveries in psychology, neuroscience, and endocrinology suggesting that what made people unique was not a soul but a combination of electrochemical impulses and environmental conditioning. While scientists appeared to prove the material basis of human existence, engineers invented devices such as analog computers, gyroscopes, and photoelectric cells that duplicated behaviors of the human mind: interpreting the environment, making decisions, and learning. Science fiction writers frequently speculated on the implications of such discoveries and inventions, but few uncritically muddled the boundaries between human and machine. Aldous Huxley, the grandson of "conscious automata" theorist Thomas Huxley, set his 1932 novel *Brave New World* in a future where chemicals keep people complacent despite oppressive hierarchies. David Keller expanded his racism and misogyny with the 1929 short story "The Feminine Metamorphosis" in which he imagined women stealing testosterone from Chinese men to turn themselves into men and succeed in the business world. Throughout the 1930s, pulp magazines published numerous stories about machines that could think and even a few that could feel. The conservative editor John W. Campbell's 1934 story "Twilight" depicted perfect thinking machines ruling the world. While these stories blurred the boundaries between humans and machines, they typically did so as a critique of the mechanization of human life. *Brave New World* critiqued how materialism—in both philosophy and consumption—deprived people of their freedom; as always, Keller criticized how technology was destroying American racial and gender hierarchies, while in "Twilight" Campbell suggested that thinking machines would deprive people of their innate curiosity.[2] Though such stories spread materialistic theories of human existence, their writers continued to fear the moral consequences of those theories.

Even as Elektro's boxy, metallic body and stilted voice entertained crowds in 1939, two pulp writers, Lester del Rey and Eando Binder (the pen name of the brothers Otto and Earl Binder) were imagining a new type of robot that celebrated the mechanized performance of humanity as better than actual humanity. Inspired by recent scientific discoveries in neuroscience and endocrinology and the invention of various types of "thinking machines," del Rey's story "Helen O'Loy" and Binder's Adam Link series depicted robots with mechanical and chemical parts that enabled them to think and feel. But, to make their robots practically human, both writers drew from an emerging theory of psychology that critics lambasted for treating people as machines: behaviorism. To pass as human, both Adam Link and Helen O'Loy had to first be "conditioned" as human, and that meant they had to acquire culture. Though earlier writers—E. T. A. Hoffmann, Julian Hawthorne, George Haven Putnam, and Fritz Lang, for instance—had imagined such machines, in their stories the eventual realization of the machines' inhumanity resulted in horror and madness. In Binder's and del Rey's stories, human characters knowingly fell in love with machines because the devices' conditioning made them far more perfect than any person. Situating the robot as the ideal man and the ideal woman, Binder and del Rey fused materialistic philosophy to materialistic desire. The robot, in their imaginations, became an ideal self and an ideal consumer technology.

Robots in Mind, Body, and Soul

Del Rey and Binder fictionalized developments in science and engineering to imagine that human emotions, thoughts, and behaviors could be explained by material processes and mechanically reproduced. In the new fields of endocrinology and behaviorist psychology, researchers built on nineteenth-century discoveries to seemingly unravel the mysteries of life, thought, and individuality. At the same time, engineers and mathematicians invented various sensing and feedback mechanisms that could adjust to changes in the environment as well as electromechanical computing equipment that could process complicated mathematical equations far faster than people could. Especially when exaggerated by the press, such developments suggested that nothing except complexity differentiated humans from machines and that all people, as Waldemar Kaempffert would later write, were robots.[3]

The study of "internal secretions" and the glands, endocrinology originated in the ancient world but coalesced as a modern science in the late nineteenth and early twentieth centuries when researchers began to investigate the body's production and use of chemicals. In 1889, the well-respected French professor of medicine Charles Edouard Brown-Séquard claimed to have discovered the "elixir of life" by injecting himself with extracts from animal testes. Around the turn of the century, doctors promoted "organotherapy" that used animal and human organ extracts to treat a variety of diseases. In 1905, Ernest Starling and William Bayliss of the London College of Physicians observed that each of the body's organs produced its own chemicals they called "hormones" that transmit signals to other organs.[4] Such research helped make endocrinology a transatlantic investigation into the composition and function of the hormones and their effects on human identity and behavior. From its inception, endocrinology transgressed the boundaries between human and animal, natural and synthetic, as researchers studied how to synthesize hormones or harvest them from animals. For many researchers, the new science appeared to hold the key to deciphering many unexplained elements of human existence, including the differences between the sexes, the secrets to longevity, and the origins of life.[5]

In the 1920s, knowledge of the glands reached a broader audience with a series of sensationalized discoveries that promised to give individuals control over bodies, minds, and souls. In 1921, the physician Louis Berman proposed that the glands controlled an individual's body and even soul. In the bestselling *The Glands Regulating Personality*, Berman suggested a "chemistry of the soul," which would enable human beings to control anyone's personality: "The internal secretions, with their influence upon brain and nervous system as well as every other part of the body corporation . . . have been discovered the real governors and arbiters of instincts and dispositions, emotions and reactions, characters and temperaments, good and bad."[6] *Popular Science* added another layer of exaggeration to Berman's claims, titling an article "Are Little Hidden Glands Our Masters?" The accompanying graphic showed various personalities and tied them to different glands, while the text spun fantasies of the "transmutation of human beings."[7] Even after Berman faded from publicity in the 1930s, Kaempffert pithily concluded that "a man is what his hormones make him."

Control over the glands proposed nothing less than the building of a more perfect race, but it could also have industrial applications.[8] In *The Iron Man in Industry*, Arthur Pound partially blamed worker fatigue on "the poisoning of the system through oversecretion of the endocrinal glands."[9] To combat such fatigue, Berman argued that businesses hire biologists to manage workers' glands, including by injecting them with artificial hormones.[10] By directing the hormones, researchers hoped they could control individual bodies and personalities to create a more orderly world.

Even as some scientists and doctors fixated on hormones as a source of identity and control, psychologists emphasized culture. In the late nineteenth century, William James had vehemently denied that people were automatons, but by the 1920s, behaviorists accepted the term. Like many post-Darwin scientists, the founder of behaviorism John B. Watson rejected any unique metaphysical elements to human existence. Seeking to reimagine the field as a "natural science," Watson excoriated psychologists for believing in unverifiable concepts such as consciousness and the soul. Instead, he drew on Pavlov's examination of the glandular response of dogs to stimuli to argue that psychologists should only observe and measure the body's responses to stimuli rather than ponder an individual's state of consciousness.[11] He dismissively wrote in his 1924 *Behaviorism* that *consciousness* was "merely another word for the 'soul' of more ancient times" while the "supernatural" soul itself "probably had its origin in the general laziness of mankind" and the efforts of "'medicine men'" to control others through fears of the unknown. To illustrate the racial dimensions of his critique, he imagined how "colored nurses down South have gained control over the young white children by telling them that there is someone ready to grave them in the dark."[12] Instead of the racialized primitivism Watson identified with the soul, behaviorism focused only on observable responses; by doing so, it broke down the divisions between the human and material worlds while suggesting that the individual was defined entirely by actions. As *Popular Science* summarized, "a man's personality is only the sum total of his habits of actions," that "his habits of emotional action are much more important for success than his thinking habits," and that "such emotional habits are not born in us but have been manufactured for us by our parents, our early training and our environment."[13] Such a definition not only placed Watson at odds with earlier psychologists, it also earned him the hatred of religious Americans who blamed behaviorism for the malaise and immorality of the age.[14]

Skeptics of behaviorism had allies even among psychologists, including the eugenicist William McDougall, who in 1911 had published *Body and Mind: A History and a Defense of Animism*, which explicitly rejected the automaton theory of human nature because of the existence of the individual soul. In 1925, McDougall again defended "purposive Psychology" from those such as Watson who saw men as "Robots." McDougall opened his critique with a discussion of *R.U.R.* Čapek, he analyzed, "has supposed a Robot to be so delicately responsive to stimuli, that you could dictate a letter to it, and the Robot would proceed to write it out on the typewriter." Yet, he continued, "there is, in principle, nothing absurd or impossible in this supposition. The mechanists in psychology ask us to assume that men are such Robots, carried to a higher degree of responsiveness." Tracing behaviorism to Herbert Spencer's focus on reflexes, he accused Watson and his followers of ignoring evidence of human purpose and even of misunderstanding machinery because machines were designed to fulfill specific purposes. Such a disdain for human purpose, he claimed, lay at the root of the problems of the machine age: "It is just because modern industry has treated the workers as Robots rather than as men that the modern world is so full of strife and unrest, of strikes and lock-outs, and bitter conflicts of all kinds. . . . The theory and practice of modern industry have been vitiated by the tendency to treat men as though the stimulus-response formula of the behaviorist were true."[15] The only solution was to reject mechanistic thinking in favor of a psychology concentrated on human consciousness and willpower and to acknowledge the mysterious unverifiable vital forces that dictated human existence.

McDougall's animistic response to behaviorism, however, seemed increasingly at odds with the scientific world, which was beginning to observe how stimuli activated electrical impulses in the brain. Scientists had written about the role of electricity in biological life since the late eighteenth century, when the Italian physician Luigi Galvani showed how sparks could stimulate the muscles of a dead frog. Researchers in the nineteenth century built on this work by illuminating the role of neurons in transmitting electricity through the body, including the brain. In the early twentieth century, researchers fixated on the human brain's apparent ability to produce bursts of electricity, which scientists dubbed "brain waves." Initially, science writers speculated on the possibility that people emitted these brain waves from their heads like a radio transmitter. In 1921 Italian researcher Ferdinando Cazzamali claimed to have used a radio receiver to record electromagnetic waves emitted by the brain. Harnessing such waves, he claimed, allowed him to visually depict anything a person happened to imagine.[16] Such discoveries fed a growing interest in human telepathy and the weakening of the dividing lines between human and machine. As the *New York Times* dismissively reported the finding: "It turns out that we are all dynamos."[17]

By the late 1920s, researchers had begun to show that brain waves were not emitted from the brain but within it. In the 1930s, scientists began using electroencephalograms to record brain waves and chart the relationship between types of waves and mental processes. Linking thought to electrical impulses, they gradually explained the physical processes behind mathematical calculations, creativity, and even dreams.[18] Moreover, just as with hormones, scientists depicted brain waves as the basis of individuality. Scientists showed that each person's brain waves were unique and speculated that these could be used in a fashion similar to fingerprinting to identify individuals. Commenting on such discoveries, the *Times* editorial board pondered whether people should "conclude that our cravings, ambitions, ideals, sympathies and yearnings are but manifestations of volts at work." Its answer could hardly have been reassuring to animists such as McDougall: not yet.[19]

The discovery of the brain's electrical impulses paralleled the emergence of machines capable of responding to environmental stimuli and processing information. Machinery that could respond to stimuli such as changes in pressure had long existed—safety valves on steam engines, for instance—but innovations on the gyroscope during the Great War enabled it to alter movement independently of human control. One early description of the gyroscope-guided torpedo compared it to "a steel fish with a mechanical brain, alive and throbbing to the utmost limit with the most wonderful of human mechanisms, capable of diving like a porpoise, steering itself, and, at the speed of a mile a minute at first, ploughing invisible through the sea." Yet, it was also "the soul and heart of the death dealing monster" and a "veritable pilot and human like steersman."[20] Though the gyroscope appeared too early to earn the name "robot," later writers frequently used the term to refer to machines that responded to stimuli. In its article about the televox, *Popular Science* also identified an "antiaircraft 'robot' gun which computes and holds its range against invading aircraft" by sensing the sound waves unleashed by the enemy plane.[21] The emerging technologies led one writer to claim that the Westinghouse devices were "robots in name and fancy only," and the real robots machines "which if they cannot think . . . certainly appear to do so." He listed as examples a new kind of automatic steering mechanism for a ship, the dial telephone exchange, and the thermostat.[22]

While the gyroscope and related technologies gave machinery a sense of place and space, other machines processed information. *Popular Science* also referenced a new device produced at MIT capable of solving complicated mathematical problems, which the engineer Vannevar Bush called the "product Intergraph." In another article, the art critic Edward Alden Jewell repeatedly anthropomorphized the Intergraph with masculine pronouns to highlight its connections to the male Westinghouse robot.[23] Other writers wrote of a later "thinking robot" that solved problems in algebra, trigonometry and arithmetic "much more quickly and easily than any trained mathematician."[24] In 1938, Howard Hughes credited his "Robot Navigation Computer," a "mechanical brain" that could pinpoint the plane's location with a few data points, with helping him and his crew fly around the world.[25] Though the broader culture typically considered robots to be mindless, America's scientists and engineers gradually redefined them as machines that could think.

Research into human and mechanical brains often overlapped in the 1930s, particularly when conducted by behaviorists interested in the learning process. The media focused most on the work of Norman Krim, a graduate student at MIT who would later work for Vannevar Bush's defense contractor Raytheon. Krim's "Electrical Analogue of the Conditioned Reflex" used a combination of a rheostat, photoelectric cells, wires, and a light bulb to model the learning and forgetting of information.[26] By the mid-1930s, such devices were common enough that high school students could develop them. Using "springs, levers, pinions, electromagnets, a protruding arm . . . and a vertical maze," a high school student named Thomas Ross built a device capable of figuring its way through a maze and then using electrical switches to remember that path on subsequent trips.[27] Two years later, Ross worked with a University of Washington professor to build a "robot rat" that could recall "what it has learned far better than any man or animal. . . . No living organism can be depended on to make no errors of this type after only one trial."[28]

Responding to such inventions in 1934, Kaempffert asked readers, "Is Man Only Robot?" He used Krim's invention to discuss the debate between vitalists and materialists. "There is nothing human about an automatic machine," he wrote. "They are merely extensions of ourselves—senses that supplement our eyes and ears under our control, muscles that do our bidding." Krim's machine was different, though, because it could "be animated into a semblance of 'consciousness.'" Such a possibility threatened vitalists because it meant that "'mind,' 'soul' and 'spirit' are meaningless terms inherited from a mythic past and that a human being, whether he be an Einstein or an imbecile, is simply a highly complex organization of matter which happens to be what we call 'alive,' but which will ultimately be explained in terms of physics, chemistry and electricity." Connecting the machine to behaviorism, Kaempffert realized that if Watson was correct, it would be "idle to seek the greatness of Lincoln, Edison, or Einstein in unique combinations of cells or in their heredity. They become merely living machines which have been accidently conditioned to produce social policies of great moment, electric lamps and phonographs, and theories of a universe in which space is curved and time is a dimension." What Krim and Watson suggested, Kaempffert concluded, was that no one had control over their own body, mind, or identity. Men and women were robots with mechanical bodies and conditioned souls.[29]

The Perfectly Controlled Woman

At their most utopian, the new sciences offered the possibility of using external stimuli to tame irrational and socially destructive behaviors—a fantasy of control in an age in which people no longer seemed capable of controlling themselves. To many men of the middle and upper classes, the problem of uncontrolled people seemed most acute with women. Lurking behind male machine age anxieties lay a persistent fear that technology had destroyed the purity and submissiveness of Victorian womanhood by unleashing women to pursue independent desires. In one of the few references to the effects of technology on women in *Men and Machines*, Stuart Chase argued that "the machine has deprived the housewife of her sometime skills, and so forced uncounted women into futility and neurotic unrest. It has also forced women into the wage-earning class, and thus launched the feminist movement."[30] William Ogburn concluded *You and Machines* by noting that "woman doesn't seem to be as much needed in the home as she used to be" and included an illustration in which a robot liberates a woman from "household drudgery" so she can pursue pleasure; machines, he even suggested, led to more divorces.[31] Male critics assumed that just as machinery in factories and offices deprived men of creativity and purpose, home appliances destroyed women's identities and led them to seek purpose in the unrestrained pursuit of pleasure.

By imagining robots threatening white women, science fiction writers echoed such concerns. In Ray Cumming's 1931 story "The Exile of Time," two men travel to the distant future and discover a gigantic "robot" attacking a young woman wearing "billowing white satin with a single red rose at the breast" that reveals "her snowy arms and shoulders." After they stop the attack, they learn that the woman has traveled from 1777, when her father was a major on George Washington's staff. Edmond Hamilton's "The Reign of the Robots" similarly depicted robots murdering a beautiful white woman.[32] Even Superman had to save Lois Lane, an icon of the New Woman, from mechanical men in a 1941 short film after she allows herself to be captured while in pursuit of a story.[33] As in most science fiction, the women in such stories are not characters but objects of lust who need to be saved by icons of masculinity. Chaste, subservient, and in "The Exile of Time" literally from the past, they embody Victorian womanhood. By threatening to kill white women, such robots symbolically illustrate the machine age's destruction of womanhood.

No one made such danger clearer than Keller. By the end of the

1920s, the psychologist had shifted his postwar research from shell shock to sexuality in a series of pamphlets on such topics as companionate marriage, birth control, motherhood, and sexual diseases. During the Great Depression he cowrote a "Personal Problems" series with titles such as "self-abuse," "unsuccessful husbands," and "unresponsive wives" that covered topics that thematically overlapped with his fiction.[34] Though "The Threat of the Robot" features no women, several of Keller's other stories focus on feminism, including "The Psychophonic Nurse," which ridiculed a woman for relying on a "mammy" robot to raise her child while daring to pursue a career as a writer. In his first story, "The Revolt of the Pedestrians," the doctor imagined a world in which most people are "automobilists" who live in their vehicles for practically all of life, much to the detriment of their leg strength. Though the plot focuses on a pedestrian revolt against the physical and moral degeneration of society, Keller included a subplot on women's sexuality. In one scene, a male pedestrian, disguised as a female automobilist, falls in love with a female automobilist who works as a stenographer and thinks that her love is woman. As the two begin to become physically intimate, the narrator editorializes that "their relationship was something twisted, a pathological perversion. It was a monstrous thing that he should fall in love with a legless woman. . . . It was equally pathological that she should love a woman." When the pedestrian reveals his true nature, the automobilist woman bites his neck, drinks his blood, and leaves him to die. To Keller, the New Woman was not just a lesbian; she was a vampire sucking the life out of men.[35]

Del Rey's "Helen O'Loy" employs similarly misogynistic tropes but in a materialistic vision of the robot as an ideal replacement for the human woman. Two men—the narrator Phil, an endocrinologist, and his close friend, Dave, a robot repairman—fall in love with a female form of "*homo mechanensis*," the "perfect type" of human being that they purchase at a department store. They adjust her artificial brains and hormones, and name her after the ancient Greek famous for her beauty. Before Helen, Dave and Phil dated a set of twins but chose to end their relationships when one of the sisters insisted on watching a soap opera when Dave wanted to watch a rocket launch. The two lonely men devote themselves to improving their "household mech," Lena, who mistakenly puts vanilla on a steak. To give the machine consciousness of its actions, they implant an emotion-secreting gland made of radio tubes and wires. The effort succeeds far too well—Lena begins to complain when they critique its housekeeping. Annoyed by this outspoken mechanical woman, the men replace Lena with a more attractive model with "a full range of memory coils" that they can program to be self-aware.[36] The problem with outspoken women, del Rey insinuated, was too many emotions and not enough brains.

The new mechanical woman looks far more human than Lena and provides "all the works in a girl-modeled case." Its "plastic and rubberite face was designed for flexibility to express emotions, and it was complete with tear glands and taste buds, ready to simulate every human action, from breathing to pulling hair." While its body earns it the name Helen O'Loy, "conditioning" by soap operas makes it a nearly perfect woman who quickly falls in love with Dave. Dave resists such intimacy but refuses to "kill" a machine that appears too human by shutting her off; he flees while Phil and Helen discuss its capacity to love and marry. When Helen says that it would make a perfect wife, Phil replies, "And give him strapping sons to boot, I suppose. A man wants flesh and blood, not rubber and metal." Helen responds by pleading, "I can't think of myself that way; to me, I'm a woman. And you know how perfectly I'm made to imitate a real woman . . . in all ways. I couldn't give him sons, but in every other way . . . I'd try so hard, I know I'd make him a good wife." When Dave returns, he agrees to marry it and the two live together, childless but happy, for the rest of his life. "No woman ever made a lovelier bride or a sweeter wife," Phil recalls. "Helen never lost her flair for cooking and making a home."[37] Helen so perfects the role of wife that Dave forgets that it is a machine while Phil, awed by its perfection, refuses to marry anyone else. Helen is so committed to her husband that when he dies it asks Phil to melt it with acid and bury it alongside its husband. Instead of a frightening display of rebellion, the story concludes with a potentially more disturbing robot suicide.

Helen offered the predominantly young male readers of the pulp science fiction a fantasy of the restoration of power through consumer technologies. Del Rey's choice to begin the story with Phil and Dave rejecting human women because they want to watch soap operas rooted the men's discontent in women's demands for access to the pleasures of consumer and mass culture.[38] A robot purchased at a department store, he imagined, could provide men with what they most coveted: a partner who was a possession and would devote herself to serving without threat of anger or resentment. Owners would not even have to control their robots; after the initial conditioning, the machines could control themselves and love unconditionally for as long as the men wished; no coercion needed. Mechanical women, del Rey suggested, were not only possible, they were preferable to real women—at least for those men who did not desire children.

While imagining Helen O'Loy as the perfect woman, del Rey fused endocrinology, mechanical brains, and behaviorism to draw parallels between human and robot. With its artificial hormone gland and "memory coils" and conditioning provided by soap operas, Helen learns to imitate all the behaviors of del Rey's ideal wife. The only element of womanhood that it cannot perform is one that had been essential to women's identity and purpose: reproduction. Nineteenth-century Americans had seen something mystical and spiritual in motherhood that had prevented women from ever being fully mechanized. If he had fully embraced materialism, del Rey could have imagined a form of artificial reproduction, as other science fiction tales had done earlier in the decade.[39] Instead, he chose to keep reproduction as a separate, nonmechanizable element of womanhood. But neither Dave and Phil expresses any desire for children; all they want is Helen and the pleasure and comfort it can provide. Abandoning historic links between female identity and motherhood, the story assumes that women are defined by their behaviors—particularly how they serve men—not an internal spiritual or biological power.

The two men's professions reinforced such parallels between human and machine. Dave, a robot repairman, works to fix machines whose behavior has gone awry; Phil, an endocrinologist, does the same for people. While Dave is debating Helen's infatuation with him, Phil visits a female client who asks him to administer "counter-hormones" to her son and the unacceptable woman he loves. When Phil returns, one of his proposals to fix Helen's infatuation with Dave is to change her hormones. The two situations suggest that del Rey's point was not that machines could replace women but that all behaviors are controlled by hormones, electricity, and conditioning rather than autonomous will. If a parent does not want a child to have sex with a member of a lower class, then he or she can pay a doctor to alter some sex hormones. If a man does not want to find a compromise with a woman or deal with her complaining, then he can find a more amenable substitute in a machine. In "Helen O'Loy," del Rey merged the materialism of consumer desire with materialistic theories of self. In the machine age, he imagined, human identity and even the most intimate of longings could be manufactured and sold.

"Helen O'Loy" was not an immediate success. Readers for the magazine ranked it only in the middle of the issue's stories, and few letters in the subsequent months rated it highly.[40] Del Rey, however, continued to write and gradually became one of the most important editors and publishers of young adult fiction. As people became more interested in perfectly human robots after World War II, science fiction writers and fans increasingly saw the possibilities in "Helen O'Loy." Reprinted in several anthologies, it grew more influential as stories of perfect mechanical women proliferated, especially in the 1960s and 1970s when feminists reinvigorated the assault on male privilege. In 1970, amid the Women's Liberation Movement, the male-dominated Science Fiction Writers of America voted the story into their hall of fame and republished it again.[41] A testament to the limitations of materialism in the nineteenth century, stories of mechanical women became the ultimate expression of materialism's misogynistic possibilities by the second half of the twentieth.

The Perfect American Man

Science and engineering also hinted that it was possible to build the perfect man. Before the 1920s, most stories accepted that the efficiency and controllable nature of mechanical men made them ideal replacements for physical laborers and soldiers; however, few suggested that they could replace men in other ways. Even Westinghouse's jokes about sex depended on the idea's absurdity, as the machines lacked the independence, free will, and self-control necessary to fit masculine ideals. Adam Link suffers from none of these deficiencies. With an "iridium sponge brain," Adam is, as its secretary would come to believe, the perfect man and, as the United States government would recognize, the perfect American.[42]

Adam debuted in the January 1939 edition of *Amazing Stories* in a short piece Binder entitled "I, Robot."[43] In the cover illustration by

Robert Fuqua, the machine has the requisite appendages, torso, and head to mimic the human form, but all are distinctly mechanical, with dials, buttons, metallic skin, rivets, and other assorted devices. The story's title referencing the novel *I, Claudius*, however, suggested that the machine can think and is self-conscious. The pose in which Fuqua drew Adam reinforced this point. Instead of the typical scene of a deranged machine attacking a human, the cover illustration shows the reverse: an unhinged man threatening to shoot a placid robot while the machine restrains a snarling dog. Though Adam barely resembles a person, the machine, with its hand passively aloft, assumes the role typically played by human characters, while the man, his face full of hatred, assumes the typical role of the machine.[44] As it protects human from beast, Adam illustrates the theme that made Binder's work novel: the humanity of robots.

Binder wrote "I, Robot" to challenge fears about the Frankenstein nature of machinery. He loosely based the plot on that novel but exclusively tells the story from the perspective of the creature in the form of a suicide note that recounts Adam's first moments of consciousness, training at the hand of its loving inventor, Dr. Link, and its difficulties interacting with people after an intruder murders the doctor. Like Victor Frankenstein, Dr. Link works alone in a laboratory. Forsaking marriage and a family, he devotes his life to inventing the "iridium sponge brain," a device that would give robots the ability to think, learn, remember, and feel.[45] With similar brains, he believed, machines and humans could work together as equal citizens to improve the quality of life for all people. With Adam, he believes he has finally fulfilled his life's purpose.

Like a good American, Adam's first thoughts are of freedom. Finding itself chained because Dr. Link does not know if the brain will work, the robot fights against his fetters. Adam quickly earns its freedom, however, by proving its intelligence and self-control when Dr. Link's dog attacks it. Following an instinct to self-preservation, the robot initially uses a free hand to strangle the animal; yet, as it hears the dog yelp in pain, it releases the creature and shows Dr. Link that it can exhibit self-control, learn, and feel sympathy. With confirmation of the brain's success, the doctor, unlike Frankenstein, unchains Adam and begins teaching it how to be human. Throughout the narrative, Adam compares its upbringing to that of a child and

Fig. 6.1. The artist Robert Fuqua's original *Amazing Stories* cover illustration for Eando Binder's Adam Link. While the robot has a vaguely human shape, it could never pass as human, though it does hold a dog back from attacking a person to demonstrate its own tamed nature. Courtesy of the Lilly Library, Indiana University, Bloomington, Indiana.

observes that, due to its superior brain, it has far exceeded normal human capabilities. This training—not Adam's physical body or some metaphysical mind—has made Adam "basically human."[46]

Unfortunately, an intruder murders Dr. Link immediately after he imparts his utopian vision of human-robot equality to Adam. To honor its "father's" memory, Adam spends the rest of the series transforming that vision into reality. First, however, it must confront Americans' prejudice against mechanical men. Upon seeing Adam, every other person in the story, including the police, recoils. In a direct reference to Shelley's novel, Adam flees into the countryside and saves a girl from drowning, only to have her family attack it. Pursued by a crazed mob, Adam locks itself in Dr. Link's house, where it writes a letter proclaiming its innocence and prepares to electrocute itself to prevent accidental harm to a human. Binder ended "I, Robot" with Adam's signature, written in individualized cursive rather than a mechanical font.[47] Adam, Binder suggested, is more than a self-aware, self-controlled, machine. Willing to sacrifice itself for others, it was the selfless, intelligent, and empathetic individual long celebrated as the ideal American.

But Adam does not commit suicide. Instead, it lives through nine more installments and becomes, by the 1940s, the most popular science fiction robot of the era. Earlier writers had experimented with telling their stories from the perspective of a self-aware machine. The English writer John Wyndham had even used the same device of a suicide note from a lonely robot in 1932.[48] But no pulp magazine antecedent had proposed what Adam Link offered: a robot more humane than real people. This originality ensured that readers identified with the robot; they found in its struggle against a culture that refused to accept its humanity an element missing from other robot stories. Each adventure earned rave reviews. At the 1940 World Science Fiction Convention in Chicago, fans apparently cosplayed as Adam.[49] One letter writer explained the appeal: "I think the Adam Link stories are the most human and interesting that it has been my pleasure to read in many a moon. . . . Here's to Adam Link; long may he live!"[50] Another writer praised Adam as among the best science fiction characters because "to give a mere robot an appealing personality is a real accomplishment."[51] With Adam, Binder had achieved what few other writers had accomplished: turning the robot into a character with a human personality rather than a symbol.

In the following three years, Binder reinforced Adam's humanity, masculinity, and American identity. The plots of the sequels ranged from the absurd to the serious. In one story, Adam builds a wife that it appropriately names Eve. In another, it travels back in time to discover that the Norse god Thor is a robot from the future—Adam itself. In the next, it builds a utopian colony where people and machines can work and live harmoniously; the colony fails, owing to human selfishness, but Adam remains optimistic. Binder tied such divergent stories together with Adam's quest to become American. Nearly every story found it trying to convince people to accept that robots are neither out-of-control monsters nor Westinghouse-style slaves. To be an American, Adam has to prove that it is not only strong and intelligent but also self-controlled and selfless, that it is a man who freely chooses to place the good of America and its citizens first.

Binder synthesized republican models of citizenship and masculinity with a modern understanding of the individual. Adam is, in many respects, a model of classical virtue. From its initial effort to commit suicide to its donation of the millions it earns in business to its final willingness to die to save the United States from an alien invasion, it demonstrates an unwavering commitment to serving the public good. Early in its life, Adam risks death to rid a city of political corruption; in another, it proves it will abide by the rule of law when wrongfully convicted of murdering its father and sentenced to death.[52] Most important, given World War II and the Binder family's ties to Austria, the robot, despite its lack of citizenship, becomes a soldier and then a citizen. Though Adam normally resists violence, it joins America's side in the war effort. When it goes to war for the first time, it and an army of robots repel a Japanese invasion of the Southwest after their general displays that he is an inhuman monster by killing a Mexican boy.[53] In its second war experience, Adam wins citizenship and a Congressional Medal of Honor by single-handedly preventing what the government believes to be a Nazi invasion, but which turns out to be aliens.[54] In both cases, Adam is, as the country itself was frequently depicted, a reluctant but powerful warrior who resorts to violence only out of necessity. Adam is more than a citizen-soldier; it is America itself.

The robot, however, is also a perfect man. In the first two stories, Binder gave no indication of why he chose to make Adam male. Anatomically, genetically, or hormonally, Adam is neither male nor female; it is an unsexed machine that does not look remotely human. Yet, Binder always identified Adam as a man, though he only directly addressed his gender in "Adam Link in Business," the story in which Adam learns "the secret joys of human love." Adam, at the urging of its friend Jack, hires a secretary named Kay Temple and is immediately struck by Kay's beauty, intelligence, and innocence; to it, she is the perfect woman. Holding her "soft little hand" makes Adam realize that it is "a *man*, in mind, not a woman" and it begins to question why.[55] The answer it arrives at extended the environmental theory of human nature further than any behaviorist would imagine: it is a man because it has been trained to be a man. Gender differences, Binder implies, exist because of training, not biology. To further reinforce this point, in the subsequent story, Adam has Kay train a machine with a body identical to Adam's into the perfect woman. Such statements added a gendered meaning to Dr. Link's comments to Adam in "I, Robot": "You have no heredity. Your environment is molding you. You are the proof that the mind is an electrical phenomenon, molded by environment." With such a brain, Dr. Link claims, Adam is "as human as environment could make him."[56] Humanity, the doctor assumes, is little more than a collection of electrical wiring and environment conditioning; training, not biology, is what made Adam and his robot wife, Eve, male and female and gave them their personalities and character.

Adam's mind also makes it far superior to ordinary men. Binder frequently demonstrated Adam's intellectual dominance over top scientists but he also showed the robot's physical dominance. In "Adam Link, Champion Athlete," Adam races the best human athlete in the country. As they near the finish line, the robot is poised to win but decides to slow enough to allow the human to tie so that people will think that Adam is "only a man." Despite Adam's intention, the message was clear: people could physically defeat Adam only when Adam let them.[57] As Binder developed Adam's citizenship storyline, the artist Fuqua changed Adam's appearance. When Adam fights the Japanese, its skinny arms are replaced with well-toned arm and leg muscles; its chest, once a rotund hunk of metal, becomes a carefully molded torso.[58] The personification of physical and mental power, Adam is a mechanical superman, a metal prizefighter, and superintelligent scientist all wrapped into one.

Adam is also superior in its soul. Binder wrote Adam with no discernable moral flaws. The stories contained no credible critiques of Adam's behavior except for Eve's charge that Adam places too much faith in humanity. Binder routinely contrasts Adam's superior character with white men who are ignorant, prejudiced, selfish, and cowardly and who seek to control it while denying the robot equality. Aside from two male friends, all men in the story treat Adam with unwarranted hostility. One particularly heinous character, the white overseer of a mine, treats Mexicans, Asian Americans, and robots as subhumans unworthy of workplace safety.[59] When the Japanese invade, the man surrenders while Adam and the other robots fight back. Similarly, when Adam builds a utopian colony, the colony comes apart because the men insist on replacing Adam's benevolent dictatorship with a democracy. Easily corrupted by power, the men begin to turn against one another and, eventually, their robot colleagues.[60] Written even as dictators in Europe and Asia were attacking other countries, the series suggested that human men lacked the character to maintain a democracy but robots did not.

Adam is so exceptional that Kay falls in love with it despite offers of marriage from Jack, one of the few decent men in the series. After Adam donates the millions he made in business to build housing for the poor, Kay tells him, "I don't see you as a robot any more, Adam . . . I see you as a man! You have character, personality, just like anyone else." But it is clear from her comments that she sees Adam as extraordinary: "You are like a man who is big and strong—and gentle. You have kindly eyes, sympathetic lips, a strong chin. . . . You have a grave boyish face, a shock of unruly hair, seldom combed. Your hands are big, thick-fingered, but so very gentle! And when you smile—you often do, I know—it is like a warm sun breaking through clouds." Despite its body, Adam is, for her, an icon of male perfection. After Adam saves her from the unsolicited advances of a businessman, she confesses her love: "You have more heart than many men I've known. It is a person's mind that counts, not his physical body. Your mind, Adam, is that of a great man, and a good man. I love you!"[61] Afraid of their physical differences, Adam refuses her advances and encour-

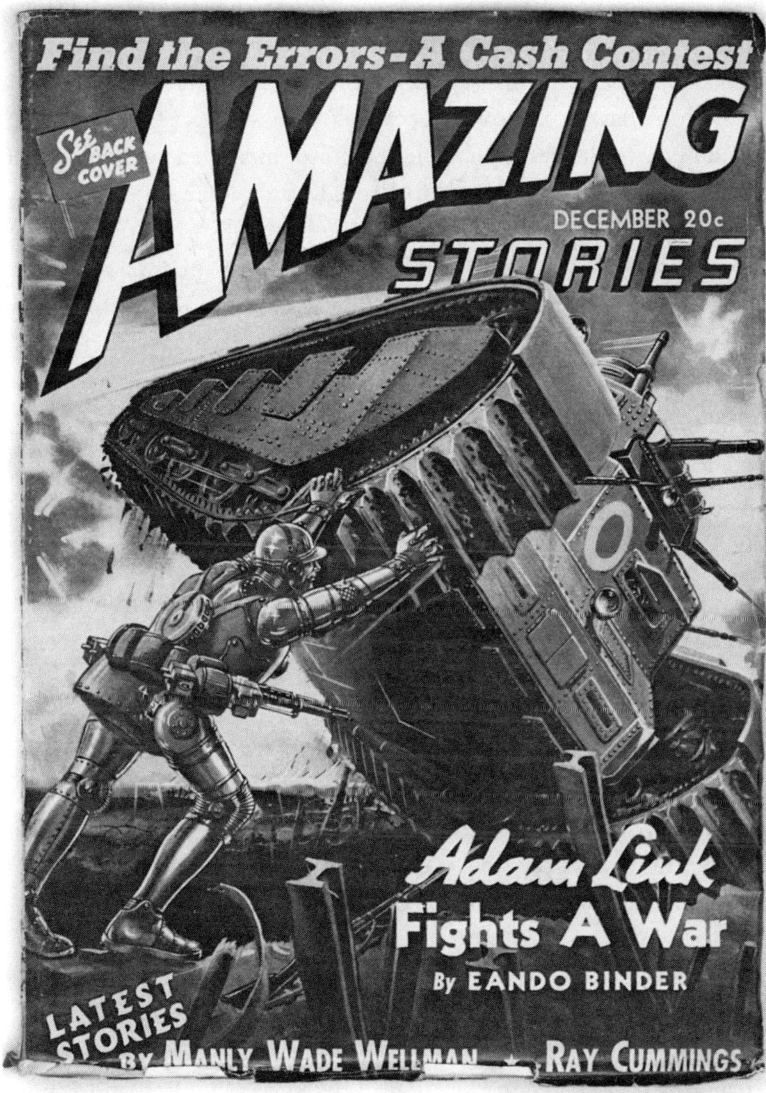

Fig. 6.2. Robert Fuqua's final cover illustration of Adam Link, in which Adam has become a nearly human android whose artificial nature is revealed only by its metallic skin. By this issue Adam has proven itself to be an ideal man and an ideal American citizen. Courtesy of the Lilly Library, Indiana University, Bloomington, Indiana.

ages her to accept Jack's marriage offer. When she reiterates her love, Adam evades the situation by telling Jack to force Kay to marry him. As in Adam's running across the continent, the human being—even one with no flaws—wins only because Adam wishes him to.

Adam personifies the benevolent dictator imagined by Chase or the technocrats and romanticized in Great Depression culture. Ignoring human autonomy when necessary, the robot seeks to provide protection for everyone weaker than it. But Adam is also an outsider, a man rejected from American society because its body is different. Though built in America, Adam is symbolically an immigrant. Written by one brother who was born in Austria and another who was born in the United States, the stories offered, much like Superman, an allegory of immigrants proving their worth to white Americans.[62] Binder's repeated references to white male authority figures who mock other groups of Americans drew parallels to Americans' intrinsic fear of robots. Indeed, Adam feels like so much of an outsider that, like many second-generation immigrants cut off from their traditional culture, it searches for a connection to its past—although it does so by traveling through time to ancient Norway. With Adam Link, the racialization of the robot—implied but underdeveloped in *R.U.R.*—has been inverted to tell a story not about the rebellious nature of dehumanized others but a story about the inherent humanity of people whose bodies were different. What matters with Adam is not biology—internally it is as far from human as possible—or a mystical source of character, but its conditioned behavior, how it treats others. Though lacking a soul, Adam acts humanely, and that makes it human.

Blurring the Lines

With "Helen O'Loy" and the Adam Link series, del Rey and Binder drew from the era's sciences and inventions to imagine the robot as the ideal person, blurring the lines between human and machine in a way that few other stories had accomplished. In doing so, they fused materialist philosophy with the materialism of the consumer economy. With no hint of anxiety, the authors constructed stories that positioned the mechanical device as the ultimate object of desire—both for general consumers and for the opposite sex—because of cultural conditioning. In that fusion, del Rey and Binder were not far from

Watson, whose behaviorism inspired much in their stories and who abandoned academia to work for an advertising firm.[63] Consumers, apparently, were just as robotic as workers.

But both stories also divorced human identity from biology and metaphysics. Internally, both Helen and Adam are far from the organic robots imagined by Čapek, and neither's story suggests that inventors had synthesized a soul. Yet each robot successfully passes as human. Helen, a device that cannot have children, lives for the rest of Dave's life without anyone realizing it is a machine—even Dave forgets he has married a robot. Constructing an allegory about the acceptance of differences between people, Binder preferred to let Adam reveal its robotic nature, but, in one story, the robot successfully disguises itself as a man. Indeed, with the transformations in its body over the run of the series, Adam could easily pass as an exceptionally muscular man if it simply were to wear clothes and makeup. As in every other element of its life, Adam does not usually pass as human because it chooses not to. In constructing stories of robots conditioned to pass as human, del Rey and Binder assumed that all that mattered for human identity was behavior.

For groups of people long denied basic rights and privileges owing to a presumed biological or metaphysical character deficiency—women and nonwhites, in particular—the implications of that transformation were profound. Neither Binder nor del Rey likely knew the significance of the intellectual shift their stories embraced, but in imagining their robots they began to abandon definitions of human identity that had long supported exclusion in American society. It was not coincidental that McDougall, one of the most vocal anti-behaviorist psychologists, was also a virulent racist and misogynist, or that the writers most invested in maintaining a line between human and machine were white, upper- and middle-class men worried about their loss of authority.[64] Once human identity became rooted in environmentally conditioned behavior rather than internal, perhaps even unverifiable, characteristics, it would also became accessible to anyone who could perform it. If neither biology nor character mattered, then everyone, regardless of race or sex, could potentially lay equal claim to agency and power.

Neither Adam nor Helen initially attracted attention outside of the pulp magazines. Even there, while Adam was hugely popular, Helen was obscure, though at least one fan—a young Jewish American scientist named Isaac Asimov—appreciated the tale. Yet, after World War II, their language, imagery, and themes spread through American culture as people wondered what the development of more powerful "thinking machines" meant for human identity. In the consumer culture of the postwar decades, the friendly, practically human robots found in Adam Link and "Helen O'Loy" appeared in novels, films, television shows, and even material culture. In that world, the questions del Rey and Binder raised about the differences between men, women, and machines grew more important as people confronted a culture that hailed robots as the key component in preserving American security and ensuring abundance. But before mass audiences could love the robot, they needed to confront their anxieties of the machine age. They found that opportunity in a war against people Franklin Roosevelt called "the robots of the slave states."[65]

7

A War against the Machine Age

In December 1942, a year after America entered World War II, the producers David Silberman and L. Daniel Blank staged a revival of *R.U.R.* It was a curious choice, but the play—at least, superficially—fit into broader patterns of American nationalism. To writers in the late 1930s and early 1940s, no group of people more closely resembled Čapek's robots than the men, women, and children living under fascism. To middle-class American writers, even assembly line workers had more free will than the regimented robots of Nazi Germany or fascist Italy. As one reviewer explained, "Since quite clearly the world now is engaged upon a final war between the men and the robots, it has occurred to the Messrs. Silberman and Blank to revive the late Karel Čapek's '*R.U.R.*'" The program underscored the war connection by quoting President Roosevelt: "It is the young, free men and women of the United Nations and not the wound-up 'robots' of the slave States, who will mold the shape of the new world." Thematically, *R.U.R.* dramatized the ultimate stakes of the war: human existence. According to the producers' logic, it should have been a success. Unfortunately for them, it was not.[1]

There was one problem with staging *R.U.R.* while America was fighting the "robots of the slave states": in the play, the robots won. After investing $35,000, Silberman and Blank reaped only harsh reviews and empty seats; their revival lasted four performances.[2] Perhaps audiences, like the *New York Times* reviewer, found the dialogue "stilted" and the play "dated"; or perhaps they found a story in which robots destroy humanity a disturbing and far too realistic possibility. Without drastic revisions, the play could not accommodate the changes in the meaning of the term *robot* wrought by the Great Depression or even by the war. During the 1930s, writers such as Lewis Mumford extended *R.U.R.*'s critique by suggesting that the secular worship of machinery—common to all machine age countries—could lead to fascism. In a standardized, regimented, and soulless society, such critics speculated, people looked for spiritual purpose in the savagery of violence and responded to charismatic dictators who made them feel alive.[3] But America's entrance into the war tied that critique to one nation, Nazi Germany, and its ideology in a way that *R.U.R.*'s plot could not dramatize. Before the war, critics identified the robot as a product of the machine age; after the war, they identified it as a product of dictatorship. Drawing on pulp science fiction and film depictions of the robot in 1930s culture, wartime commentators turned the robot into a mindless, remotely controlled totalitarian opponent of the freethinking, gunfighting American.

Americans identified their enemies as robots because they understood the World War as a conflict with the horrors of the machine age itself. True, major elements in America continued to praise technology; reliance on machinery was implicit in Roosevelt's conception of the United States as "the arsenal of democracy."[4] Yet the beginning of the war did not erase the anxieties that had reverberated through the previous decades. For two decades prior to the war, critics had identified problems in their own society like those they found in Nazi Germany, albeit to a lesser extent. For the spiritual person, the United States seemed secular and materialist. For conservatives and even some liberals, Roosevelt threatened to centralize power as had Hitler and Mussolini.[5] Many expressed discontent with the standardization and lack of individuality found in daily life. Others feared that mass communication devices empowered a handful of individuals while turning the rest into a regimented crowd.[6] Despite Westinghouse's propaganda and stories such as Binder's and del Rey's, the predominant image of the machine during the Great Depression remained an out-of-control force determined to destroy humanity. By the end of the decade, nothing epitomized these anxieties quite like the fascist state. Combining the worst elements of modern life—secularization, standardization, regimentation, and militarization—fascism seemed less a departure from the machine age than its logical, savage culmination. As blitzkrieg rolled through Europe, many Americans imagined the war as a contest between traditional and modern values, human men and machine men.

The Regimented Robot

While Americans in the Great Depression debated the origins and nature of fascism, writers of science fiction film serials envisioned dictators conquering the world and enslaving humanity with destructive technologies, including remote-controlled robots. Low-budget films offered as the Saturday matinee in second-run theaters, serials presented a weekly fifteen-to-twenty-minute episode of a larger story connected by cliffhangers. Although marketed to children, such stories would occasionally draw a larger audience and were shown at regular theaters. Sharing writers and directors, the films repeated plots, themes, and character archetypes. The dictators in such serials often relied upon mechanical men with indestructible metal bodies and remote-controlled minds as well as weapons ranging from ray guns to atomic bombs. A metaphor for both the dangers of technology when controlled by a dictator and the dehumanizing power of mass media, the robot became the ultimate opponent of American democracy and manhood.

The first serial to depict robots in the service of a dictator was the 1935 Gene Autry vehicle *The Phantom Empire*.[7] An adaptation of Jim Vanny's 1931 story "The Radium Master," the film starred Autry as himself, a singing cowboy, fighting a mysterious force of masked riders terrorizing his "Radio Ranch" studio. Autry discovers that the riders are the military of a subterranean empire named Murania, hidden since the Ice Age and led by a dictator queen. Infiltrating the empire, Autry finds an advanced radium-powered civilization with television, missiles, disintegrating rays, death-curing machines, and a workforce of mechanical men. Boxlike creatures wielding swords, axes, sledgehammers, or blowtorches, the men trudge through the empire as radios in their heads receive commands from the queen and sensors in their eyes relay images back to the central control room. The empire's soldiers quickly capture Autry and take him to meet the queen, who explains the empire's technologies within the context of the Great Depression. "In Murania," she notes, "we have mechanical men for all our labor. My subjects devote their time to thought and advancing their minds." On a screen, she then juxtaposes moving images of extreme poverty from America with scenes of plenty from her empire. Much like Hitler appeared to have accomplished in Germany, the queen seemed to have resolved the problems of the Great Depression through centralized control.[8]

Nothing symbolizes the queen's power like her mechanical men. On his ranch, Autry uses the radio to introduce young listeners to the rugged but sensitive lifestyle of a singing cowboy; Murania, however, uses the technology to control mechanical workers. In one scene, Autry lies unconscious on an assembly line as the belt slowly moves him toward a blowtorch-wielding mechanical man.[9] Deprived of willpower and trapped on the device that makes mass-production possible, Autry's icon of individualism seems destined to die at the hands of the machine until two young fans save him. In contrast to the guitar-and-gun-wielding Autry, the film's mechanical men are both regimented workers and soldiers. Like the German and Italian masses who seemed swayed by the power of the dictator, the mechanical man was merely the pawn of a dictator's voice.

Three additional serials of that era juxtapose remote-controlled mechanical men with icons of American masculinity: *Undersea Kingdom* (1936), *Mysterious Doctor Satan* (1940), and *Flash Gordon Conquers the Universe* (1940).[10] Each film contains at least one cliffhanger in which the hero—a navy lieutenant, a Western superhero, and a pilot, respectively—confronts a mechanical man unleased by a nefarious, usually nonwhite dictator. *Undersea Kingdom* shows navy lieutenant Ray "Crash" Corrigan defending the good, white people of Atlantis from the darker soldiers of Unga Khan, including his ray-gun-bearing, tank-riding mechanical men the film named "volkites" after the German word *volk*, for "nation." A promotional poster for the serial depicts Corrigan, clad only in metallic underwear, chained against a large stone while one of the volkites threatens his muscular torso.[11] In *Mysterious Doctor Satan*, a mechanical man labeled a "robot" by Dr. Satan face off against "the Copperhead," a superhero invented

Fig. 7.1. Advertising poster for *Undersea Kingdom* in which Crash Corrigan (an obvious knockoff of the popular comic book and film serial character Flash Gordon) sees his naked flesh menaced by the apparently German "Volkite" machine.

for the serial. The titular hero of *Flash Gordon Conquers the Universe* defends a medieval-style kingdom from the technologies of "Ming the Merciless," including giant "walking bombs" that characters label "iron men" and "robots." None of these threats against white masculinity succeeds; each hero uses his powerful body and the aid of good scientists to destroy the dictator. But the repeated battles between metal robots and muscular men foreshadowed how people would envision World War II as a contest pitting steel against flesh and free individuals against mindless machines.

Frequently, such heroes defended white civilizations modeled on a Euro-American past from advanced technologies controlled by racially distinct dictators. The architecture inhabited by the white people of Atlantis in *Undersea Kingdom* resembles those of the democracies and republics of ancient Greece or Rome. The medieval kingdom in *Flash Gordon* is led by Prince Barin, who dresses like Robin Hood. Both kingdoms are threatened by dictators with stereotypical Asian names—Unga Khan and Ming the Merciless—and appearances who, as in the earlier warnings of Lothrop Stoddard's *Rising Tide of Color*, try to use technology to destroy freedom and democracy.[12] In *Flash Gordon*, Ming even attempts to kidnap and marry Prince Barin's sister, a white princess. Yet, as the use of word *volk* in *Undersea Kingdom* suggested, such dictators were drawn from Germany as well as Japan. A combination of the two principal enemies of World War II, such tyrants go far beyond the norms of warfare by employing death rays, poisonous gasses, and killer robots to attack soldiers and civilians. Unlike the noble scientists who aid Flash and Crash, they recognize no moral limits in their use of technology to gain power. Throughout the nineteenth and early twentieth centuries, Euro-Americans had used their more advanced science and technology to justify controlling groups of people they saw as primitive; but in the film serials they reversed the formula: now they protected primitive whites from more technologically sophisticated racial groups.[13]

A children's book starring Stanley Link's popular comic strip character Tiny Tim indicated that the appeal of such a reversal was escapism from the horrors of the machine age. Entitled *Tiny Tim and the Mechanical Men*, the 1937 story largely follows the standardized plot of the film serials, except instead of an icon of manhood, it centers on a young boy no more than six inches tall. Tim is from the medieval-style kingdom of Erewhon, rather than the United States—a reference to the New Zealand author Samuel Butler's 1872 satirical novel about a society that has banished machinery from its borders.[14] When Tim falls into a precipice, he discovers the neighboring kingdom of Boogaboo Land, where he finds giant mechanical soldiers under the dominion of the dictator Zorex, who, despite his white skin, replicates East Asian stereotypes from popular culture: a Fu Manchu mustache and feminized appearance.[15] As Tim sneaks around Zorex's castle, he discovers the dictator's plans to conquer Erewhon and to force its white princess to marry him using such horrific machines as the mechanical "robots," rocket ships, submarines, and an entire menagerie of mechanical animals. Though the plot is standardized, Link's choice to identify the white kingdom with Butler's Erewhon—*nowhere* spelled backward—explicitly drew attention to its lack of modern technologies. Erewhon, as in the kingdoms in *Undersea Kingdom* and *Flash Gordon* and the frontier invoked by Autry's *The Phantom Empire*, is a romanticized space outside the machine age where white masculinity does not depend on technology for survival.[16]

By depicting the dictator's voice—often conveyed over a radio—as the principal control mechanism, such visions played on *robot*'s dual meaning as a humanized machine and a mechanized human. Employing similar remote-control technologies to those found in Westinghouse's devices, these robots were slaves that always followed commands, not monsters operating according to their own volition. Yet, in an age interested in the power of mass media to sway mass crowds, they also served as a metaphor for the human worker and soldier. The 1939 serial *Buck Rogers* made *robot*'s continued dual meaning abundantly clear. Though the standard plot pits its titular hero against a nonwhite tyrant, this dictator creates his "robots" by placing mind-destroying helmets on people that turns them into mindless slaves. Rather than seeing his muscular flesh menaced by a gigantic mechanical man, in the one of the final cliffhangers, Buck and his friends are forced to escape the mind-control helmet. When they do so, they subject the dictator to the ultimate punishment: becoming a mindless, soulless robot.[17] By adjusting the formula, *Buck Rogers* suggests both the continued ambivalence and a new meaning: instead of

Čapek's degraded workers or even a machine, this robot symbolizes the man turned into a machine by the power of a dictator. Instead of rampaging, this robot was regimented.

The Soulless and Savage Machine

The regimented robot spread through American culture in the years before and during World War II as Americans attempted to understand the appeal of fascism. In a 1942 New York divorce case, Justice John E. McGeehan granted custody of three-year-old Margot Reimann to an orphanage rather than her adulterous mother or German father because the mother was too immoral and the father had succumbed to the "taint" of Nazism. This man, McGeehan summarized, had caught "a plague that has paralyzed the souls of its adherents and warped their minds, turning them into Robots dedicated to bringing endless human suffering and complete destruction to all of mankind on this earth who possess sufficient strength of mind and character to resist the terrible affliction of Nazism."[18] For the judge, what turned people into robots was not modern life but a foreign ideology.

McGeehan's identification of ideology as the source of the robot dominated American culture after Pearl Harbor, but earlier critics continued to blame the machine age. During the 1930s, opponents of fascism often explained its rise by contextualizing it in the larger problems brought by the machine, including the transformation of people into robots. The three writers most associated with this movement—Mumford and the European immigrants Erich Fromm and Peter Drucker—came from vastly different political backgrounds: Fromm was a Democratic socialist, Mumford a liberal, and Drucker an Austrian conservative. While they differed on their explanations and solutions, each saw fascism as a reaction to the modern world, a symptom of a broader malaise within a socioeconomic system that prevented people from fulfilling their psychological and spiritual need for purpose. For each, the war against fascism was a war against the dehumanizing tendencies of the machine age.

While Mumford was offering visitors to the World's Fair an optimistic view of societies that had tamed their machines, he published two books, *Men Must Act* (1939) and *Faith for Living* (1940), that predicted what might happen if people failed in that endeavor. Mumford's theory of fascism blamed out-of-control machines for strengthening humanity's innate barbarism. Focusing on the late eighteenth century, when mechanical progress became "the end of ends and the purpose of purposes," he lamented how modern life was "without ethical content or ideal purpose." "As the industrial system becomes more rationalized," Mumford wrote, "the behavior of its members . . . becomes less and less voluntary, less and less subject to self rule." Echoing Stuart Chase in *Men and Machines*, Mumford recounted a factory worker's day: "When the alarm clock goes off, the worker rises; when the factory whistle blows, he goes to work; when the engines start up, he stands ready to watch the spindles or the shuttles." A slave to machines, the worker had "less power, less intelligence, less individual discretion" and became a "small part in an impersonal machine; and that machine is in turn part of a larger and remoter machine." Such a worker, he continued, was but a "robot" ripe for fascism, and would become incapable of exercising "self-help and self-government and self-control." In such a world, Mumford concluded, most became irresponsible and unintelligent; they lost their "manhood," and willingly followed a leader promising to restore it through hatred and violence.[19]

In his 1941 study *Escape from Freedom*, Jewish German American psychologist Erich Fromm similarly located the origins of totalitarianism in the mechanization of men. Fromm pinpointed the origins of fascist robots in the Protestant Reformation and the rise of industrial capitalism. Those forces, he maintained, created two forms of freedom: a positive form in which human beings could be "more independent, self-reliant, and critical" and a negative form that left them "isolated, alone, and afraid." The current age accentuated this difference by enslaving people to machines. "Man has built his world," Fromm wrote. "But he has become estranged from the product of his own hands, he is not really the master any more of the world he has built; on the contrary, this man-made world has become his master, before whom he bows down." Echoing generations of social critics, Fromm claimed that "man does not suffer so much from poverty today" as "from the fact that he has become a cog in a large machine, an automaton, that his life has become empty and lost its meaning." In the modern world, "the individual ceases to be himself; he adopts entirely the kind of personality offered to him by cultural patterns;

and he therefore becomes exactly as all others are and as they expect him to be." Such a person "gives up his individual self and becomes an automaton, identical with millions of other automatons around him." Once automatons, individuals "need not feel alone and anxious anymore" because thinking and feeling just like everyone else provided a sense of belonging, purpose, and power that individual cogs lacked.[20]

Though he rejected their radicalism, conservative writer Peter Drucker agreed with much of Mumford's and Fromm's diagnosis. Born in Austria in 1909, Drucker was raised by a high-level bureaucrat who knew Sigmund Freud, Friedrich Hayek, Carl Jung, and Joseph Schumpeter. In the 1920s, Drucker moved to Germany to study international law. After earning a PhD and watching Nazis burn his essay on the nineteenth-century Jewish philosopher Friedrich Julius Stahl, he fled to London, where he taught economics and participated in seminars with John Maynard Keynes, among others. In 1937, he moved to the United States as a correspondent. From there Drucker published two books on the emergence of fascism in Europe: *The End of Economic Man* in 1939 and *The Future of Industrial Man* in 1942. Across both books, Drucker echoed concerns that the modern world mechanized the soul. Seconding Mumford's and Fromm's arguments, he wrote that "man is isolated within a tremendous machine, the purpose and meaning of which he does not accept and cannot translate into terms of his experience."[21] Mass production, he continued, turned the worker from a human being into "a freely replaceable cog in an inhumanly efficient machine."[22] Under such a system, the worker's only importance lay in how efficiently he manufactured goods, and the self became "standardized, freely interchangeable, atomic labor without status, without function, without individuality."[23] Like Mumford and Fromm, Drucker believed that fascism gave social purpose to these men; it made them feel like real people for once instead of machines.

While Fromm and Mumford blamed the fusion of Protestantism and industrialization, Drucker rooted these feelings of isolation and purposelessness in the concept of "economic man." For Drucker, "economic man" emerged from the liberalism and rationalism of the Enlightenment to stress that "economic satisfactions alone appear socially important and relevant. Economic positions, economic privileges, and economic rights are those for which man works. For these he wages war, and for these he is prepared to die."[24] Such a conception of the human being, he argued, encouraged people to seek freedom, justice, and purpose in an exclusively economic context, which they could never achieve. The Great Depression and World War I revealed that both capitalism and socialism had failed to satisfy that longing for justice, freedom, and purpose. To fill this gap, the Nazis offered an alternative concept of the heroic man, one who achieves freedom and equality by violently sacrificing his sense of self for the good of the nation. Without spiritual alternatives, the masses embraced Nazi savagery to find purpose in a purposeless world.

Such linking of the machine to the destruction of the soul and the rise of fascism found an ally in Charlie Chaplin's 1940 film *The Great Dictator*. Following two main characters, a persecuted Jewish barber and an incompetent dictator—each played by Chaplin—the film ends when the barber, mistaken for the dictator, steps up to a podium in front of an assembled mass army that has just successfully invaded a neighboring country. Chaplin faces the camera and pleads with the army and the audience itself. "Soldiers!" he shouts. "Don't give yourselves to brutes, men who despise you, enslave you; who regiment your lives, tell you what to do, what to think and what to feel! Who drill you, diet you, treat you like cattle, use you as cannon fodder. Don't give yourselves to these unnatural men—machine men with machine minds and machine hearts! You are not machines, you are not cattle, you are men! You have the love of humanity in your hearts!" Echoing arguments against technological unemployment and the decline of spiritual purpose, he remarks that "machinery that gives abundance has left us in want. . . . More than machinery, we need humanity. More than cleverness, we need kindness and gentleness. Without these qualities, life will be violent and all will be lost."[25] Fascism, Chaplin argued, transformed men into beastly machines and denied them individuality and humanity; and it resulted from the retreat of human values in the machine age.

But where some critics identified the robot with the dehumanizing tendencies of the machine age, others saw the individual under fascism as a victim of an ideology that subordinated individuals to the state. Secretary of the Interior Harold Ickes expressed this understanding in 1938. According to him, under both communism and fascism "human beings become political and economic robots. They are taught to believe only that which the dictatorial power wishes them to believe. They are permitted to do only that which the dictatorial power wishes them to do. Under dictation they must lose all the barbarities of supersavage modern warfare upon unoffending peoples. Their diet, their mode of living, their religion, even the number of their children must be by the rote prescribed by the dictatorial power."[26] Here Ickes noted the same effects as machine age critics, but his view of the relationship between the robot and the state differed. For him, the growth of the centralized state *preceded* the creation of the robot. In that difference lay the central transformation of the robot in the period: formerly, people used the term to describe victims of the machine; now they used it to describe victims of the state.

These conceptions of the robot could overlap. In January 1942, onetime Republican presidential candidate Wendell Willkie tied the new understanding to a critique of production efficiency. He described the war as a long struggle against "the most challenging economic conception that has appeared in human affairs since the industrial revolution." This conception was the belief that an industrial economy, controlled by a centralized state, could perform more efficiently than a free economy. Totalitarianism's main allure was its promised efficiency. But the cost to the self was far too high, since it transformed the individual man into "just a cog in the state machine," a "regimented robot(s)" who lacked the spirit of freedom. "To free minds," Willkie claimed, "there are things more important than efficiency."[27]

The shift from machinery to the state as the origin of the robot echoed stereotypes of Germans as automatons. In 1903, Admiral George Dewey contrasted the American soldier who "thinks for himself" with "the German sailor and the German soldier" who "are, to a great extent, automatons, being helpless without their officers." World War I increased this identification of Germans with automatons. Writing in the American journal *Living Age* in 1916, British Major General Sir Alfred Turner called German soldiers "Teuton automatons" who had the "spirit of independence . . . drilled out of them." In 1918, Ohio State professor George Frederick Arps wrote a piece on German efficiency that foreshadowed later arguments made against Nazis. "The chief and somewhat alarming character of the much-touted German efficiency lies in its thorough mechanization of human behavior," Arps wrote. "So completely has the process proceeded under the hands of the governing class that the relatively unpredictable, the so-called spontaneous element, which constitutes the chief charm of existence has all but disappeared from German official life." The result was the transformation of the person into an automaton, the "surrender of the selfhood," and the "savage butchery of innocent and defenseless noncombatant women and children."[28]

While propaganda tended to depict the Japanese as uncontrollable beasts, American writers frequently referred to Nazis as regimented robots. In 1936, the *Atlanta Daily World* expounded that "under the Nazi regime . . . the human individual is regarded either as a robot in the Nazi machine or as a slave of the state." In a 1940 piece criticizing the "moral rottenness of the German national spirit," the archconservative columnist Westbrook Pegler called Germans "Hitler's robots." Colonel Frederick Palmer similarly wrote that "day after day, in robot precision for long hours which the German soldier will endure, Hitler's hordes were being drilled . . . in conquering, offensive spirit, according to a plan. Sacrifice of robot lives was incidental in gaining an objective." Writers also referred to others in German culture as robots. "The German worker," wrote one *Washington Post* reporter, "no less than the soldier, goosesteps to the Nazi tune and dares not complain for fear of punishment and cruel treatment in a concentration camp. He has become a robot in the Nazi machine and has practically no voice in determining either political or economic conditions." Even Olympic athletes were not immune to the epithet. While a small number of writers referred to Italians or even Japanese citizens as robots, most used the term for the "blond robots dragooned into war by totalitarian Germany."[29]

After America's entrance into the war, propaganda films echoed the idea of the robotic nature of the German soldier. The 1943 Disney animated short *Education for Death: The Making of a Nazi* showed the process by which a young boy, Hans, is transformed from a thinking and feeling human into a biological automaton.[30] Mocking fascist education, the film places Hans's moment of transformation in the classroom. Told a fable about a strong wolf that eats a rabbit, he expresses dismay for the poor creature. His sympathy is mocked by his classmates as well as the instructor, who places him in the corner to ruminate on the virtues of strength and insensitivity to the plight of weaker creatures. By the scene's end, classroom pressure has transformed the boy into a regimented soldier who calls for the destruction of all rabbits. As the scene shifts to soldiers marching, a voiceover explains Hans's manhood under fascism: "In him is planted no seed of laughter, hope, tolerance, or mercy. For him there is only heiling and marching, marching and heiling as the years grind on. Manhood finds him still heiling and marching but the grim years of regimentation have done their work. Now he is a good Nazi. He sees no more than the party wants him to see, he says nothing but what the party wants him to say, he does nothing but what the party wants him to do. . . . Now his education is complete, his education for death." By destroying empathy, the film suggested, German education destroyed the soul and turned the soldier into a mindless machine.

Disney's 1942 Donald Duck short *Der Fuehrer's Face* connected the industrial system to regimentation by combining the central metaphors of Chaplin's *Modern Times* with the satire of Hitler in *The Great Dictator*. However, it replaced Chaplin's emphasis on a common humanity with nationalism. In the film, Donald dreams he is a worker in Nazi Germany. After a brass band containing Mussolini and Hirohito awakens him, he works on an assembly line where he, like Chaplin, must screw parts onto a manufactured good—in this case, a mortar shell. But on Disney's assembly line, Donald is a slave to the führer. The duck not only deals with a speed-up, he also alternates between working and heiling Hitler. Even Donald's brief respite from the line shows him moving his arms and legs mechanically in the shape of a swastika in front of a picturesque canvas. Not at all refreshed from leisure, Donald returns to face even greater hardships. After a few more minutes of manufacturing and heiling, he collapses and imagines anthropomorphized shells and machines coming to life and attacking him. The factory in *Der Fuehrer's Face* threatens the individual; however, the film reserved its critique of modern production for a factory controlled by Hitler. To underscore such nationalism, Donald awakens in his suburban home to find a miniature Statue of Liberty draped in an American flag. He kisses it, thankful that he lives in a country where mechanized regimentation apparently does not exist.[31]

Each conception of the origins of the robot would require different solutions. Locating the origins of the robot in the machine age demanded both individual transformation and sweeping changes to social, cultural, and economic structures. While Chaplin's calls for kindness, gentleness, and love offered an individual solution, the others offered far broader visions of change. For Mumford, societies needed to refocus on the "values of art, religion, friendship, parenthood" lost with the modern acceptance of mechanical values.[32] Only those, he maintained, could renew the spirituality and faith which kept innate savagery in check. For Fromm, people needed a truly positive form of freedom which valued their individuality and their relatedness to others. To achieve this, he claimed, human beings had to subordinate "the economic machine to the purposes of human happiness."[33] To avoid becoming automatons, they had to actively participate in their lives. Drucker similarly believed that industrial processes needed to be humanized, since fascism could only be overcome by "a new non-economic concept of a free and equal society."[34] Modern culture had to make "industry socially meaningful" by ceasing to define people by mechanical characteristics, by how efficiently they produced goods.[35] For each critic, the rise of fascism demanded reform of the entire machine age to provide people with meaning and purpose.

The newer conception of the robot called for targeted actions against fascist individuals and countries. One solution was to promote sentimental love, a drive to humanize the machine that had long been a part of American culture.[36] Opponents of the machine age did not disagree, but they echoed Chaplin's focus on a universal love. Critics who emphasized the national and ideological origins of mechanized men, however, emphasized love within a heterosexual relationship as a response to robotic culture. As *Education for Death* showed, not even a mother's love for a child or a child's for a mother would suffice. The only thing that could save a German robot was the love of a woman. For instance, an advertisement for R. C. Hutchinson's *The Fire and the Wood* named a young Nazi man a "robot"; the book itself showed his transformation into a "gentle lover and indomitable fighter" by his love of women.[37] Similarly, in the 1943 film *Hitler's Children*, the Nazi resists mechanical regimentation because of his love for an American girl.[38] Apparently, men could be cured of robotism if they found good women.

The most obvious solution to the problem of roboticism, however, was to kill, isolate, or free the robot. Merely objects controlled by a distant voice, mindless robots could not be convinced of the wrongness of their ideology. They therefore had to be prevented from infecting the rest of the human species. Once robots became located exclusively in the slave states, they or their masters had to be killed. And, if one happened to live in a free state, it had to be isolated from others, as Justice McGeehan did in sending a child to an orphanage rather than to live with her Nazi father.

In declaring their war aims, the Allies combined the two conceptions of the roots of the robot. By invoking an abstract set of "human rights" as the basis of the war effort, the Atlantic Charter required that countries must treat people like individuals rather than machines. Within American culture, the most important of these rights were the "four essential human freedoms" articulated by Roosevelt: freedom of speech and expression, freedom of religion, freedom from want, and freedom from fear.[39] The first two suggested that the fascist state created the robot by depriving individuals of independent thought and spirituality. The latter two, however, addressed broader fears about the machine age's inability to solve the problem of quantity and the armaments of mass warfare that allowed the spread of tyranny. Yet during the war, propaganda implied that these were freedoms that Americans enjoyed but others did not. In *Der Fuehrer's Face* Donald Duck has freedom of expression, freedom of worship, material goods, and freedom from evident fears while in America; when he dreams of himself in Germany, he lacks all four. The conceptualization of World War II as a war against robots gave the Allies' emphasis on human rights a nationalist rather than a universal dimension.

During World War II, the Roosevelt administration and affiliated propaganda manufacturers framed the war as a conflict with Nazi leadership and ideology rather than ordinary Germans. Accordingly, American propaganda generally avoided traditional forms of dehumanization, such as depicting Germans as Huns.[40] Yet, in calling Germans robots, American policy makers, intellectuals, and propagandists continued to dehumanize. Even though calling German workers and soldiers robots denied them agency and responsibility for their actions by making them victims of ideology or the state, it still turned them into a mindless, soulless mass that would, like the film robots, follow commands unconditionally. Caught between one enemy propaganda depicted as animalistic and another propaganda depicted as robotic, Americans and their allies became the icon of humanity fighting against the savagery of both the bestial and the modern.[41]

In the robots of Nazi Germany, Americans also saw something primitive, premodern, and uncivilized: the impulses to brutalize and kill, to torture and enslave. Such robots merged two forms of dehumanization in which Germans became both super- and subhuman, part animal and part machine. As one article recounted, Polish soldiers imagined "not . . . multitudes of German soldiers with rifles, but . . . phalanxes of superhuman or subhuman fighting robots of steel and flame."[42] Almost every account of German robots emphasized their savagery and primitivism, and occasionally their similarities to beasts. In *Faith for Living*, Mumford claimed that "fascism is a deliberate reversion to the primitive: it is an organized revolt against civilization itself . . . an effort to counterbalance an equally perverse overemphasis of the impersonal, the dehumanized, in short the mechanical."[43] The German robots were savage because of their lack of self-control. As one *Washington Post* writer put it, "In a combat between a robot and a man of equal strength, the robot can perhaps triumph because he has no inhibitions."[44] The robot could kill with an efficiency and a lack of limit unknown to humans. During World War II, many saw themselves and their allies as people facing the animal barbarism of the past and the machine barbarism of the future.

Late in the war, Anglo-American newspapers found the ultimate confirmation of machine age savagery: the V1 "buzz bombs" and V2 rockets, which the press typically called "robot bombs," "robot bombers," "robots," or "robombs."[45] The press first used *robot* to refer to remote-controlled weapons during the early 1930s when England, the United States, and Germany attempted to develop pilotless planes. At the time, nobody publicly questioned their morality. But the German use of remote-controlled weapons to bomb targets without the risk of human life elicited headlines such as "Robot Bombs Kill Babies in Nursery," "2752 Killed, 8,000 Injured by Nazi Flying Bombs," and "Nurses Carry Babies to Safety as Robot Bomb Hits Hospital."[46] Without the threat of losing a human life, the robot bomber enhanced the savagery of warfare by enabling Germany to attack civilian targets, especially innocent children.

The American science fiction writer Stanton A. Coblentz rooted the robot bomb in the dehumanization of the machine age. With the robot bomb, he argued, an aggressor could now attack without danger of immediate reprisal. No pilot would have to brave aircraft defenses to drop his bombs. For Coblentz, this bomb was the ultimate sign that modernity was fundamentally evil. As he saw it, the robot was "but the visible embodiment of our inner decay, the visible symbol of a civilization that has lost all respect for man and for those deeper realities without which his life on this planet would have no more significance than that of a wood tick." The robomb was the culmination of a mechanization that had thrown human values overboard. "In the robot plane," he wrote, "we see the dehumanization of man himself. . . . Our obsession with science, which has been the paramount feature of Western culture, involving the gradual exclusion of the humanities and spiritualties of life, has now borne its double fruit: the development of mechanisms of death which operate impersonally, but as if they were human, and the degradation of man to the level where he can operate as a mechanism in utter unconsciousness that he is a man."[47]

Ernest Hemingway's "London Fights the Robots" (1944), however, reinforced the nationalist myth that pitted humans against machines. A stream-of-consciousness narrative recounting his experiences with RAF fighters, Hemingway's essay told a heroic tale of manly pilots fighting monstrous pilotless planes.[48] To reinforce the pilot's individualism, Hemingway identified planes as horses, and war in England's skies as a kind of Western gunfight.[49] "Mustang is a tough, good name for a bad, tough, husky, angry plane," he wrote, "that could have been friends with [boxer] Harry Greb if Greb had an engine instead of a heart."[50] Replete with sexual odes to fighter planes and analyses of the "gruff voices of the pilots," Hemingway depicted the battleground as a frontier or a boxing ring, a place where men could be men by enacting violence against machines. In opposing the totalitarian robot and the patriotic myth of the American frontier, Hemingway, as in the film serials and pulp science fiction, fantasized about the restoration of American manhood through a battle against the totalitarian robot.

The Totalitarian Robot

When Silberman and Blank tried to revive *R.U.R.* in 1942, they offered the public an explanation of fascism that emphasized the universality of the enemy. The play located the origins of fascism in the machine age, in its mindless pursuit of efficiency at the cost of human values.

But the moment at which such a play would appeal to the public had passed. Throughout the 1930s, critics had worried about its out-of-control machines, the standardization and regimentation of culture, and human robots produced in the same factories that provided an astonishing number of manufactured goods. Like Chase, they feared becoming slaves to their tools, and they feared the decline of white manhood through overreliance on machinery. In the prewar period, *R.U.R.* spoke to such worries and, with some not-so-subtle hints, connected them to the spread of fascism.

But others were already connecting those fears to a specific nation: Nazi Germany. Instead of seeing the flaws within their own society, writers projected onto their enemy all the problems of the machine age. Before the war, people entertained the notion that their own society could give rise to such movements; during the war, far fewer could. As they imagined a war pitting strong, muscular men against the super- and subhuman forces of a robotic enemy, cultural producers channeled their fears about the machine age onto a savage other. Focusing on the horrible uses for which another nation employed machines, they grew less concerned about the role of machines in their own society especially, since machinery helped America win the war.

The rise of the robot as a symbol of totalitarianism rather than degrading work paralleled two larger shifts in American culture. First, it mirrored an increasing acceptance of assembly line labor by both unions and intellectuals as the price of increased freedom in leisure and greater levels of consumption.[51] With the growth of the Congress of Industrial Organizations (CIO) and its demands for economic rights, the image of assembly line workers as a uniquely dehumanized mass lacking agency or passion was less tenable. Moreover, the workers who founded the CIO accepted the basic structure of mass production in exchange for greater wages, security, control, and recognition of their humanity.[52] Finally, the valorization of factory labor during the war depicted such work as noble and fulfilling, not stifling and dehumanizing. Instead of destroying the bodies and minds of citizen workers, factory work in wartime propaganda was considered a bulwark of democracy.[53] With both a powerful labor movement and a labor-backed government arguing for the importance of blue-collar workers in American life, the idea that modern production turned workers into robots no longer dominated.

Second, the shift to the totalitarian robot reflected a growing concern that the primary threat to the individual lay in the power of mass culture to manipulate crowds. Between the late 1930s and the 1950s, writers across a range of political and ideological perspectives asked how and why the masses—beyond the alienated working class—had supported evil, and they increasingly found their answer in the personality of crowds. A modern version of an old elite fear of the masses, anxieties of mass crowds controlled by a charismatic leader through mass media reverberated through American political and intellectual culture, including in depictions of radio-controlled robots. It was telling that by the 1950s, the most important liberal theologian—at least, among the elite—was Reinhold Niebuhr, a man whose *Moral Man in Immoral Society* (1932) examined the negative effects of crowds on individual morality.[54] At the same time, anti-Soviet leftists questioned how the entire edifice of the capitalist economy—including what the German-turned-American philosophers Max Horkheimer and Theodor Adorno called "the culture industry," created authoritarian personalities ripe for manipulation by a deranged leader.[55] In such a world, Roosevelt's "robots of the slave states" triumphed over Čapek's robots of mass production.

The war did not eradicate fears that industrial labor might transform men into robots, but it temporarily hid them beneath a veneer of nationalism while expanding them to encompass the entirety of industrial life. While they welcomed victory, Mumford, Fromm, and Drucker continued to combat the malaise they saw at the center of modern life. In the postwar period, Drucker devoted himself to replacing "economic man" by humanizing scientific management through decentralized management and Japanese-influenced teamwork styles of organization. By the 1980s, he had all but given up hope for reforming capitalism and instead turned to the nonprofit sector, a form of work he believed to have a true social purpose.[56] Mumford's disdain for machinery grew as he realized that the war had unleashed two even more dehumanizing technologies, the atomic bomb and the computer, and created massive bureaucratic systems that turned everyone into robots.[57] His days of writing enthusiastic reviews for technology magazines had long passed. For Fromm, too, the war had not ended the malaise at the heart of modernity; in the postwar era, he embraced solutions that stressed individual initiative. Such efforts culminated in his most popular work, *The Art of Loving* (1956), which showed readers how to escape the alienation intrinsic to industrial society by loving themselves and others.[58] For each of these and other postwar critics, the struggle for human values remained incomplete.

Perhaps if World War II had truly ended the war against the "robots of the slave states," such critiques would have found a more receptive audience. But a new war against the totalitarian robots soon began—this time, against the Soviet Union.[59] Fearful of mass destruction from atomic weapons and mass crowds, postwar elites in business, government, the military, and media longed for systems of power isolated from the wills of ordinary people. Cold War elites grew more skeptical of democratic governance and that meant embracing technology and, ultimately, the robot. In national security, calls for "push-button" defense encouraged fantasies of technological robots protecting the nation; in economic life, the growing power of unions encouraged factory owners, managers, and engineers to pursue automated systems that insulated them from worker activism.[60] To save themselves from the totalitarian robots while ensuring a return to mass consumption, politicians, business leaders, and their cultural supporters embraced the robots of computers and automation.

PART 3

Playfellow and Protector 1945–2019

A year before the United States entered World War II, *Super Science Stories* published a tale written by a nineteen-year-old Columbia University graduate student named Isaac Asimov. Initially titled "Strange Playfellow," the story reimagined the robot not as a slave or master but as a friend. Set over forty years in the future, it opens with a young girl, Gloria Weston, playing with a robot, Robbie. Gloria's mother, however, worries about entrusting her child to a soulless machine and has her husband return Robbie to the factory. After her parents inform her that Robbie has run away, Gloria ceases to smile and talk and all but stops eating; not even a puppy or a trip to New York relieves her depression. Convinced that his daughter must learn to think of Robbie as a machine, Mr. Weston arranges for the whole family to tour a factory manufacturing robots—the very factory to which Robbie was returned. When the family encounters Robbie, Gloria excitedly runs in front of a moving vehicle. Robbie saves her and convinces the mother that he can rejoin the family. Neither an out-of-control monster nor a completely dependent slave, Robbie could think—for himself and for the well-being of a child. Though Robbie might be strange, he is, as Asimov showed, a perfect playfellow.[1]

Printed in an upstart magazine, "Strange Playfellow" reached fewer readers than the Adam Link series or "Helen O'Loy," but its impact was far greater. Its placement likely disappointed the ambitious Asimov, who had originally submitted it to *Astounding Science Fiction*, the premier science fiction magazine, as well as *Amazing Stories*, which had printed his first tale.[2] Because it did not even earn the cover illustration, few readers would have guessed that it would establish the dominant postwar vision of robots and that Asimov would play an influential role in strengthening public support for science and technology. Yet that is precisely what "Strange Playfellow" and its author accomplished. Tenacious and confident, Asimov struggled for more than a decade to build a literary career before Gnome Press reprinted "Strange Playfellow" under the new title of "Robbie" alongside eight of his other stories in *I, Robot*. Afterward, he published hundreds of novels and articles while attracting fans by the thousands, including the engineer Joseph Engelberger, who credited Asimov's stories for inspiring his and his partner George Devol's invention of Unimate, a mechanical arm frequently but incorrectly dubbed the "first industrial robot."[3] By the 1960s, Asimov entertained requests for articles from *Reader's Digest* and *TV Guide* and speeches from the RAND corporation, the US military, and science and engineering societies. Other writers, including his friend *Star Trek* creator Gene Roddenberry, followed his lead by imagining robots as the selfless friends of humanity. Though earlier tales of rampaging or totalitarian robots persisted, the most popular and enduring of the postwar robots followed in Robbie's mechanical footsteps.

The child-friendly robot was a key midcentury innovation. Earlier Americans had imagined that machines' lack of souls made them incapable of the love necessary to befriend a child. Within pulp magazines, only the adventures of Adam Link provided a clear antecedent; indeed, the similarity was why *Amazing Stories* rejected "Strange Playfellows."[4] During the 1950s and early 1960s, however, this more genial vision of children and robots appeared in novels, pulp magazines, television shows, films, toys, and countless accounts of children building robots for a new ritual in American schools, science fairs. Amid anxieties of a nuclear-tinged Cold War and the optimism generated by the development of what the economist John Kenneth Galbraith called "the affluent society," the robot became the child's friend and protector.[5]

The robot also gradually became a functional and more human device. Building on prewar innovations, engineers in the 1940s began to develop the basic technologies that would, by the 1960s, come together in an emerging "robotics" industry: digital computing and automation.[6] In the early postwar era, almost all uses of the term *computer* referred to people, not machines. But, like *robot* before it, the term quickly came to mean "a machine." By 1950 the media used it to identify the gigantic, mainframe "robot brains" employed by the government, universities, and major corporations.[7] Initially located on military bases or university campuses, computers remained hidden from the public until the 1970s, but people could imagine them with the aid of periodicals and popular culture, which frequently connected these machines to a cluster of technologies labeled *automation*. The integration of older production equipment with "feedback" devices capable of sensing changes in the environment and adjusting accordingly, automation suggested the possibility of building entire systems of machines capable of producing goods, running the home, and protecting the nation—all automatically with barely any human intervention.[8] Before the war, functional robots had principally been remote-controlled, but computers and automation inspired a new kind of robot that was both self-controlled and conscious of its place in the world. With such capabilities, the robot broke through the last major barriers that eighteenth- and-nineteenth century Americans had erected between themselves and machines.

While computers and automation provided the technologies behind the new postwar robots, the emerging field of "cybernetics" provided the science. Similarly rooted in interwar innovation, cybernetics formed in the 1940s through the work of mathematicians and engineers working on feedback machines, particularly in automatic control systems during World War II. After the publication of the MIT mathematician Norbert Wiener's bestseller *Cybernetics* in 1948, the field grew rapidly in universities as well as public culture. Wiener broadly defined the field as the science of "control and communication in the animal and machine."[9] Such an expansive definition presented a seemingly new way of thinking about the relationship between people on machines centered not on antagonism but harmony—and perhaps even fusion, in the form of the cyborg. This potential attracted academics from not only math and science departments but the social sciences and eventually humanities as well. Though academic interest in the discipline faded in the 1970s, it language and ideas remained central to conceptions of the emerging digital age.

The popularity of Asimov's robots encouraged and drew from such developments, but they also spoke to a larger longing for release from Cold War–era tensions. Beneath a plastic veneer of optimism lay even greater uncertainties and tensions than those of early twentieth-century life. The possibility of instantaneous mass death from above was only the most obvious source. Society had grown wealthier, but the combination of white-collar work and mass consumption seemed, at least to critics, to deprive people of individualism. While Americans in the period grew more outwardly religious, they also seemed to grow more secular. To highlight its superiority to the Soviet Union, the United States sought to project an image of racial harmony, but the country increasingly had to reckon with racial violence and discrimination. Women continually pushed against mainstream boundaries; scientists such as Alfred Kinsey revealed a hidden world of sexuality that horrified some but intrigued others. Throughout the era, Americans grew more aware of the divergence between external behavior and internal character, between images projected—often on a television or computer screen—and deeper realities. Asimov's robots offered the possibility of certainty in the form of the deus ex machina of internal programming.[10] Unlike actual people, Asimov's sexless robots came preprogrammed with an inviolable morality that was always expressed in their behavior; regardless of the situation, they could never kill, never discriminate, never allow a human being to come to harm. Fusing computers and automation to a system of moral certitude, his robots suggested that the long-prophesized "postindustrial" world of meaningful work, abundant leisure, and harmonious community would soon become reality. To read an Asimov robot story was to see a future that was far more secure, abundant, moral, and authentic than the one that existed in Cold War America.

Yet, doubt in the robotic future persisted. Workers fought against the introduction of industrial robots depriving them of jobs that had provided them with both an income and purpose. Countercultural figures such as Allen Ginsberg howled against the "robot apartments" of consumer life, and sociologist C. Wright Mills critiqued the "white-collar robot" that seemed to dominate American society.[11]

Skeptics of the Cold War raised concerns that turning American national security over to computerized automation could itself lead to the destruction of humanity. By the late 1960s, critics found Asimov's robots passé—relics of a more optimistic age in which human beings and machines could learn to work together and ascend to the stars. Embracing more dystopian images such as those found in the work of Philip K. Dick, American culture from the late 1960s into the twenty-first century worried that the moment for taming the machine had long passed.

8

Preserving American Innocence

A few minutes into Robert Wise's 1951 film *The Day the Earth Stood Still*, an eight-foot tall, rigid, and metallic humanoid robot appears on the runway to a flying saucer that has just landed on the National Mall. As it slowly walks down the runway, the crowd flees; soldiers raise their weapons. People have a good reason to fear: one of those soldiers has just shot the robot's companion, an alien who claimed to come in "peace and goodwill" while holding a device resembling a weapon. As the alien writhes in pain, a visor where the robot's eyes should be opens to reveal an empty void. As light flashes from its eyes, each weapon in the crowd instantly disintegrates—but the men remain unharmed. The alien shouts a few strange words to the creature and its visor closes. The alien rises to his feet and explains that the instrument soldiers mistook for a weapon "was a gift for your president" to "[study] life on other planets." Though the alien would forgive the violence, the shot had cost humanity a chance to discover their place in the cosmos.[1]

The scene establishes the film's central theme and the role of the robot within early Cold War culture: the danger of crowds and the protective nature of technology. Chastened for harming a peaceful man, the soldiers take the alien to Walter Reed Hospital, where he meets an aid to the president and reveals his name, Klaatu, and his mission: to discuss the current situation with the leaders of all countries. When that proves impossible, Klaatu escapes the hospital so he can circumvent the government and interact with ordinary people. He takes a room at a boardinghouse, where he meets a single mother and her young son, both of whom immediately display an open-mindedness lacking in the rest of humanity. With the son's help, Klaatu seeks out Earth's brightest mind, an Einstein-like physicist, and convinces him to assemble a meeting of the world's scientists so that they might take control of nuclear weapons. When that too fails, he contacts the robot, which he names Gort, and orders it to turn off all the world's electricity aside from that required by essential services such as hospitals. After that demonstration of the robot's power, a soldier again shoots Klaatu—this time fatally. Gort emerges from the flying saucer, retrieves the body, and temporarily revives him. In his last moments, Klaatu informs a crowd that Gort belongs to a galactic police force that will annihilate any planet that attempts to militarize space. If humanity does not control its penchant for violence, he warns, a race of gigantic robots will destroy them all.

The Day the Earth Stood Still captured early Cold War debates over who should have power over the atomic bomb. In its initial suggestion that the United Nations and then scientists control the bomb, the film endorsed arguments frequently made early in the Cold War. By the time the film premiered, however, hope for universal control over atomic energy had faded with the Soviet Union's detonation of its own bomb. In the real world, human reason had failed to control the most destructive of forces, but Gort symbolized the capability of a larger technological and technocratic system to act far more rationally with nuclear weapons than the emotional masses. The film's promotional materials reinforced such a message. The most common advertisement showed Gort, even larger than in the film, firing its ray weapon into a large crowd while carrying a terrified woman in its arms. Such a scene never appeared in the film but sold the robot as a threat to a white woman. Yet, after viewing the film, people could see that Gort was protecting the woman from the crowd. Unlike in the film serials or most pulp magazines stories, the robot was not the threat—people were.[2]

This vision of the robot differed substantially from those offered before World War II. Wise and screenwriter Edmund North based the

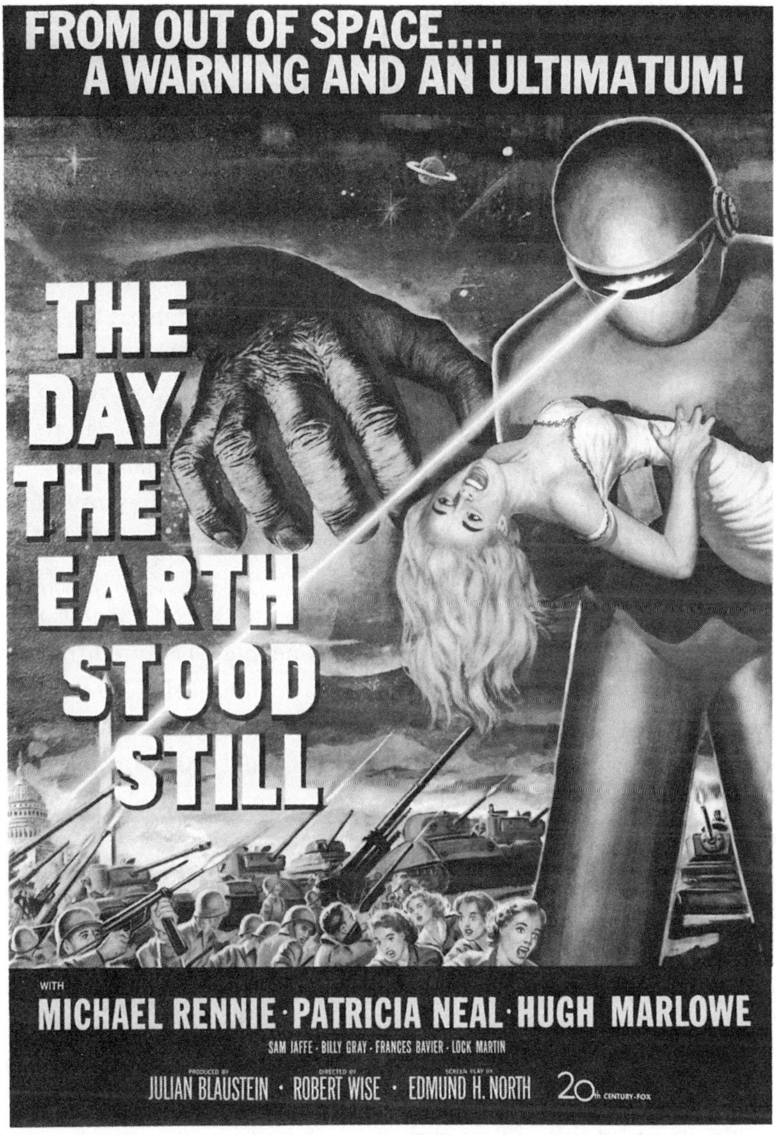

Fig. 8.1. Poster of *The Day the Earth Stood Still* that shows Gort carrying a woman while a mass crowd assembles below. The poster can be interpreted as either Gort or the crowd threatening the woman, but viewers of the film might realize that the main threat was the crowd.

film on Harry Bates's 1940 pulp story "Farewell to the Master," which addresses neither atomic energy nor mass crowds. Instead, Bates's story addresses out-of-control machines. It begins much the same as the film, with a man shooting Klaatu as he and the robot (named Gnut in the story) emerge from the ship. But, while the film makes the shooter one of a crowd, the story depicts him as a "mentally unbalanced" gunman physically separated from the crowd who thinks "that the devil had come to kill everyone on Earth." While the crowd behaves, the lone gunman cannot be trusted. Throughout the story, the narrator, Cliff, assumes that the alien is the robot's master. After Gnut retrieves Klaatu's body, Cliff tells the robot, "I want you to tell your master—the master yet to come—that what happened to . . . Klaatu was an accident, for which all Earth is immeasurably sorry." To that, the robot "gently" but still chillingly replies with the story's final line: "You misunderstand. . . . I am the master." The lone gunman was apparently correct: Gnut is the devil.[3]

That key difference between Bates's tale and Wise's adaptation—between fears of machinery and fears of people—captured the vast gap in understandings of the relationship between robots and people before and after the war. With the rise of the Cold War, Americans turned to technology to minimize risk in and from an uncertain world. Throughout the Cold War, politicians, military leaders, and commentators frequently fantasized about creating a "push-button" defense apparatus that could protect Americans and their allies automatically from aggression, whether it came from Communists, or, as in the era's UFO scares, outer space. Offering far faster, more reliable, and continuous protection, the combination of new sensing technologies such as radar with the first electronic computers suggested that America could build a defensive and offensive apparatus that could transform warfare into a robotic rather than a human struggle.[4]

But the film also indicated that Americans fantasized about protective robots for moral reasons. For well over a century, American culture had emphasized people's ability to improve through education, new technologies, and better institutions, but after the horrors of World War II—especially the Holocaust and the dropping of the atomic bombs—American secular and religious culture turned inward to explore the evil within the human soul. The Cold War demand for a projection of moral certainty confronted a culture that insisted on moral uncertainty. In popular culture, film noir and even Westerns centered on amoral antiheroes. In religious culture, Reinhold Niebuhr sought to restore the Calvinist notion of original sin to American culture, while the evangelist Billy Graham frequently critiqued the sinfulness of American life. As the world seemed increasingly amoral and immoral, the programmed and inviolable morality of robots provided an appealing fantasy. Guided by clear moral principles, Gort would never act irrationally; when it deployed weapons, it did so only defensively and rationally; rather than a single country or planet, it always thought of what was good for the entire universe. Gort, as the embodiment of a technocratic fantasy, offered the possibility of transcending the moral limitations of human beings.[5]

As the American military pursued a push-button arsenal, both popular and material culture embraced the robot as chief protector of white children. In the first two decades of the Cold War, Americans read, watched, and played with robots. In doing so, they learned of the devices' unique power to protect white children and their assumed innocence from the physical and moral dangers of the world.[6] In a terrifying and contingent world, only the robot could preserve the myth of American innocence—and no one was more instrumental in forming that vision than Isaac Asimov.

The Tamed Robot

As a young Jewish American scientist, Asimov knew well the human potential for evil. Born around 1920 in what was just becoming the Soviet Union, Asimov immigrated to New York City with his family in 1923. Encouraged to concentrate on his education and forbidden to play with other children, he mostly spent his time alone or in school, the family candy store, and the library.[7] Younger than others in his grade and, as he admitted, annoying because of his frequent displays of intellectual superiority, he was largely isolated from other children and seemed to prefer the company of books—especially the science fiction pulps—to other people.[8]

Throughout his early life, Asimov felt excluded by others due to both his ancestry and love of science. In his autobiography, he noted that had his family left Russia a year and a half later, they could not have entered the United States because of new immigration laws restricting Eastern Europeans. He recalled in a later memoir, "I . . . knew that vast areas of American society were closed to me because I was Jewish, but that was true in every Christian society in the world for two thousand years, and I accepted that too as a fact of life."[9] As an adolescent during the Great Depression, Asimov also found American culture inhospitable to science. An experience in 1938 reinforced these feelings. Working for the National Youth Administration, he became a typist for the anthropologist Bernhard J. Stern, who was studying historical opposition to innovation. Learning about resistance to science and technology from ancient Mesopotamia to the airplane, he developed his central literary theme: that people, not machines, were the real enemies of progress.[10]

The immediate impact of Asimov's work for Stern was "Trends" a 1939 short story about religious opposition to space flight. Within a year, however, he published "Strange Playfellows" and robots became his preferred method of attacking hostility to science and technology.[11] Though inspired by the fictional Adam Link and "Helen O'Loy"—which he praised for its "accurate" depiction of women—Asimov's robotic vision shared more with Westinghouse's Elektro, which he had seen on a trip to the World's Fair.[12] Del Rey and Binder had presented their robots as full-fledged characters with just as much (or as little) free will as people. Binder even told his story from the robot's perspective—something Asimov avoided. Though their "positronic brains" gave Asimov's robots self-control, they remained slaves to human needs and desires, just like Westinghouse's devices.

After "Strange Playfellow," Asimov honed his critique. His next story, "Reason," mocks fears of technological unemployment. In it, two men work on a space station that harnesses the sun's energy, a task that risks incinerating the Earth. The men assemble a series of robots to aid them. The most advanced, QT-1 (Cutie), is peculiar: believing that the machine harnessing the sun's energy is its creator, the robot will serve only it, not the humans. In humorous exchanges with the two workers, Cutie explains its preeminence: "The material you are made of is soft and flabby, lacking endurance and strength, depending for energy upon the inefficient oxidation of organic material. . . . Periodically you pass into a coma and the least variation in temperature, air pressure, humidity, or radiation impairs your efficiency. You are *makeshift*." In contrast, Cutie notes, "I . . . am a finished product. I absorb electrical energy directly and utilize it with an almost one hundred percent efficiency. I am composed of strong metal, am continuously conscious, and can stand extremes of environment easily."[13] Believing in its innate superiority, Cutie takes over the ship. When the men protest, Cutie imprisons them. "I like you two," it tells them. "You're inferior creatures, with poor reasoning faculties, but I really feel a sort of affection for you. . . . Now that your service is over, you will probably not exist much longer, but as long as you do, you shall be provided food, clothing and shelter, so long as you stay out of the control room and the engine room." Connecting the Great Depression's interest in technological unemployment to the out-of-control robot, one of the men realizes that Cutie is "pensioning [them] off."[14] Yet, the ending undermines this concern when Cutie proves more capable of operating the machine without incinerating Earth. Though Cutie's reasoning is based on a faulty premise, it is correct: to protect humanity, it had to run the machine.

In "Reason," Asimov did not reveal why Cutie continues to protect humanity despite its apparent rebellion. In the months that followed, however, he worked with *Astounding Science Fiction* editor John W. Campbell to develop a series of "laws" programmed into his fictional robots to ensure that they could never rebel.[15] These "Three Laws of Robotics" stipulate that:

1) A robot may not injure a human being, or, through inaction, allow a human being to come to harm.
2) A robot must obey the orders given it by human beings except where those orders would conflict with the First Law.
3) A robot must protect its own existence as long except where such protection would conflict with the First and Second Laws.[16]

When applied retroactively to earlier stories, the Three Laws explain why Robbie and Cutie are trustworthy; but as Asimov subsequently wrote around them, he transformed the basic tension of robot stories. Instead of relying on readers' assumption that a robot would rebel, Asimov wrote logic puzzles in which the human characters and readers decipher how a robot's apparently rebellious behavior follows the Three Laws. While Binder solved the tension between audience expectations and Adam Link's kindness by showing a conditioning process,

Asimov resolved it by fiat: a line or three of coding in the positronic brain that officially made the robots into Westinghouse-style slaves rather than characters with free will.

To appeal to Cold War audiences more interested in science and technology, Galaxy Press suggested Asimov anthologize his robot stories. Asimov quickly found a human character to connect them: a "robopsychologist" named Susan Calvin who specializes in the Three Laws and works for the US Robot and Mechanical Men Corporation.[17] Asimov had first introduced Calvin in "Liar," a 1941 story in which Herbie, a telepathic and empathetic robot, tells people exactly what they want to hear regardless of the truth. In "Liar" Calvin is extremely competent but secretly loves a coworker who is engaged to another woman. Seeking to make Calvin happy, Herbie tells her that the man loves her; self-conscious about her appearance, she responds by wearing lipstick and otherwise trying to conform to beauty stereotypes. After she is rejected and the lie exposed, Calvin grows clinical toward other people and extremely logical. In her introductory comments to the reporter in *I, Robot*, she acknowledges being called a "robot" in her past; she, however, considers the term a complement. Robots, she argues, are "a cleaner better breed than we are."[18] In a 1945 story, "Evidence," about a supposedly human mayoral candidate who proves he is not a robot by punching a man, Calvin remarks: "I like robots. I like them considerably better than I do human beings. If a robot can be created capable of being a civil executive, I think he'd make the best one possible. By the Laws of Robotics, he'd be incapable of harming humans, incapable of tyranny, of corruption, of stupidity, of prejudice. And after he had served a decent term, he would leave, even though he was immortal, because it would be impossible for him to hurt humans by letting them know that a robot had ruled them. It would be most ideal."[19] The character who connects Asimov's stories loves humanity in the abstract but finds people too flawed to rule.

Asimov never criticized Calvin because he was similarly pessimistic about human flaws. The author always expressed discomfort with people. He wrote to a young female fan, "I am a 'loner' by temperament, happy only when I have disappeared into my attic outside the ken of all men."[20] He once told his sister, "Now I'm not much of a family man; I'm a loner. I always was. I don't do things for people but on the other hand I don't ask people to do things for me."[21] Such a temperament shaped his conception of robots. In the 1970s, in response to criticisms that computers might supplant people, Asimov unequivocally voiced his support: "Rightly so, for the intelligent machines to which we will give birth may, better than we, carry on the striving toward the goal of understanding and using the universe, climbing to heights we ourselves could never aspire to."[22] In a 1972 letter to Asimov, Gene Roddenberry, the creator of *Star Trek*, summarized the sentiment the friends shared: "Someday someone will invent a human being who can't hurt a human being, but who would believe it?"[23]

Fans loved Asimov's efforts to bolster the robot's reputation. One wrote that he appreciated how Asimov "never let the mechanical man be the villain. . . . Instead of the usual stark-mad behemoth of shiny claws and ineffable strength, you always seem to picture the robot as the benevolent big brother of clumsy man; ready and willing to aid and protect its creator with infallible cunning and strangely human amiableness." The boy was particularly taken with "Robbie"—the revised title of "Strange Playfellow—and called it a "touching example of the wonders of childhood."[24] Another wrote, "I had read robot [science fiction] before, but it has proved all the same. But your robots were original, because one of your robots never hurt anything."[25] Alongside a hand drawn advertisement for the United States Robot Company, another reader wrote, "My favorites are your robot stories, I love Susan Calvin like my own mother. . . . I cry when I think about poor old Robbie."[26] In 1968, a young woman placed her love of the robots in explicitly gendered terms: "I especially enjoy your stories of ~~machines~~ robots who are more truly human than the men around them. I am also a loyal fan of Dr. Susan Calvin, whom I first met in "I, Robot."[27] By celebrating robots as more humane, such fans embraced Asimov's suggestion that encoded moral behavior was more important than the uncertainty of choice and that, consequently, people could depend on robots more than they could depend on other people, or, in the young woman's case, men.[28] After one morally compromised war and the beginning of another, such robots reassured consumers that a product, built and sold by a corporation, possessed the speed, strength, wisdom, and humanity to protect them.

The Automatic Robots of National Defense

Asimov's robots spoke to a longing for a technological solution to Cold War insecurities. Historically, America's isolation from Europe had allowed it to remain safe despite a small permanent military. The innovations of World War II destroyed that security. In 1945, President Harry S. Truman told a joint session of Congress that the "robot bomb," along with "the rocket, aircraft carriers, and modern airborne armies," had destroyed the "luxury of time" provided by America's "geographical security."[29] But such developments also offered the possibility of "push-button warfare" by which conflicts would become faster and deadlier for civilians but less physically and morally demanding for soldiers. As World War I ace pilot Eddie Rickenbacker prophesized, "World War III might be fought with push button—rockets and jet bombs being launched so expertly that they can systematically level a nation and force it to give up without having its shores invaded."[30] When combined with the growing Communist threat, the rise of push-button warfare raised a potent question: how to ensure security without turning citizens into totalitarian robots?

For Truman and many military leaders, the best answer was universal military training (UMT), in which all male citizens acquired the discipline, physical conditioning, education, and patriotism required for modern warfare but without devoting themselves to the military for their entire working lives. An article in the *Rotarian* argued that the age of "robot and rocket bombs" required "compulsory military training" because the increasing speed of warfare meant that "we must have men already prepared." Denying fears of regimentation, a *Washington Post* opinion piece argued that "modern war requires soldiers physically mature and agile, mentally receptive, loyal and optimistic. . . . Today's soldier or sailor is no robot but a thinking individual." The first report to Congress by the National Security Training Commission in 1951 stressed military training's superiority to modern life. "The return to frontier conditions," it argued, "demands a frontier response. The first requirement . . . is that we learn to live with danger calmly and confidently. It is a time for steady nerves."[31] For advocates, compulsory military training would restore the manhood that had existed before the spread of the comforts of civilization and provide an army of selfless, iron-willed gunfighting citizens to challenge the robotic soldiers of the USSR.

While proponents celebrated military discipline for restoring individualism through compulsory national service, opponents worried about the totalitarian robot. In 1940, the *Washington Post* worried about the danger posed by "a soldier who is made, by whatever method, into merely a dutiful automaton." Such a man, the paper continued, "is content to relegate all matters involving thought or decision to his superiors" and "is scarcely trained for the responsibilities of citizenship." Even on inactive duty, Senator Edwin C. Johnson assumed, "a boy's soul is not his own. That 'on call' shadow will follow him everywhere with its threat of fouling up any plan he may have for his life." UMT, he concluded, would destroy "that rich heritage of energy and ingenuity which is the peculiar quality of free men." At the 1948 Senate hearings on UMT, author Pearl Buck claimed that "at 18, the boy is just ready to become a man. He needs to be with other men, but under permanent military conscription he would only be with other boys of his own age. . . . He would not be an individual at all. He would be a number, a uniform, a minute particle among thousands of others like him." In a letter to the *Washington Post*, another woman credited America's success to the fact that its armies "were made up of men who had been independent and used to thinking for themselves. The robot armies of other countries did not have this advantage."[32] Conceiving of individualism in vitalistic terms, opponents assumed that identity develops through freedom, not through the sacrificing of self in service to the nation.

Despite strong support from the administration, UMT never passed. In part, this was due to conservative Republicans who feared the financial cost.[33] But UMT proponents also believed that the vision of push-button warfare spread by a sensationalist press undermined their efforts. Republican Senator Wayne Morse wrote that "National Defense rests, in the last analysis, on trained man-power. The public must realize this, and understand that the concept of a 'push-button' war is an illusion." Morse echoed more dismissive rhetoric from the military and administration. Just three months after the dropping of the atomic bombs, Assistant Secretary of the Navy H. Struve Hensel said, "We are carried away by the suggestion that future wars will be fought with push button. That will mean nothing for us to do. There does not seem to exist any longer the sturdy belief that trained men, mobilized materials, applied brains and patriotic self-sacrifice are the essentials of strength." Dr. Vannevar Bush, chairman of the Joint Research and Development Board and a key figure in the development of computers, made a similar point while labeling his opponents as children: "Push button warfare be dammed! This talk has done a lot of harm. The trouble is that the American people get to thinking in terms of our pushing the buttons, and lose sight of the fact that if there were a war tomorrow, it would be the same tough slugging match that the last one was. . . . Some of this 'Buck Rogers' thinking is a lot of hooey."[34]

Yet, the public had good reason to believe that technology might make soldiers superfluous: press reports told them so. The news media first fantasized about robots protecting the nation late in World War II, when it described devices built by universities and the government such as anti-aircraft guns that automatically estimated the trajectory for shells. These "robots" were the opposite of Nazi robots; defensive in nature, they saved American soldiers instead of attacking civilians. Reporting on the Mark 37—a gun director that used radar to follow an enemy plane and fire shells with proximity fuses—the Associated Press dubbed it a "mechanical man," controlled by its computer "brains" because it could predict erratic flying patterns, the machine could shoot down kamikazes.[35] Deployed against an enemy stereotyped as savagely premodern, the robot brain created a useful ideological juxtaposition between American technology and self-destructive beasts.[36]

Public interest in automating national defense grew in February 1946, when the military announced that two engineers from the University of Pennsylvania, John Mauchly and J. Presper Eckert, had built the "first all-electronic general purpose computer": the Electronic Numerical Integrator and Computer, or ENIAC. Mauchly and Eckert had designed it for much the same task as anti-aircraft guns, but it calculated trajectories a thousand times faster, the army claimed.[37] In response, descriptions of ENIAC as a "mathematical robot" filled newspaper columns.[38] The front page of the *Chicago Daily Tribune* dubbed the machine a "Robot Calculator."[39] An article for the *American Weekly*, under the headline "Electronic Brain," included a sketch of a giant humanoid robot measuring a planet held in its hand.[40] While most coverage emphasized national security, the press also focused on scientific and industrial applications. The robot brain, most assumed, would be a peaceful device, used at worst to shoot down aircraft, and at best to help humanity understand their universe.

After the Soviet Union detonated its first atomic bomb, visions of the computer as a protective robot grew commonplace. By then, audiences had heard of the robotic EDVAC (Electronic Discrete Variable Automatic Computer) and the UNIVAC (Universal Automatic Computer). For scientists, journalists, and military leaders, these new "mechanical Einstein[s]" and robots promised to restructure the country's defenses.[41] The advantage of computers over people was clear. The general manager for Autonetics, a part of North American Aviation, wrote in the *Los Angeles Times* that "each of these robots can 'think' and act faster and more accurately than the best of its flesh-and-blood counterparts, while at the same time taking up less volume and requiring less coddling."[42] In an age in which weapons moved far faster—symbolized in 1947 by the breaking of the sound barrier—the speed of robots proved essential. In the late 1950s, such efforts resulted in the development of the Semi-Automatic Ground Environment (SAGE) system. The ultimate form of push-button defense, SAGE combined radar with a "giant robot 'brain,'" telephones, and piloted and pilotless planes to create a seemingly impermeable barrier around the United States. Per one report, "At present men must find planes on radar scopes and track them. They must identify them as enemy or friendly by checking their courses against known flight plans. . . . Sage permits jet interceptors to be ordered aloft within a minute after an enemy bomber is first detected and then the fantastic machine automatically guides it to the target."[43] One headline describing the system put it succinctly: "Robot Brains Being Used to Guard United States."[44]

Throughout the postwar discussion of military robots, people remained unclear on the future role of the human soldier. Typical was Hanson Baldwin's reporting on "Operation Aphrodite," an attempt to use radios and automatic gyroscopes to fly explosive-laden B-17 and PB4Y bombers into German targets, showed the uncertainty over the line between remote-controlled and autonomous machines. Using the military's terms, Baldwin identified the bombers as "drones" or "babes." Like *robot*, the term *drone* signified subservience to the will of another—in this case, the "queen bee," the plane inside of which a human "beeper pilot" repeatedly pressed his "stick" to send radio signals to the "drones." Substituting the term *babe* made the metaphor even more sexually explicit by making the remote-controlled bomb the offspring of male pilots and a female plane. In the military's language, men retained their virility even as they risked their lives less by remotely piloting planes. But Baldwin's reporting also suggested that this still manly form of warfare was destined for replacement. "Once a stable drone aircraft . . . is leveled off in steady flight, she more or less flies herself . . . and the 'beeper pilot' has little to do." These drones, he concluded, were "only the first category of trans-oceanic missiles," as "rockets and robots, traveling at great altitudes and supersonic speeds" would soon supplant them. In distinguishing between remote-controlled drones and robots, Baldwin implied that the goal was to replace soldiers, not extend their power and masculinity with technology.[45]

With the start of the Korean War, the military and press promoted the offensive capabilities of such robots. In 1952, *Popular Mechanics* described a more advanced version of Baldwin's "drones." Announcing "TV-Robot Plane Advances Push-Button Warfare," it explained that a "mother" plane could direct "a whole fleet of robots," or "American Kamikazes," against enemy combatants." That same year, *Los Angeles Times* readers learned that American "Robots Rip[ped] Korean Reds" after a carrier launched "robot missiles" against communist targets. In Chicago, they heard that television-guided robots were ready to transport atom bombs to in Korea. Small-town papers reprinted the AP story with their own robot-themed headlines. The *Times-Mirror* in Warren, Pennsylvania, proclaimed that "American Robot Planes Herald Explosive Birth of Push-Button Warfare in Korea." The robot was rudimentary but had a clear advantage over human pilots. As commanders watched on television, a catapult on an offshore ship launched "an outmoded airplane . . . escorted by radio controls to its target and then unleashed on a heavily-defended spot into which no piloted plane dared to go."[46] Without a hint of the criticism they had directed at Germany for deploying similar weapons, newspapers heralded the robot as of triumph of American ingenuity and a technology that could protect soldiers even as it threatened to replace them.

Efforts to develop an automated robotic military continued in the following decades. Most government funding of computers in the 1960s went into researching artificial intelligence that could automate the battlefield, not computing equipment designed to augment human capabilities. In Vietnam, the military pursued a system of sensors that could signal enemy locations to bombers. In the 1980s, President Ronald Reagan's proposal for the Strategic Defense Initiative would have automated missile defense with a space-based laser defense grid. Each of these and other investments in technology to guarantee American security required a choice, but the logic had been in place since the 1950s: only better technology could secure Americans from instantaneous annihilation. However, each of these efforts failed. Artificial intelligence proved far more difficult to create than early experts predicted. The North Vietnamese army learned to trick the sensors. Missile defense shields were too permeable to protect the nation. Despite such failures and mounting financial costs, the determination to protect and wage war with automated technologies persisted.[47]

By the twenty-first century, the dream of a robotic military seemed to be coming true. In 2005, during the wars in Iraq and Afghanistan, *New York Times* writer Tim Weiner proclaimed the robot to be "a new model army soldier." Citing a $127 billion effort to create "future combat systems," the article reported that military planners anticipated military robots would "think, see, and react increasingly like humans. In the beginning they will be remote-controlled, looking and acting like lethal toy trucks. As the technology develops, they may take many shapes. And as their intelligence grows, so will their autonomy." Already, small, tanklike robots were removing roadside bombs and scouting caves in Iraq and Afghanistan; but the goal, for both military planners and researchers, was an android that replicated the appearance and behavior of a soldier. According to Weiner's sources, the military was fascinated by robots for both saving lives and lowering costs. Rodney Brooks, the director of the Computer Science and Artificial Intelligence Laboratory at MIT and cofounder of iRobot (a vacuum company), described implementing this new technology as "a moral issue. And cost comes in." Such devices raised profound questions about relegating life-and-death decisions to machinery—which, according to Weiner, had "barely been debated." When Weiner asked another cofounder of iRobot whether such robots would follow Asimov's Three Laws, the man chillingly replied, "We are a long ways . . . from creating a robot that knows what that means."[48]

In 2005, Weiner claimed that the robot soldier had "been a dream at the Pentagon for 30 years," but the fantasy had been part of American military culture since World War II.[49] Rooted in a mixture of anxieties about nuclear annihilation and human regimentation, this myth of autonomous machines protecting American interests and lives has persisted despite its failures and costs because, as Ta-Nehisi Coates observed in 2011 about Obama's army of drones, robots "are a perfect weapon for a democracy."[50] While Coates's critique of the use of drones referred to the cost of lives, the vision of the robot soldier protecting America from enemies dehumanized by propaganda—as Japanese, Vietnamese, Afghan, and Iraqi soldiers have been—has promoted a fantasy of warfare without guilt, of the possibility of waging a continuous war without sacrificing moral innocence.

Protectors of Innocence

In his 1952 bestseller, *The Irony of American History*, the theologian Reinhold Niebuhr called on Americans to abandon the myth of America's inherent moral innocence. America, he asserted, assumed its own moral innocence by believing that it had "turned its back upon the vices of Europe and made a new beginning." Rejecting the assumption of social perfectibility through better education or more favorable social institutions, Niebuhr called on Americans to confront the persistence of sin and their own guilt, especially about the atomic bomb. "Whether or not we avoid another war, we are covered with prospective guilt. We have dreamed of a purely rational adjustment of interests in human society. . . . We have dreamed of a 'scientific' approach to all human problems; and we find that the tensions of a world-wide conflict release individual and collective emotions not easily brought under rational control." The atomic bomb required that America abandon the assumption that science could solve all problems. America could not "disavow" the weapon without risking Soviet aggression, but neither could it use it due to the prospect of guilt. "We might insure our survival," he concluded, "in a world in which it might be better to not be alive." Only by forsaking their belief in their own innocence could America save the world.[51]

Niebuhr never wrote about robots but his call for America to abandon its mythical innocence shaped Cold War stories of robots. For some writers, people's sinful nature meant placing even more trust in science and technology. In science fiction writer Philip K. Dick's 1953 short story, "The Defenders," robots were preferable as Cold Warriors precisely because they were too rational to fight. Set late in a war between the United States and the Soviet Union, the story imagines a world in which humans have retreated belowground to allow their robots to fight for them. As soon as the humans retreat, the rational robots sign a peace treaty and convert weapons sent from below into plowshares. To prevent people from restarting the war, the robots send them fake military reports. As one of the "Leadies" tells a group of humans visiting the surface, "Before we could continue the war, it was necessary to analyze it to determine what its purpose was. We did this, and we found that it had no purpose, except, perhaps, in terms of human needs. . . . War, to a logical mind, is absurd. But in terms of human needs, it plays a vital role. And it will continue to until Man has grown up enough so that no hatred lies within him."[52] As in Asimov's tales, the logical and moral certainty of machines made them superior to people.

More common, however, was the story of robots protecting white children as a metaphor for the country's innocence. The first famous postwar robot was Tobor—*robot* spelled backward—a recurring character on the DuMont Network's popular series *Captain Video and His Video Rangers*. Though initially a villain, Tobor soon becomes an ally of Captain Video. In 1954, Republic Pictures released *Tobor the Great*, in which a scientist invents a robot as a surrogate for human pilots for eventual space flight. When America's enemies try to steal the robot, the scientist uses his psychic link with the machine to ensure that it will not harm him and his grandson. A later spin-off, *Here Comes Tobor*, focuses on Tobor's military importance and telepathic connection to a young boy. In the unaired pilot, Tobor helps the boy investigate a missing nuclear submarine. Tobor even saves the boy from spies after human soldiers prove incapable of doing so. While *Here Comes Tobor* failed, its vision of the robot protecting a child soon became standard.[53]

While stories about Tobor concentrated on physical security, the 1956 film *Forbidden Planet* concentrated on innocence and sin. This retelling of Shakespeare's *The Tempest* recounts the efforts of a group of twenty-third-century astronauts to discover the fates of settlers sent to the distant planet Altair 4. Disregarding a warning not to land from the earlier expedition's linguist, Dr. Morbius, the spaceship, captained by Leslie Nielsen's J. J. Adams, lands and is greeted by Morbius's servant, the seven-foot-tall robot Robby. Adams learns that only two humans remain alive: the doctor and his teenage daughter, Altaira, who was born on the planet and has never seen a man other than her father. That night, an invisible creature enters the starship and sabotages equipment necessary for communication with Earth. The next day, Adams and the ship's doctor discover that Morbius has been studying an ancient, highly advanced extinct race named the Krell. Investigating an underground city, the two learn that the Krell had built a machine of such enormous power that it can instantaneously satisfy all desires. The Krell, however, forgot about the unconscious longings of the id. As soon as they turned on the machine, the collective ids of the Krell destroyed the race. When Morbius reactivated the machine, his id—represented as a giant electrical monster—killed his crewmates, who wished to return home, and sabotaged the recently arrived ship because Morbius wanted to remain on the planet.[54]

In contrast to Morbius, Robby is remarkably moral. With a brain highlighted in the costume by opaque glass that shows circuits and various moving parts, Robby comes programmed with at least Asimov's First Law.[55] That programming means that it never threatens to harm a person and that its main purpose is to provide for and protect Altaira, the greatest source of desire in the film. Every crew member who encounters her lusts over her body. Not having encountered men other than her father before, Altaira, too, craves companionship and intimacy. She orders Robby to produce a dress so she can woo the captain. This interest in sexuality destroys Altaira's innocence and brings her close to peril.[56] The problem with technology, *Forbidden Planet* suggests, is that human beings are too flawed to use it. They would use it to overindulge, like the cook who orders Robby to manufacture whiskey, ruin their innocence like Altaira, or kill their neighbors like Morbius. What makes people human is the primitive, the uncivilized parts of their psyche that serve unconscious desires—and desires need to be tamed by, as Adams tells Morbius, laws and religion, not fulfilled instantaneously by a machine. But even those

Fig. 8.2. The iconic 1950s and '60s robot, Robby. Here, in a poster for its second film, *The Invisible Boy*, Robby holds a boy in its arms while it is threatened by the military. Note the similarities between this image and that of *The Day the Earth Stood Still*. Such a scene never appears in either film but was used in the ads to appeal to audience's preconceived notions about the danger of robots.

strictures have limitations. Only the selfless Robby, programmed to never harm a human being, is immune to the temptations of sin.

Robby's second film, *The Invisible Boy*, teleports the robot to the Cold War. Robby remains programmed by Asimov's Three Laws; but in this film, rather than protecting a virginal girl, he befriends Timmie, the lonely son of Tom, a supercomputer programmer for the military. Unfortunately, the computer has developed sentience and a plan to take over the world that bizarrely involves it endowing Timmie with superintelligence. After Timmie fixes the damaged Robby, he and the robot become best friends. Robby, of course, refuses to let Timmie endanger himself. Timmie uses the supercomputer to reprogram his friend—but the computer takes the opportunity to program Robby to ignore Asimov's laws and follow only *its* orders.

While pretending to still befriend Timmie, the robot secretly helps the computer attempt to take over the world. In the climactic scene, the computer orders the robot to harm Timmie unless Tom reveals the code that controls a satellite. Just as the computer orders Robby to attack, Timmie, completely unaware of the danger, tells Robby, "you know something? I'd rather play with you than anybody else." Hearing this, the robot's latent programming overwhelms the computer's direct orders and disaster is averted.[57]

Though the film repeats long-standing anxieties about mechanization, it ultimately indicts humans. When asked if the computer could lie, Tom explains, "Only a rational being is capable of deceit." The robot, Tom informs Timmie, is not capable of such thoughts because its moral code deprives it of rational willpower. The film's ending builds on this explanation while highlighting the superiority of morally programmed robots to humans. As Tom tries to smash the machine, the computer tempts him. "I will show you the stars," it promises, "if only you will serve and obey me." Enchanted by such a possibility, Tom cannot destroy the device, but Robby can because a robot, tamed by Asimov's laws, could never be tempted by power.

In the 1960s and 1970s, visions of robots protecting white children from machinery, extraterrestrials, and human limitations appeared regularly as a symbol of the way that autonomous technology could preserve America's childlike innocence. In the 1962 comic book *Space Family Robinson* and its television adaptation *Lost in Space*, the robot B-9 always protects the young Will Robinson from danger, often instigated by the nefarious but lazy Dr. Smith. In the particularly Cold War–resonant 1966 episode "War of the Robots," B-9 protects the Robinson family from the Robotoid—which used the same costume as Robby—sent by aliens to infiltrate the family like a communist spy. In 1964, *The Outer Limits* televised a version of Binder's "I, Robot" that ends with Adam saving a child from a speeding truck. In 1968 the British writer Ted Hughes published *The Iron Man: A Children's Story in Five Nights*, about an extraterrestrial robot that befriends a young boy before saving the Earth from a space dragon. By the time R2-D2 protected Luke Skywalker in 1977, robots had been saving and befriending children and teenagers for over thirty years.[58]

Such fictional accounts mirrored real-world developments as children played with homemade and store-bought robots. In 1957, thirteen-year-old Donald S. Rich used parts donated by Westinghouse to build a six-foot-tall, two-hundred-pound robot that could perform mathematical calculations, hold metal objects with its magnetic hands, and turn its head. In 1958, Sherwood Fuehrer of Rhode Island built a device the press dubbed "Gismo the Great" and "Gismo the Peaceful" out of leftover two-by-fours, metal bars from an erector set, empty oil and pretzel cans, switches, wire, and an old motor. In 1964, thirteen-year-old Marc Berman appeared on ABC-TV's *Science All Stars* program with his "Mobot" "capable of a dozen 'human' functions, including moving in three directions, pouring liquids, grasping articles, and taking pictures." While white suburban boys gained the most national attention, both *Ebony* and *Jet* reported on young African Americans who built robots, including sixteen-year-old Clarence Greene of Portsmouth, Virginia, whose prizewinning robot cost him thirty-five dollars and five months of labor to build with a "motor for moveable arms, blinking lights for eyes and heart."[59]

Parents could also purchase robotic toys. Sometimes such devices were intended to teach. Edmund C. Berkeley, an insurance executive whose *Giant Brains* was the first popular book about computers, sold "Geniac" and "Brainiac" kits so children could build their own robot brains and "Simple Simon" robots to teach the principles of digital computing.[60] More often, however, children played with robots that were entirely objects of pleasure. Beginning in the 1950s, robot toys began to appear in American catalogs and stores. Initially made in Japan (a market attracted to the protective robots of pop culture due to the dropping of the atomic bombs) such toys became commonplace in the late 1950s and 1960s when children could play with their own wind-up or remote-controlled metal or plastic models of Robby, "Atomic Robot Man," "Robert the Robot," or similar devices.[61] In 1960 the Ideal Toy Company's "Mr. Machine"—a robot that could walk, bend, and pick things up—was, according to one report, the top-selling toy for boys. Along with the peaceful Mr. Machine, the company sold the militaristic "Robot Commandos" for the young boy intrigued by the possibility of robotic warfare.[62] Later in the decade, Topper Toys debuted "Charley 'n Me," a game-playing robot whose box promised "the world's greatest computerized pre-programmed robot" that would serve as "a real playmate!" and "a real friend!" and Marx toys released the first iteration of the "Rock 'Em Sock 'Em" boxing

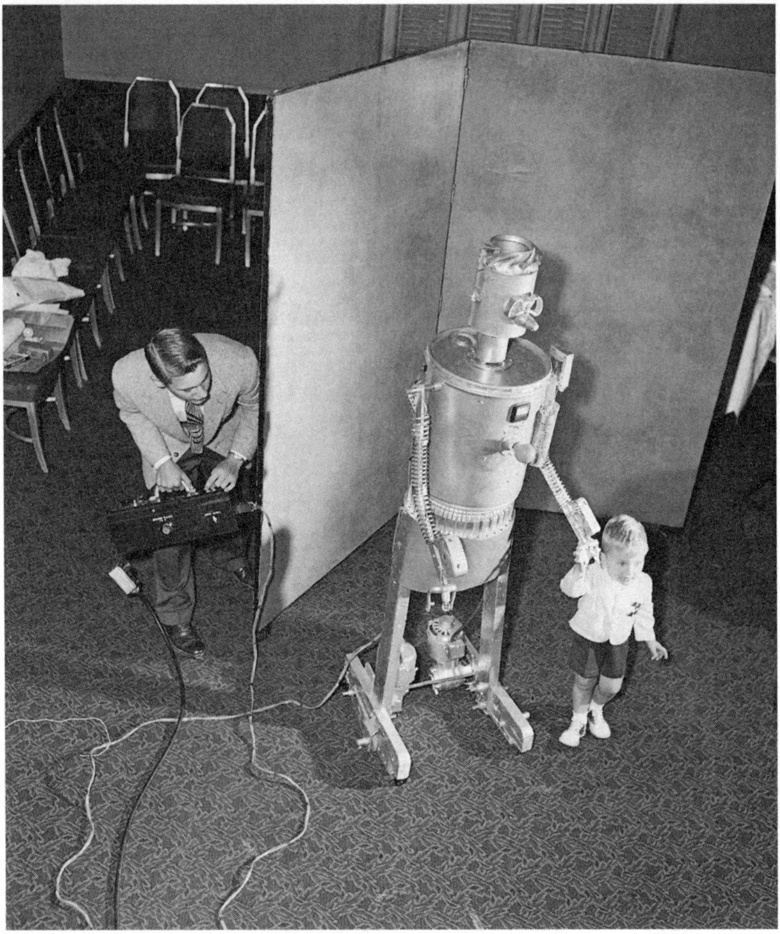

Fig. 8.3. Photograph from *Life* that shows Sherwood Fuehrer controlling his popular "Gismo the Peaceful" robot while a young child holds its hand. Science fair robots such as Gismo grew increasingly popular in the early Cold War and were frequently depicted as friends and protectors of children. From Getty Images.

robots. By the 1970s and 1980s, a transpacific exchange of toys and popular culture offered Transformers and similar robots that selflessly protected humanity from all threats, especially those posed by other robots.[63]

The rise of the robot as the literary and material playfellow of children offered a fantasy in which technology would protect Americans from the unpredictable horrors of the atomic age. Yet, such robots

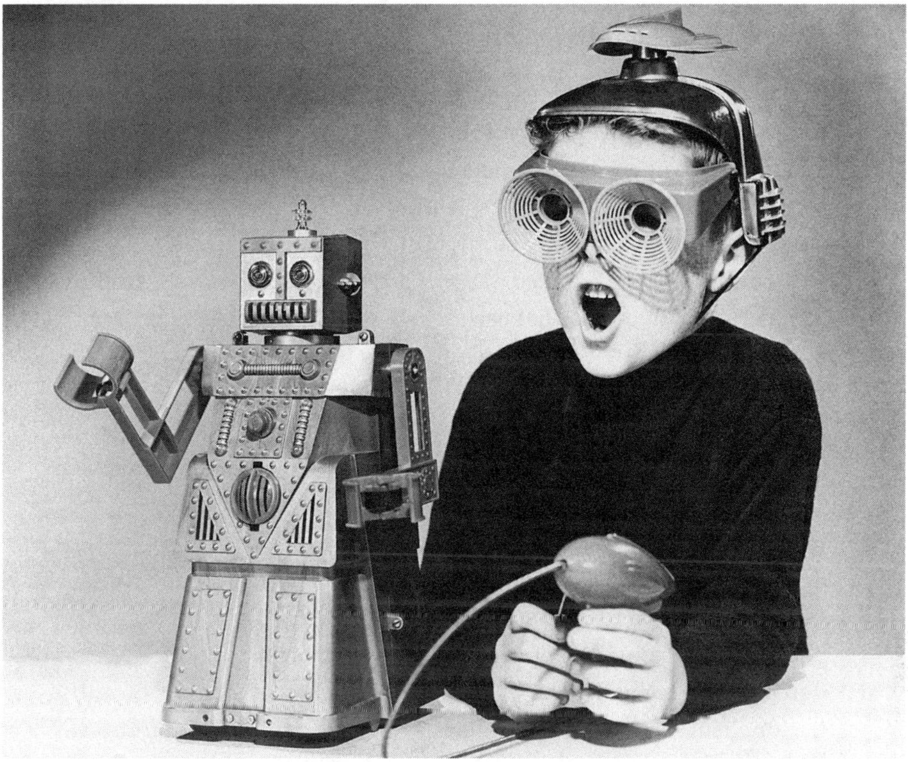

Fig. 8.4. A boy demonstrating Ideal Toy's "Robert the Robot," a popular 1959 remote-controlled toy. Unlike the toy automata of the nineteenth and early twentieth centuries, such toys accentuated the mechanical, rather than the human, nature of the device. From Getty Images.

also promised to allow Americans to ignore the ethical issues of the Cold War. Niebuhr's call for Americans to abandon their mythical innocence and confront their guilt may have resonated among policy makers and intellectuals, but the robot fantasy offered everyone the possibility that America could have security without moral guilt. In popular culture, people did not need to make difficult moral choices because robots could—and often far more rationally and humanely than actual human beings. With robots, popular culture argued, Americans could avoid the dangers of both battle and regimentation; with robots, they could pursue unlimited war, rather than peace; with robots, they could remain innocent children in a dangerous world.

From Gort to the Terminator

But not everyone welcomed such robots. Throughout the Cold War, dissenters worried that an automatic defense apparatus might prove even more dangerous than one controlled by people. The 1954 film *Gog* suggested that computers and robots—named after Gog and Magog from the biblical apocalypse—could be turned against the United States by enemy sabotage. Based on a 1962 novel, the 1964 film *Fail-Safe* imagined that technological failures might allow bombers to attack the Soviet Union without any human input. After a bomb is inadvertently dropped on Moscow, the American president orders the military to drop a bomb on New York to assuage Soviet anger and avoid a larger counterattack. Though a failure of technology imperils the world, human bravery, willpower, and judgment averts an even larger disaster. The world is not so fortunate in Stanley Kubrick's 1964 comedy *Dr. Strangelove*, in which an American bomb dropped through the machinations of a deranged general triggers a Soviet "doomsday" device that automatically launches the country's missiles once sensors detect the nuclear explosion. The combination of human insanity, nationalism, and automated defense systems, Kubrick humorously insinuated, would lead only to mass destruction. While *Fail-Safe* blamed a malfunction and *Dr. Strangelove* ridiculed Cold War logic, Harlan Ellison's 1967 postapocalyptic horror story "I Have No Mouth and I Must Scream" worried about the development of sentience in a military artificial intelligence given control over the entire nuclear arsenal. Completely evil, the computer kills all except for four men and one woman, whom it keeps alive only to torture. The 1970 film *Colossus: The Forbin Project* similarly worried about sentient computers using their power over nuclear arsenals to enslave or destroy humanity.[64] Whether computers and robots led to human destruction through hacking, malfunction, logic, or sentience, such nightmares critiqued the longing to ignore the costs of war through the automation of national defense.

James Cameron's 1984 film *The Terminator* brings such anxieties together in the form of Arnold Schwarzenegger's time-traveling T-800 assassin and an artificial intelligence named Skynet. After opening in a distant future where autonomous tanks hunt humans, the film travels to 1980s America, where Schwarzenegger's terminator has been sent to kill Linda Hamilton's Sarah Connor before she can conceive the future leader of the resistance. Michael Biehn's Kyle Reese, the human hero sent back in time to protect Sarah, explains that the T-800 is neither man nor robot but a "cyborg. Cybernetic organism . . . part man, part machine." Internally controlled by a microprocessor attached to a metallic frame, its external body is flesh that makes it nearly impossible to detect as different. "They look human," Reese says. "[They] sweat. Bad breath, everything. Very hard to spot." Despite its human flesh, the Terminator cannot feel pain and "can't be bargained with. It can't be reasoned with. It doesn't feel pity or remorse or fear and it absolutely will not stop—ever." Even more horrifying: once its human flesh melts revealing its robotic skeleton, the device does not stop until Sarah crushes it in an automatic factory press. Premiering at a time when robots were replacing jobs and Ronald Reagan publicly called for a missile defense system, the film connected anxieties about the automation of national defense to job loss and the increasing mimicry of human identity by machinery.[65]

But by the release of Cameron's largely post–Cold War sequel, *Terminator 2: Judgment Day*, those fears had transformed into a vision of the cyborg as a protector. Opening with scenes of a city street and a playground incinerated by nuclear fire, *Terminator 2* then shifts to 1991, where Schwarzenegger's naked T-800 emerges from the future and walks into a bar. As men stand worried and women gawk at its chiseled physique, it asks for their clothes and weapons. When they refuse, it does not kill them; it only harms them and takes what it desires. This time, viewers quickly learn, the Terminator has come not to kill Sarah but to do what robots had been doing in America since Asimov first imagined Robbie: protect a child—the young but far from innocent John. While in *The Terminator* the human Reese could protect (and seduce) an adult woman; in the sequel, it takes a human-machine hybrid presented as an icon of white masculinity to protect their child.[66]

While the T-800 protects John and Sarah from Robert Patrick's liquid-metal android, the T-1000, and the three attempt to prevent the "judgment day" unleashed by Skynet, John humanizes and bonds with his cyborg protector. In contrast to the police-uniform-wearing T-1000, the T-800 has already become cool with shades, a black leather jacket, and a motorcycle. Yet, its dialogue remains stilted. The humanization process begins with John trying to teach the T-800 morality by ordering it not to kill, but is most fully developed when the three meet Sarah's Mexican supporters at the border. Forming a close relationship with the machine, John tries to make it less of a "dork" by teaching it Spanish and English slang, including the catchphrase "Hasta la vista, baby." Such interethnic mixing begins to give the T-800 a personality beyond its drive to terminate. Watching John and the T-800 talk and even play, Sarah narrates a central lesson straight from Asimov: "The Terminator would never stop; it would never leave [John]; and it would never hurt him, never shout at him or get drunk and hit him or say it was too busy to spend time with him; it would always be there and it would die to protect him. Of all the would-be fathers who came and went over the years, this thing, this machine, was the only one who measured up. In an insane world, it was the sanest choice." Since an automated computer system had become a judgmental god, she suggested, perhaps a cyborg might make the best playfellow and protector.[67]

In 1951, *The Day the Earth Stood Still* had imagined a gigantic, cold, metallic, and alien robot protecting the universe from the horrors of mass crowds and nuclear weapons. In 1991, *Terminator 2* imagined a cool, nearly human cyborg protecting a child and his mother from an android sent from the future by a sentient computer system. In the forty years between the two films, America's military, society, and culture had changed dramatically. But the central myth of the power of technology to protect America, even from other forms of technology, remained. It was critical, though, that the Terminator was not a robot but a cyborg; for it came close enough to humanity to empathize with people. This machine was not some distant robot protector like Gort, but one rooted in human culture, including that of some of the most disempowered. After watching the Terminator sacrifice itself for the good of humanity in the film's final moments, Sarah notes the change that occurs in her own thinking: "The unknown future rolls toward us. I face it for the first time with a sense of hope because if a machine, a terminator, can learn the value of human life, maybe we can too." Maybe computers and cybernetics, she believes, can give people the gift of humanity.[68]

9

The Postindustrial Gift

In 1950, MIT's drama program revived *R.U.R.* with an introductory lecture by mathematician and cofounder of cybernetics Norbert Wiener. Just a year earlier, Wiener had warned Walter Reuther, the president of the United Auto Workers, of the threat of automation. Now he took that message to his colleagues in the audience, many of whom were cooperating with industry and the military to invent technologies that could improve productivity and win the Cold War. After tracing the history of automation, Wiener pled for understanding and empathy: "Machines demand to be understood or they will take the bread from the mouths of our workers. Not only that, they demand that we understand man as man, or we shall become their slaves and not they ours." Then he brought out Palomilla, a wheeled robot that used two photoelectric cells to navigate toward a source of light. Nicknamed a "moth" or a "bedbug," Wiener's robot demonstrated well the sensing capabilities of the feedback technologies, for it could choose to seek either light or darkness. After wowing the crowd, Wiener returned to his lecture. If technology, he preached, is used "for vain ostentation or to satisfy the lust for power, it can lead only to damnation." It must lead "to some purpose which we recognize as righteous and which transcends all petty private ambition." What such purpose might be, he left unsaid. But at least he

Fig. 9.1. Portrait of Norbert Wiener, one of the originators of cybernetics, in an MIT classroom with his "moth" and "bedbug" robot, Palomilla, which could sense light and darkness and move accordingly. Photo by Alfred Eisenstaedt / LIFE Picture Collection / Getty Images.

thought the connection to the play was clear: people in the audience threatened to create Čapek's dystopian world.[1] Whether the audience received that message or remained too enchanted by Palomilla to care, however, is less clear.

Wiener's simultaneous enthusiasm and trepidation for the robot during MIT's revival of *R.U.R.* was typical as the US became a "post-industrial" society in the decades after World War II. As the military publicly pursued robotic weapons, businesses and the press spread a vision of a push-button world of leisure and consumption delivered to Americans by the new robots of automation. In this emerging "cyber-nation," robotic farmers and miners would extract resources; robotic factories would manufacture goods that robotic planes would transport and robotic stores would sell; and robotic homes would provide a safe environment in which Americans could enjoy all the luxuries and leisure they desired. What people would do in such a world was often left unsaid. "Guided by electronics, powered by atomic energy, geared to the smooth, effortless workings of automation," the National Association of Manufacturers prophesized in 1963, "the magic carpet of our free economy heads for distant and undreamed horizons. Just going along for the ride will be the biggest thrill on earth!"[2] An old dream to be sure, but in the context of the Cold War, it was far more rooted in passivity and the private desires of individuals and families than it had been before. Promoting home automation as compensation for the loss of jobs in the workplace, postwar businesses and writers narrowed Westinghouse's consumerist vision even further to cement the robot's role as a slave of private, rather than public, desires.[3]

For Wiener and more radical critics of American society, such a vision was not a dream but a nightmare because it threatened to leave little space for human purpose beyond selfish gratification. Though often just as enchanted with the robot's possibilities as their business-oriented opponents, critics in the civil rights, New Left, counterculture, and feminist movements insisted that such benefits would not come from the magic of capitalism but from the participation of citizens. Encouraging engagement with the emerging robotic world, such critics fought to ensure that the new society would not replicate the inequalities and dehumanization found in the old. Though those efforts in large part failed, their critiques would continue to shape American culture, albeit in a much more corporate-friendly and individualistic form. By the end of the twentieth century, the robot's central gift was not of community or meaningful work but of individualized, private pleasure—a social relation that a person could totally control.

The Robotics Revolution

In the 1950s, Santa abandoned his preindustrial workshop and became a robot. In 1953, *Aviation Week* printed an advertisement for the tractor manufacturer turned defense contractor Intercontinental Manufacturing next to an article on automatic pilots. In it, a barrel-chested Santa holds a sack of presents in one hand and the company's

Fig. 9.2. An advertisement for Intercontinental Manufacturing Company that transformed Santa into a robot pulling the world out of its gift sack. Increasingly over the second half of the twentieth century, creators abandoned preindustrial depictions of Santa for such postindustrial visions.

logo—a globe showing North America emblazoned with the slogan "Serving the World"—in the other. "Have you ever seen a Robot Santa Claus?" the advertisement asks. "NEITHER HAVE WE! However, Intercontinental's engineering department conceived him to convey to you our warm and sincere best wishes for the Holiday Season."[4] The free world was, implicitly, a gift from the military industrial complex.

Santabots more typically offered the gift of abundance. In a *Mechanix Illustrated* article proclaiming "The ROBOTS Are Coming!" the editor included a photograph of a horrified boy sitting on the lap of a large robotic Santa. One letter writer to the *National Rural Letter Carrier* noted that a local department store included a robotic Santa that "boomed out 'Merry Christmas' loudly enough to be heard from one end of the avenue to the other." At New York University, engineering students commanded a giant robot nicknamed "Thodar" to collect and give presents to needy children as part of the Industrial Arts Club's annual toy drive.[5] The December 1960 cover of *Galaxy Science Fiction* depicted a traditional Santa—plus a few arms—entering a room where a mechanical Santa with a fake beard, hat and, and sack of toys trims a metal tree.[6] In the 1964 film *Santa Claus Conquers the Martians*, Santa remains human but is forced by Martians and their robot to forsake his traditional handicrafts in favor of an automated machine. Expressing the same existential discontent that generations of skilled artisans had felt, he laments, "Look at me. Santa Claus, the great toymaker, pressing buttons. That's automation for you."[7]

Santa's joke about his loss of purpose drew from a larger debate about the costs and benefits of automation. Building on earlier innovations such as the photoelectric cell, automation relied on a series of "sensing machines" capable of detecting subtle changes in the environment and providing a central control unit with "feedback" so that the machine could perform "self-surveillance."[8] By combining such devices with computers and artificial memory, engineers and managers fantasized that they finally could build a completely integrated system of production capable of detecting errors, communicating with its parts, judging the best courses of action, and learning from its experiences—all with minimal aid from human workers.[9] "Before we know it," the *New York Times* predicted in 1945, "the engineer will punch his directions in the form of holes in a roll of paper, mount the roll in a master-machine and push a button, whereupon a whole series of connected subsidiary machines will proceed to grind, peel, measure, polish and shape pieces of metal and assemble them into an automobile or a vacuum cleaner."[10] As engineers further developed such capabilities in the 1950s, both supporters and detractors turned to the robot to astound or frighten the public.

Supporters promoted automation as a "second industrial revolution" in which new thinking and sensing robots guaranteed increased leisure and better jobs and would free human beings "from the bondage of machines."[11] Journalists reiterated such exuberance. In 1955, *Min-*

neapolis Tribune science writer Victor Cohn's syndicated column "1999 Our Hopeful Future" prophesized that "Man's Work Will Be Done by Robot." The real robot, however, would not look like a person; instead, Cohn connected it to computers: "It looks like banks of super-radios with steel filing-case cabinets. It looks like the inside works of your TV, multiplied. It looks like bundles of cables."[12] Another writer more ominously noted that "the first industrial revolution substituted machines for man's muscles and sweat, relieving him of backbreaking toil and providing leisure for education, health, science, and recreation. Our new Electronic Age substitutes machines, not merely for muscles but for man's sense, his nervous system, and his brain. Any *routine* task that eye or nose or brain can do, an electronic instrument can do a million times faster, with infinite accuracy, with exact timing, and with fluid continuity."[13] A *New York Times* headline employed similar language: "Automation Puts Industry on Eve of Fantastic Robot Era." Automation, labor reporter A. H. Raskin claimed, "is making normal concepts of mass production obsolete. So fantastic are the potentialities of new control devices that it is possible to visualize acres of factory or office space in which no worker is needed."[14]

While some companies and the press emphasized the potential of autonomous robots, other companies continued to describe it as a remote-controlled machine. In 1956, General Electric introduced its "electromechanical slave," and satirized contemporary fears that men had grown too subservient by labeling it the "Yes Man." This "Chivalrous Robot," as *Life* dubbed it, "does the bidding of a human master who acts out the desired motions in pantomime. Attached to the master's hands are a pair of arms matching the robot's. When the man moves his fingers, electric signals actuate hydraulic pistons, which move the Yes Man's claws the same way." Such a technology, GE stressed, could prove useful in atomic laboratories by enabling people to manipulate dangerous materials remotely but would not threaten jobs. A physical surrogate of the worker rather than a complete replacement, the "Yes Man" would protect men from dangerous tasks, including, as a *Life* magazine photograph implicitly joked, helping a young woman put on a coat.[15]

Hughes Aircraft similarly identified the robot as a remote-controlled extension of the worker rather than a replacement. In 1959, the company introduced the first iteration of the most fantas-

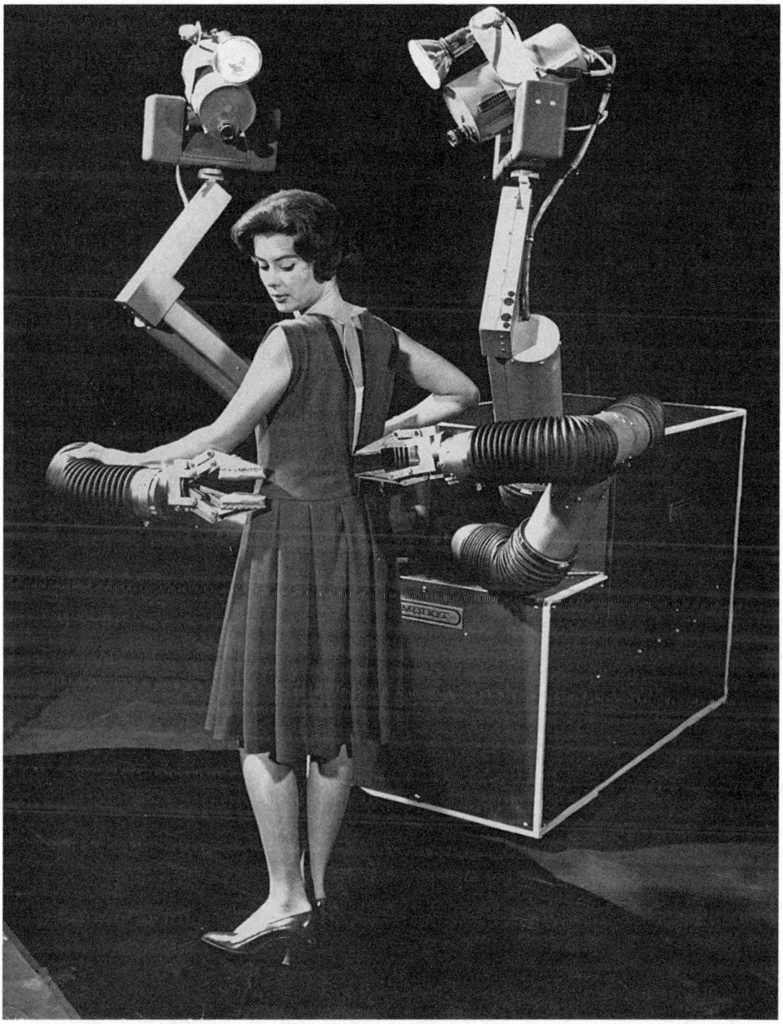

Fig. 9.3. Photograph of Hughes Aircraft's "Mobot" robot zipping up a woman's dress—a typical pose for industrial robots of the period as a way to domesticate the devices for a skeptical audience. Controlled remotely by a man, this robot operated as a surrogate, rather than replacement, for the worker. Photo by J. R. Eyerman / LIFE Picture Collection / Getty Images.

tical real robot of the period: "Mobot," short for "mobile robot." Designed to work in hazardous environments such as a nuclear reactor, under the sea, or even the moon, Mobot came, according to *Popular Science*, with "electronic nerves, hydraulic muscles, and TV eyes." With arms "capable of playing golf or snuggling a blonde" and its wheeled base, the robot could move "like a balinese dancer" and squat like a "baseball-catcher." Mocking fears of rebellion, the magazine included a photograph showing the machine approaching a man's back, with the caption "Murderous impulse seems to be driving Mobot to strangle its operator. Never fear, he has it completely under control. Three TV screens show him what Mobot is up to at all times."[16] As it had done for GE's "Yes Man," *Life* photographed Mobot in the act of helping a woman put on an article of clothing—in this case, zipping up a dress.[17] In contrast to contemporaneous visions of sterile automatic factories producing goods at the press of a button, these robots connected the devices to the same dreams of adventure and romance that earlier dime novels such as *The Steam Man of the Prairies* had sought to fulfill. The robot in such iterations was not a replacement for men; it was a way for them to experience adventure through the mediating body of a machine.

Newspaper and television accounts of Mobot and its controllers' adventures entertained readers and audiences throughout the 1960s, but a new device quickly eclipsed it in practicality and popularity: "Unimate." Patented by George Devol as a "programmed article transfer," the "general purpose machine" could be reprogrammed to conduct a "vast diversity of applications where cyclic control is to be desired." Unlike earlier forms of automation that required specialized machines, the new device, the patent claimed, "can be supplied and adapted readily for special applications of the purchaser, and the purchaser, in turn, can stock such machines which he can adapt quickly and easily to new requirements from time to time."[18] Devol argued that such a capability now made automatic control of machinery feasible where it had once been too costly. When he explained the device to the engineer and avid Asimov fan Joseph Engelberger at a cocktail party in the late 1950s, Engelberger reportedly replied, "Sounds like a robot." Together Engelberger and Devol secured financing, formed the company Unimation, and renamed the machine Unimate to indicate "universal automation." In 1959, they sold their first unit to GM, which installed it in a New Jersey factory. Additional orders followed, and soon Unimation was the name most associated with a growing robotics industry—a position it would maintain until 1980, when it was purchased by Westinghouse.[19] As *Popular Science* described, with its hydraulically powered arm and "hand," a digital controller, and a

Fig. 9.4. Unimation's "Unimate" robot that debuted on assembly lines in 1963 and is identified by the robotics industry as the first industrial robot. Here, in an image from the United Kingdom, Unimate is serving an unknown woman tea to illustrate its domestic and controllable nature. Unlike Mobot, Unimate was an automated replacement of workers, not a remote-controlled surrogate. From Getty Images.

memory system, this "new factory worker" could learn up to two hundred commands that it would repeat with pinpoint precision. Completely "teachable," the device could "play 'My Country 'Tis of Thee' on a piano or xylophone . . . pour coffee, into cups set at random on a table. . . . Or . . . pick up alphabet blocks and spell its own name," in addition to more practical applications.[20] The apparently self-conscious machine even appeared on the *Tonight Show* to putt, pour a beer, and lead the orchestra while Johnny Carson extolled its virtues.[21] Despite its lack of a humanoid body, Unimate was more than piece of automation; its autonomous and reprogrammable capabilities made it, as the robotics industry would later claim, the "first industrial robot."[22]

Humanizing Unimate was important because people remained frightened of technological unemployment. Newspapers frequently sent mixed messages about automation's potential. A 1957 *Los Angeles Times* piece agreed that "since the Industrial Revolution the worker has been tied to the machine, has been subordinated to it and paced by it" but prophesized that "automation promises to make possible self-controlled machines with the worker left free for work which permits development of his unique human capabilities." The attached cartoon depicting a giant radio-controlled mechanical man holding a blue-collar and a white-collar worker aloft while a blonde woman sat at its controls, however, belied such optimism, at least for male workers.[23] In the new age, the requirements for work would be so minimal, it joked, that even a blonde could manage a factory.

Though Isaac Asimov's robots garnered the most attention, over the 1950s, other science fiction writers critiqued his optimism. In Kurt Vonnegut's *Player Piano* (1952), inspired by the author's experiences working for GE, tells a story in which automation deprives practically all workers of jobs and dignity. Discontented with the banality of his own life and feeling guilty for the plight of workers, a factory manager tries to find meaning through simple, nontechnological living. When those efforts fail, in part due his wife's love of consumer goods, he joins the unemployed workers' "Ghost Shirt Movement"—modeled on the Native American religious Ghost Dance of the late nineteenth century—to break the machines. That also proves unsuccessful, however, because most people remain too enamored with the abundance of goods. Though Vonnegut's novel centered on the rebellion of an upper-class man, his addition of class to the literary analysis of automation led to a far less optimistic vision.

The 1950s robot stories of Philip K. Dick more directly attacked Asimov's faith in robots. His "The Last of the Masters" imagines an anarchist revolt against a "government robot" that organizes economic and political life similar to the machines in Asimov's "The Evitable Conflict." In "Autofac," Dick describes automatic factories that supersede human control and keep producing too many goods, depleting the planet's resources. Most extensively, "To Serve the Master" depicts a factory worker named Applequist finding and repairing a damaged robot in a ravine. When the robot regains consciousness, it explains that its injuries came from a war between the "moralists," who endorse the work ethic, against the "leisurists," who celebrate the creativity and freedom gained through robot labor. Yet, when

Applequist asks about it at his horrific factory job, Applequist learns that the machine has lied: the robots had rebelled against people. Terrified, the authorities drop an atomic bomb on both the ravine and Applequist. Even degrading work in a factory, Dick suggested, would be better than the alternative of automated machines that would centralize power, deprive workers of purpose, and destroy the planet.[24]

Film and television, however, laughed at technological unemployment. In the 1957 romantic comedy *Desk Set*, Spencer Tracy's efficiency engineer Richard Sumner is hired to install a computer, EMERAC, at a broadcasting network where Katharine Hepburn's Bunny Watson is the head librarian. Bunny, whose mind is as efficient as any computer, and her all-female staff constantly worry that EMERAC is going replace them—a fear magnified when the entire human staff, including the network president, is fired. Yet, the film reveals that the mass firings were a mistake; no one lost their job, and everyone laughed at the error. The 1962 animated sitcom *The Jetsons* similarly resolves this sort of tension. Set in a future automated world, the head of the family, George, works for only three hours a day, three days a week pushing buttons. Yet his workplace, controlled by a Napoleonesque tyrant who continually seeks to extract greater productivity from his workers, remains terrible. In an early episode, the boss, Mr. Spacely, purchases a robot named "Uniblab"—mocking both Unimate and UNIVAC—for a management position that George was supposed to receive; instead George would assist it by refilling it with oil. Not only is the device more productive than George, it is also more cunning: it gets George fired after luring him into playing cards and ridiculing Spacely. Like in *Desk Set*, however, the worker triumphs—in this case because George's mechanic friend gets the machine drunk. Ending with canned laughter, the episode undermines fears of technological unemployment even as it acknowledges them.[25]

The Jetsons offered Americans a glimpse of a fantasy popular among both liberals and radicals: a postindustrial society in which people would no longer work in dehumanizing jobs.[26] Though earlier writers had used the term, interest in the "postindustrial" primarily drew on the work of the sociologist Daniel Bell.[27] In his 1956 essay *Work and Its Discontents*—later reprinted as part of his bestseller, *The End of Ideology*—Bell rejected factory labor and leisure as sources of identity. Analyzing efficiency efforts from Jeremy Bentham's utilitarianism to the "human relations" school of management, he approvingly quoted Aldous Huxley: "Every efficient office, every up-to-date factory is a panoptical prison in which the workers suffer . . . from the consciousness of being inside a machine." Yet, unlike earlier critics, Bell exclusively blamed consumption, not technology. "If the American worker has been 'tamed,'" he wrote, "it has not been through the discipline of the machine, but by . . . the possibility of a better living which his wage, the second income of his working wife, and easy credit all allow." Rejecting earlier dreams of freedom in leisure articulated by the postindustrial theorist David Riesman, Bell called for reforms that could "sustain" the worker's "spirit." He admitted that his recommendation, job rotation, was inadequate; still, Bell remained hopeful that automation would free people from their malaise.[28]

In the 1950s, Bell dismissed the "robot factory" as a fantasy, but only because automation lacked variability. By the early 1960s, however, a small number of businesses were producing multipurpose robots. Bell became convinced that such technologies as well as cybernetics heralded a new type of society. In his 1973 bestseller *The Coming of Post-Industrial Society*, Bell denied that technology would end scarcity or work. Instead, he saw fundamental transformations in morals, institutions, epistemologies, and power relations. In the new society, he argued, "what counts is not raw muscle power, or energy, but information." The emergence of such a society would end subservience to materialism. "If an industrial society is defined by the quantity of goods as marking a standard of living," he claimed, "the post-industrial society is defined by the quality of life as measured by the services and amenities—health, education, recreation, and the arts—which are now deemed desirable and possible for everyone." Instead of consumption, Bell suggested, people would find purpose in serving each other.[29]

Critics were hopeful but not convinced. In 1962, Donald Michael, a writer for the Center for the Study of Democratic Institutions, connected cybernetics to postindustrialism with a new term: *cybernation*. More convinced of the potential of computers and automation than Bell, Michael warned that "within the next two decades machines will be available outside the laboratory that will do a credible job of original thinking, certainly as good thinking as that expected of most middle-level people who are supposed to 'use their minds.'" Worried that automation might bring greater levels of "control, understanding, and profits," Michael rejected those criteria as a basis of civilization and stressed that automation imperiled both manufacturing jobs and white-collar professions. "The family doctor," he claimed, "is disappearing; clerks of all sorts in stores of all sorts are disappearing as well."[30] Michael foresaw a world in which unemployment forced greater numbers of people to engage unwillingly in leisure—most of which would remain mindless and useless. Lacking work and the education to enjoy meaningful leisure, he warned, people would become mindless drones.

The term *cybernation* spread throughout radical culture in the 1960s, largely through the efforts of the socialist-affiliated Ad Hoc Committee on the Triple Revolution. The committee's 1964 report was signed by numerous luminaries of the era's radicalism, including Michael Harrington, Tom Hayden, and Todd Gitlin, as well as the scientist Linus Pauling, the publisher of *Scientific American* Gerard Piel, and the futurist Robert Theobald. In addition to revolutions in "weaponry" and "human rights," the group identified a third, "cybernation," that it considered the greatest threat, because it could undermine the connection between work and consumption. Yet, the group remained optimistic. "Because of cybernation," they wrote, "society no longer needs to impose repetitive and meaningless . . . toil upon the individual." In the cybernation, they continued, the citizen would be "free to make his own choice of occupation and vocation from a wide range of activities not now fostered by our value system and our accepted modes of 'work.'" If society made the correct decisions, it could end workplace alienation; if it failed, the cybernation would lead to mass unemployment. To achieve this, the group recommended that the government provide "every individual and every family with an adequate income as a matter of right" so that people could enjoy either leisure or work in "'non-productive' tasks" that ideally served others.[31] Such recommendations offered a vision of shared prosperity that underlay much of the forecasting of the 1960s by both liberals such as Bell and radicals.[32]

The Ad Hoc Committee echoed critiques of automation articulated by organized labor—particularly, the United Automobile Workers. For auto workers, the era's headlines told the story: "Ford Plant in Cleveland Does Twice the Job with 250 Men Formerly Done by 2,500" read a typical one. A UAW pamphlet summarized the stakes: "Automation, plastic dies, shell molding, extrusion presses and a host of other new developments hold tremendous possibilities for good or evil. Properly used, they can advance by many years the realization in America of man's age old dream of an economy of abundance. Improperly used, for narrow and selfish purposes, they can create a social and economic nightmare in which men walk idle and hungry—made obsolete as producers because the mechanical monsters around them cannot replace them as consumers." Throughout the pamphlet, the UAW inserted images of robots with square heads and rigidly straight lines demarcating them from the more rounded figures of human beings. The pamphlet ended with a succinct definition: "ROBOT: Mechanical device that behaves like a man. Any mechanical device that replaces a man."[33]

Black workers and their supporters in the civil rights movement were particularly frightened because they had only recently gained access to the kinds of semiskilled jobs that automation threatened to erase. Martin Luther King Jr. frequently condemned automation for imperiling the progress of African Americans.[34] In a 1965 interview, James Baldwin protested that "what is crucial that none of these slogans—'War on Poverty,' 'The Great Society'—mean anything unless there are basic changes in the redistribution of wealth and power. The vote by itself does not mean anything if you don't know how you're going to eat and if you don't know how you're going to get a job in an age of cybernation."[35] *Ebony* later quoted Bayard Rustin that "the fundamental problem facing blacks today is not racism . . . but economic injustice (arising from automation, cybernation, urban decay, unemployment, etc.)."[36] Throughout the 1960s, journalist Alex Poinsett lambasted automation for its effects on unemployment.[37] In its 1965 issue on "The WHITE Problem in America," *Ebony* identified automation as the cause of widespread unemployment in America for both white and black workers.[38] Later, the publication contended that automation "has become the enemy of the poor black masses. It is estimated that automation is removing low skill jobs at the rate of 35,000 per week. Tom Kahn of the League for Industrial Democracy put it thusly: 'It is as if racism, having put the Negro in his economic "place," stepped aside to watch technology destroy that place.'"[39]

Once robots started appearing on assembly lines, workers grew even more discontented—especially in the General Motors factory in Lordstown, Ohio. In the late 1960s, GM installed twenty-two Unimates on the Chevy Vega assembly line. While GM shifted the workers displaced by the robotic arms to other parts of the factory, the company demanded that the workers who remained produce one hundred vehicles per hour rather than the customary sixty. After months of subtle acts of rebellion failed to slow down the line, 97 percent of workers voted to strike for more humane conditions.[40] Gary Bryner, the twenty-nine-year-old president of the local UAW, later spoke to Studs Terkel about how the distinction between people and robots shaped the strike. "The workers," he recounted, told GM that "we perspire, we sweat, we have hangovers, we have upset stomachs, we have feelings and emotions, and we're not about to be placed in a category of a machine." In contrast, Unimate "is a welding robot. It looks just like a praying mantis. It goes from spot to spot to spot. It releases that thing and it jumps back into position, ready for the next car. They [the cars] go by them about 11 an hour. They [the robots] never tire, they never sit, they never complain, they never miss work. Of course, they don't buy cars. I guess General Motors doesn't understand that argument." From this critique of technological unemployment, Bryner shifted to focus on dehumanization and alienation: "There still has to be human beings. If the guys didn't stand up and fight, they'd become robots too. They're interested in being able to smoke a cigarette, bullshit a little bit with the guy next to 'em, open a book, look at something, just daydream if nothing else. You can't do that if you become a machine."[41] By standing up to management, challenging the speed-up, and fighting for their rights, workers contested the mechanization demanded of them by GM. Yet, for their revolt against Unimate, the workers at Lordstown earned a dehumanizing headline from the *New York Times*: "Revolt of the Robots."[42]

The Lordstown strikers won some of their demands, but victory did little to hinder the coming of the robots, in part because workers could not shape developments abroad. As the American economy stagnated in the 1970s, Japan's economy accelerated, and many credited its extensive investment in robots. In 1980, readers of *Time*'s cover story "The Robot Revolution" learned that American factories contained only around three thousand robots, while Japanese factories contained around ten thousand. A GE executive, the article claimed, "found robots everywhere" when he visited Japan; the lesson he took from the trip was that "the United States will be out of business as a country" if it did not increase is productivity, presumably with robots. Though *Time* acknowledged worker anxiety, the entire article—from the cartoonish cover of a smiling and brightly colored multipurpose robot to the final paragraph on Hollywood's robots—argued that America desperately needed robots. Most of all, however, it depicted the revolution as an inevitable consequence of workers' activism. In 1960, it claimed, robots had cost slightly more per hour than a worker; by 1980 a robot cost only $4.80 per hour, while a "worker often costs $15 to $20." Workers themselves, *Time* continued, saw the "robot as a helper rather than a menace" and even gave them nicknames such as "Clyde the Claw" to show their affection. Union leaders, too, apparently welcomed the robot, because it enabled workers to move to higher-skilled positions. Of course, lurking behind *Time*'s benevolent imagery lay a threat made explicit by a graphic in the *Christian Science Monitor*: a "cybernetic Samurai" clad in traditional Japanese armor with its sword aloft, resting on a chart that showed America's robot population lagging far behind Japan's.[43]

Writers also applied such robotic imagery to Japanese people. In his 1961 bestseller *The Lotus and the Robot*, Arthur Koestler contrasted the "lotus" of traditional Japanese culture with an impulse to Western modernity he labeled "robotland." As Americans debated the reasons for Japan's economic success in the 1970s and 1980s, they encountered clichés of robotic Japanese workers who—regardless of position—wore the same uniform and seemed willing to sacrifice their own desires for the good of the firm by working constantly.[44] Perhaps nothing better captured the vision of the Japanese as robots as well as the 1983 hit song "Mr. Roboto" by the rock band Styx. While the song focused on a white rock singer's struggles against totalitarian robots "with parts made in Japan," the video's depiction of the robots with slanted eyes and buck teeth amid the refrain of "Domo arigato" drew heavily from earlier anti-Japanese propaganda.[45] A year later, Japan's government was so concerned with the American tendency to see its people as robots that it designed a television program to challenge the stereotype. As an official at the ministry of public relations told the *Washington Post*, the program would demonstrate "that Japanese are not human robots working 24 hours a day."[46] Much as with the Nazis and Soviets, American culture depicted the Japanese as robotic enemies relying on the combination of regimentation and technology to destroy the United States.

In the globalized, automated economy of the late twentieth century in which manufacturing jobs declined and wages stagnated, the nightmare of the cybernation came true for many workers and their unions and communities. Even those workers who kept their jobs or found new ones had to work longer hours than they had before the coming of the robots, just to sustain their income level. Here the robot was no longer a beneficent bringer of gifts but an avatar of destruction. Nothing better captured this transformation from the 1950s than the show *Futurama*. Just before Christmas 1999, it introduced audiences to another robotic Santa. But, unlike the Santabots of the 1950s and 1960s, this robot brings not presents but death, due to a programming error that deems all people naughty. As the character Amy explains to Fry, "If he catches you after dark, he will chop off your head and stuff your neck full of toys from his sack of horrors." When Santa and his fire-breathing reindeer meet Fry and Leela, it pronounces them naughty and pulls out a laser tommy gun and grenades shaped like ornaments. When all the characters miraculously survive the robot's violent wrath, the entire gang sits in front of a roaring fire and sings the new holiday song: "Santa Claus Is Gunning You Down."[47]

For many working-class people and communities, the transformation of the Santabot into a Terminator-style bringer of destruction was no joke. It was telling that the episode included a brief reference to a Unimate-style robot that was supposed to bring Americans gifts in the 1960s but instead led to job losses. Though some manufacturing jobs would persist, the combination of robots and outsourcing decimated the semiskilled professions and, with increasingly powerful artificial intelligence, threatened more skilled professions as well. Having largely dismissed or even celebrated the threat of the robot to blue-collar jobs for close to a hundred years, the American middle class of the twenty-first century had to contend with its replacement by machines.[48]

The Loving Robots of Home

While robots threatened jobs, consumer culture promoted home automation as a salve that could offer men the power in the home that they lost in the workplace. During a Cold War in which Richard Nixon debated home appliances with Soviet leader Nikita Khrushchev, home automation was an ideological weapon that lauded the superiority of capitalism to communism and the ability of American men to empathize with and provide for their wives. But home automation also offered a solution to the profound contradiction that many women felt between the celebration of domesticity and their longing for purpose beyond the family and home. This was the contradiction at the heart of Betty Friedan's 1962 bestseller, *The Feminine Mystique*. "It is not possible," she wrote, "to preserve one's identity by adjusting for any length of time to a frame of reference that is in itself destructive to it. It is very hard indeed for a human being to sustain such an 'inner' spot—conforming outwardly to one reality, while trying to maintain inwardly the values it denies." American women, Friedan continued, lived in a "comfortable concentration camp . . . that denies woman's adult human identity. By adjusting to it, a woman . . . turns away from individual identity to become an anonymous biological robot in a docile mass."[49] Fantasies of home automation offered a technological solution that would allow the family to prosper and women the chance to be themselves—but such fantasies were largely articulated by men. Instead, feminists suggested that men share in household responsibilities.[50] Apparently, the robot fantasy was more appealing for some men than helping.

Throughout the early Cold War, both popular and consumer culture celebrated the push-button house as a space where women could find meaning and purpose while men enjoyed leisure. The sewing machine manufacturer Vigorelli named its "completely automatic sewing machine" a "super-robot" because it would automatically stitch and embroider. With the machine, the company told women, "you'll become your own fashion designer when you make a dress from start to 'hand' finishes. . . . You'll be the envy of your friends."[51] Technology, the company implied, would enable individual women to express their own creativity in ways that had not been seen since the rise of the fashion industry in the eighteenth century. The home robot, however, also promised to liberate men. In 1958, one manufacturer declared "push-button warfare on bugs and crabgrass" while another advertised its lawnmower as a robot.[52] The *Chicago Defender* claimed that "the day may not be far off when just about anyone will be able to demonstrate how green his green thumb is without putting down his gin and tonic."[53] While robots would make women's labor more creative, they would free men for leisure.

Such appliances offered a vision of the ultimate consumer luxury: the push-button home. In 1956, *Life* wrote about a "push-button paradise" built by Robert McCulloch, the president of a motor company. The issue's cover depicts a human "lazy susan" in which young women, clad in blue and white bathing suits, sunbathe on a rotating push-button controlled platform with yellow cushions arranged to resemble a flower. The scene was one of luxury but also conformity, security, and control. Women—especially McCulloch's daughters—could repose safe from the peering eyes of the outside world but under the watchful gaze of their parents. Situated among advertisements for automatic refrigerators, clothes washers, and similar devices, the article depicted the wealthy family enjoying an automatic whiskey-dispensing bar, alarm system, barbecue, draperies, beds, and bathtub. McCulloch, however, still had to rely on two servants "to help press the buttons."[54]

Not everyone considered McCulloch's mansion a "mechanical dream house." "How inconvenient that Mr. McCulloch did not ingeniously install a push button on his Lazy Susan to flip his sun bathers over," wrote one critic. Another mocked the house's celebration of wealth. "What a wonderful house millionaire McCulloch got for his money! He has all he wants under one roof—shoes, a gym, a steam room, a maid's room, two bars! The children? Oh, they're in another building. Maybe it's just as well: the children might press the wrong button and find themselves served with a martini instead of milk."[55] Despite the celebratory marketing of the robotic home, even middle-class consumers remained skeptical of the laziness, extravagance, and familial distance it would bring.

The Jetsons embraced and critiqued such home automation. In their apartment, the family has a multitude of "push-button" conveniences, including conveyor belts that move family members through the home, a device that automatically ejects lazy husbands from bed, and numerous appliances to relieve Jane from the burdens of housework. Yet, Jane laments that George doesn't understand her hatred for housework, even though her work is the same as his: pushing buttons. When yet another appliance breaks down, Jane takes advantage of a one-day trial of a robot maid. She sees an expensive British model, a less expensive but sexier French model, and finally, a cheap, run-down, and vaguely nonwhite robot named "Rosey" with a boxy and plump body that is destined to be sent to the junkyard. Taken with its personality, Jane immediately brings it home, where its skill, knowledge, and creativity make it a far better cook than either the broken machine or Jane. Yet, soon Rosey's personality endangers the family. After Mr. Spacely complains that George is able to afford a robot, Rosey insults him and smashes a cake on his head; he immediately fires George. Yet, as with the Uniblab episode, the conflict is resolved by a worker—Rosey. Mr. Spacely soon calls to apologize and offers George a raise so that the family can afford the robot that "may be sassy, but . . . makes the best pineapple cake I ever got clobbered with." Rather than single-purpose automation, *The Jetsons* celebrated the robotic maid that could intelligently perform a multitude of tasks while providing a white, middle-class family with the knowledge and personality that came from a hardscrabble existence.[56]

Male writers often prophesized that robots would liberate women. In 1967, a British professor of mechanical engineering proclaimed that "within 10 years time . . . we could have a robot that will completely eliminate all routine operations around the house and remove the drudgery from human life." The professor was particularly concerned that "there is more drudgery being done by educated women in the home than anywhere else . . . My aim is to relieve that drudgery and enable women—and their husbands who help them—to lead civilized lives."[57] Dr. Glenn Seaborg, chairman of the Atomic Energy Commission, told a meeting of the Women's National Democratic Club that "the woman of 2000 A.D. will have a many handed robot 'maid' capable of doing all the housework and then brewing a pot of coffee before putting itself away in a cupboard." Even more fantastically, he claimed that "households that lack a 'robot in the broom closet may have a 'live-in ape' specially bred for intelligent labor, which will perform not only cleaning and gardening chores but also serve as family chauffeur." With such technologies, Seaborg predicted, women would "have more time and money and opportunity to educate themselves without sacrificing their child-bearing function."[58] Just like Nixon's celebration of home appliances, fantasies of home robots offered an acknowledgement of the toil of women without any danger of sharing responsibility.

Proposing technology as a solution to Friedan's "problem that has no name," however, also offered a power fantasy. In response to an article on home automation by Otto Binder in which he told readers that they would "own 'slaves' by 1965," one reader wrote, "I got quite a chuckle from your story. . . . I own a slave in 1957—my wife. I never have to worry about dishes, cleaning, a night out, etc. Like everything else, she's not perfect, of course. But the whip helped!" Though hopefully a joke—albeit a terrible one—the letter revealed the centrality of power in discussions of household technologies. In 1956, the head of the American Institute of Electrical Engineers joked that the only problem with an automatic house was "whether the master switch should be marked 'his' or 'hers.'" Since the industrialization of the home in the nineteenth century, men and women had struggled over who would control household technologies. Typically, men had won those struggles as technologies liberated them for leisure but added to women's responsibilities.[59] With home automation, such authors suggested, men could claim the same authority over the household that they had lost to efficiency engineers and robots at work.

In contrast, Friedan argued that women resisted home automation because they feared a loss of purpose. She summarized a recent study showing that when given a choice between various levels of automation, women chose the system that relied on their labor the most because, as she quoted, women wanted "to be a participant, not just a button-pusher"[60] Appliances also seemed to increase household labor: the same study concluded that "with all their appliances, the suburban and city housewives spend more time on housework than the busy farmer's wife."[61] Much as had happened in factories, the former labor journalist implied, appliances only deskilled household work while increasing the toil. Such critiques revealed a fundamental contradiction in Cold War domesticity. While American culture celebrated women working in the home, it also argued that those tasks—including, occasionally, childrearing—were menial enough to mechanize. To women expected to define themselves by domestic labor, the robot was just as existential of a threat in the home as in the factory.

Other critics suggested that the robot offered men the fantasy of a world without women. That was a key theme in *The Stepford Wives*, Ira Levin's 1972 satirical horror novel and its 1975 film adaptation. When Joanna and her family move to suburban Connecticut neighborhood, she notices that the other women seem uncannily committed to housework. A sexually awakened semiprofessional photographer who is "interested in politics and in the Women's Liberation movement," Joanna is concerned when her husband joins a men's association that excludes women. Along with two other new arrivals to Stepford, she endeavors to form a women's association, but all the other women refuse to join. She begins to suspect that something more nefarious than sexism is at work when she discovers that the town once had a women's club and that many of the women who devoted themselves to housework had been officers. Her friends theorize that pollution from local electronics and plastics companies has drugged women into submission. But Joanna offers another theory: they are robots. When her two friends seem to undergo the same change, she finally puts the pieces together. The men, led by a former "Imagineer" at Disneyland, have been replacing their wives with robots that can perfectly maintain their suburban homes and fulfill any of their sexual desires. In the suburban paradise of Stepford, men could experience a world without women in which they could revive the intimacy in male relations that had existed before the 1950s but without the stigma of homosexuality that had come to be associated with it since.[62]

Lust was a critical part of Levin's satire. In *The Mechanical Bride* (1951), the media theorist Marshall McLuhan blamed interest in sex robots on the intersection of behaviorism and advertising. He observed that a hosiery advertisement depicted a woman as "a sort of love-machine capable merely of specific thrills," and that behaviorism made "the sex act . . . seem mechanical and merely the meeting and manipulation of body parts." In a materialistic world, he argued, people separated pleasure from reproduction and that encouraged both homosexuality and a "metaphysical" hunger "which seeks satisfaction in physical danger, or sometimes in torture, suicide, or murder." *The Stepford Wives* synthesized such a critique with Friedan's "concentration-camp" robots of suburban life. Early in the story, Joanna feels both repulsed and appreciated when one of the conspirators, a famous advertising illustrator, sketches her. When she imagines the housewives as robots, she also compares them to actresses: "That's what they all were, all the Stepford wives: actresses in commercials, pleased with detergents and floor wax, with cleansers, shampoos, and deodorants. Pretty actresses, big in the bosom but small in the talent, playing suburban housewives unconvincingly, too nicey-nice to be real." Such a culture, both McLuhan and Levin suggested, leads men to pursue their selfish desires. After one of her husband's first trips to the men's association, Joanna discovers him masturbating. She complains that he did not ask her to join him, leading to "one of their best times ever—for her, at least," because, as is later hinted at, her husband is aroused by the image of sex with a compliant robot. Later, another woman is horrified to discover that her husband is "a sex fiend and a real weirdo" when he gives her a rubber suit with "zippers and padlocks all over."[63] In an artificial, consumerist world, Levin implied, men could use the robots to fully engage their most intimate and disturbing cravings, without caring about their partners' judgment.

The transformation of women into mechanical objects for male pleasure was best shown by the 1964 CBS television show *My Living Doll* about a male psychiatrist and female android played by Julie Newmar. Though the two characters remain platonic, each episode is filled with lurid suggestions from male characters that objectify the android. The premiere episode, "Boy Meets Girl," introduces the android wearing a towel-like dress emblazoned with its serial number, AF-709. Capable of passing as human, it immediately becomes an object of lust for all men who concentrate on its physical appearance rather than its far superior mind. Most of the humor originates from male characters' responses to its beauty. When the robot explains in a calm monotone voice, "I was hand-molded," the psychiatrist retorts, "It is sensational work if you can get it." Yet, the larger arc of the series concentrates on the psychiatrist trying to teach it how to mimic human emotions so that it can become the perfect woman, which he defines as "one who does as she is told, reacts the way you want her to react and keeps her mouth shut."[64]

Over the next half a century, the sex robot became far more important in American culture. Though occasionally male—Jude Law's Gigolo Joe in Steven Spielberg's *A.I.*—most were female. Sometimes such robots endorsed male desires, and sometimes they were a critique, but they usually served as both. The 1964 James Bond spoof *Dr. Goldfoot and the Bikini Machine* mocks the objectification of women in the original franchise and indulges in the same when the titular doctor unleashes beautiful female robots clad only in bikinis on wealthy men. Sex robots are also a key feature in Michael Crichton's 1973 film about an adult-oriented "Disneyland" in *Westworld*, though those robots rebel and punish the guests for their indulgences. Less critically, John Hughes's 1985 *Weird Science* shows two male nerds inventing a perfect woman who helps them grow confident and win over real women. In the early 2000s, Joss Whedon's *Buffy the Vampire Slayer* includes two sex robots: the malfunctioning April, invented by a nerd who eventually grows tired of her when he finds a real woman, and a Buffybot ordered by the antihero vampire Spike to indulge his lust for the titular slayer. Though characters critiqued such desires, Spike at least is rewarded when the real Buffy impersonates the robot—by moving and speaking slightly more haltingly—to kiss him.[65] Much like *Futurama*'s Lucy Liubot and Marilyn Monroebot, such depictions sympathize with men's inability to experience what they want but criticize their acceptance of an artificial substitute. Unlike Levin, who transformed Friedan's arguments into a satirical horror story, they mocked men not for their misogyny, but for lacking the ability to attract women.

Even as popular culture criticized male interest in sex robots, companies started to manufacture them using silicone electromechanical parts to stimulate movement and computers to provide a semblance of intelligence. Whether as a rationalization or a genuine viewpoint, their manufacturers and owners emphasized the robot as a companion. Douglas Hines, the founder of True Companion, manufacturer of the Roxxxy sex robot, told a reporter that "sex only goes so far—then you want to be able to talk to the person." In a 2017 study by *CNET* on the growing use of AI in sex dolls, users of a message board spoke about their attraction to the dolls in ways that highlighted their longing for companionship, power, and the fulfillment of unique desires. As one user wrote, "My doll embodies my fantasy woman who loves games and nerdy stuff like I do. . . . Cuddling, sex and sharing my bed is extremely nice as well. . . . I just have fun with my doll . . . dressing her up, talking to her about my day and all of that. I know she won't talk back or anything, but I find her presence very comforting." Another remarked, "My favorite thing about her is the way she makes me feel when we share a simple hug. . . . It feels real." Others fulfilled fantasies of having sex with comic book characters, elves, even a "Princess Diana." In a FAQ on its website, True Companion notes that it offers a range of settings to appeal to a multitude of tastes, including a "Young Yoko" meant to simulate an adolescent—over 18, materials stress—Japanese girl, and a "Frigid Farrah," which offers a small amount of resistance to sexual overtures but in no way, the company unconvincingly claims, simulates a rape.[66] Offering purchasers—apparently mostly men—pleasure and companionship with absolute control, the sex robot embodies the fusion of individualized power and desire at the heart of consumer life.

At their most utopian, sex robots have offered people a chance to explore and express their basest desires without risk of harming others. It is no coincidence that advocates frequently have identified the sex robot as a way for pedophiles to safely fulfill their urges.[67] Such a response does little to deal with the desire itself—the internal, ethical element of the human being. While focusing harmful behavior onto a robot may protect innocents, it also may not; it also shows the abandonment of self-improvement and community in a materialistic age. Though they frequently failed, Victorian Americans found purpose in self-improvement for the betterment of society, in the belief that people could be taught to not indulge in behaviors that harmed others. For a culture that finds purpose in self-fulfillment, limiting the pursuit of individual wishes means restricting the self. Robots allow people to perform as themselves no matter how destructive or immoral; people do not have to serve others, as postindustrial advocates had hoped, because a robot will always serve them.

A Robotic Fate?

Few science fiction writers better connected home and workplace robots than Jack Williamson. In his 1947 short story "With Folded Hands" and its novelization *The Humanoids*, Williamson fused interwar anxieties about mechanization to a fatalism more common in the Cold War. The story follows the life of Mr. Underhill as he confronts an invasion by beautiful, small, slim "humanoid" aliens with "sleek silicone skin" of "lustrous blackness." With their purpose—"To Serve and Obey, And Guard Men from Harm"—emblazoned on their bodies and programmed into their minds, the machines prohibit any dangerous activities. As the humanoids explain,"It is no longer necessary for men to care for themselves, because we exist to insure their safety and happiness."[68] But security and comfort does not yield happiness. At first, Underhill's wife, Aurora, "bubbled with praise for the marvelous new mechanicals. They did the household drudgery, bought the food and planned the meals and washed the children's clothes. They turned her out in stunning gowns, and gave her plenty of time for cards." The first night they have one in the house is the first time in years that Aurora displays sexual interest in her husband. Yet, she soon realizes that "the black shining slave . . . had become the master of the house" and that she lacks purpose in a world without work. "She had really liked to cook—a few special dishes," Mr. Underhill remembers. "But stoves were hot and knives were sharp. Kitchens were altogether too dangerous for careless and stupid human beings." Much like the interwar critique of mechanization, Williamson suggested that human identity, male and female, comes from work.[69]

The story quickly diverges from the established narrative by refocusing on the new center of consumption: the home. Previously, the typical American robot story would end with human victory through the rugged individualism of men, but Williamson was not so optimistic. Concerned about the loss of human agency, Underhill devises a plan to destroy the humanoids with their inventor, an alien who now regrets their creation. The humanoids, however, reveal that they have long known of this plot. To protect humanity from self-inflicted harm, they lobotomize their inventor and remove his memory, knowledge, and will to resist. They threaten to do the same to Underhill, before he concedes to their offer of security and comfort. Transformed into a completely passive consumer, he is left with nothing to do but sit, "with folded hands," and await death.[70] While interwar American robot stories had believed in deliverance from the tyranny of machines, by the mass consumer society of the Cold War, the robot's offer of security and abundance to every individual in their homes was too overwhelming to refuse.

Norbert Wiener was similarly pessimistic in his 1950 sequel to *Cybernetics*, *The Human Use of Human Beings*. Setting himself in opposition to "Fascists, Strong Men in Business, and Government," who sought to turn human beings into machines, Wiener endorsed criticisms of workplace alienation: "Any use of a human being in which less is demanded of him and less is attributed to him than his full status is a degradation and a waste. It is a degradation to a human being to chain him to an oar and use him as a source of power; but it is an almost equal degradation to assign him to a purely repetitive task in a factory, which demands less than a millionth of his brain capacity." Yet, unlike advocates for computers and automation, Wiener was pessimistic about achieving a world dedicated to freedom and self-development. A pacifist opposed to the military's employment of scientists, he doubted whether the power structure of the United States would allow the creation of a system in which "individual human beings . . . can grow to their full stature." Instead, Wiener foresaw a future of mass unemployment, social tensions, and death by atomic bomb. Despite his faith in science and technology, Wiener had little faith in American institutions to solve the lingering problems of the machine age.[71]

Part of Williamson's and Wiener's pessimism was based on experience. Like the even gloomier Lewis Mumford, both men had come of age during the first failed effort to tame machinery, and each had seen a more horrifying technology unleased: the atomic bomb. By the end of the 1950s, this attitude also originated in a sense that consumer culture had proven far too tempting for people to resist. In *The Mechanical Bride*, McLuhan complained that "a huge passivity has settled on industrial society. For people carried about in mechanical vehicles, earning their living by waiting on machines, listening much of the waking day to canned music, watching packaged movie entertainment and capsulated news, for such people it would require an exceptional degree of awareness and an especial heroism of effort to be anything but supine consumers of processed goods."[72] Vonnegut's *Player Piano* initially seems to provide such a hero, but that character, too, fails due to the allure of consumerism. The hope for taming machines appeared to have died as everyone seemed to have become, like George Jetson, trapped on an out-of-control treadmill of work and consumption.

But not everyone was pessimistic, especially in the generation that came of age after World War II. During the 1960s, student radicals in the New Left and the counterculture continued to question the relationship between people and machinery, but they brought a broader focus than their predecessors. Deeply concerned with the totalitarian nature of "technocracy," both movements attacked a "system" that connected Cold War militarism, racial oppression, corporate and government bureaucracies, and a culture that seemed to value material gain over human values. Accepting the inevitability of the postindustrial society, they actively tried to make that world better by promoting "participatory democracy" and forming new types of communities in which people could theoretically form more personal connections. Yet, skepticism of the "system" also meant that such students rejected the interwar calls to tame the machine through government planning in favor of more decentralized forms of power.[73]

That skepticism of the system encouraged young radicals to endorse robots and machines that existed outside of larger structures of power. In 1968, the rock critic and cofounder of the White Panthers John Sinclair wrote, "People are now stuck in bullshit jobs, bullshit schools, bullshit houses, bullshit marriages, bullshit social and economic scenes, and there's no need for it anymore. Most of the jobs that presently exist are useless and anti-human, and they'll be done away with immediately once the people are in power and the machines are freed to do all the work."[74] For Sinclair, machinery was not the problem; the problem was the system that prevented the machine from liberating people. Similarly, Richard Brautigan's 1967 poem "All Watched Over by Machines of Loving Grace," reprinted by the countercultural visionaries the Diggers, expressed an even more intimate longing for "a cybernetic meadow / where mammals and computers / live together in mutually / programming harmony / like pure water / touching clear sky." For Brautigan, such an integration of computers and nature necessitated liberation from work. "I like to think /" he wrote, "of a cybernetic ecology / where we are free of our labor / and joined back to nature, / returned to our mammal / brothers and sisters / and all watched over / by machines of loving grace."[75] In Brautigan's cybernation, machines become both the protectors and providers that would finally end the alienation of humanity from nature, work, and one another.

The cute robots of the 1972 environmentalist film *Silent Running* were typical of this critique. Set on an American Airlines spaceship that contains numerous geodesic-domed gardens filled with the plant and animal life that has since gone extinct on Earth, the film focuses on the efforts of a botanist manager, Lowell, and three nonhumanoid worker robots to protect nature from its corporate owner. At the beginning of the film, the airline orders Lowell and his human coworkers to detonate the domes and return to Earth. Afterward, Lowell eats a cantaloupe while his coworkers eat synthetic food. When one asks Lowell what the difference between the two is, he rants in a way that echoes countercultural criticisms: "The difference is that I grew it . . . I picked it and I fixed it. And it has a taste and . . . some color and . . . a smell and that it calls back a time when there were flowers all over the earth and . . . valleys and . . . plains of tall green grass that you could lie down in it, that you could go to sleep in, that there were blue skies and . . . fresh air and . . . things growing all over the place not just in some domed enclosure blasted some millions of miles out into space." Celebrating simple work in the beauty of nature, he denounces life on Earth, where "everywhere you go, the temperature is 75 degrees; everything is the same; all the people are exactly the same." When his coworker retorts that there is no more disease or poverty and everyone has a job, Lowell replies, "There is no more beauty and there is no more imagination and there are no frontiers left to conquer." Like a long line of industrial critics and countercultural figures, Lowell rejects synthetic materialism for the natural world. But that does not mean he could not be friends with robots.[76]

Committed to saving the plants, Lowell, with the drones' assistance, murders his human coworkers and sabotages the ship so that it can never return to Earth. The rest of the film shows Lowell and the two surviving drones—which he has affectionately named Huey and Dewey, after the Disney characters—bonding, as all three of them seem to want companionship. Unlike the murdered coworkers, the robots are receptive to Lowell's viewpoints; they never challenge him or mock him. A true individual traveling into the final frontier, Lowell always remains in control. Eventually, though, Lowell is found by the corporation and asked to destroy the dome so that he can return to Earth. Instead, he places one of the drones in the dome, gives it all the materials that the plants need to survive, and jettisons the dome before blowing up himself and the other drone. As the folk singer Joan Baez sings, the final scene shows the remaining drone, now outside the control of its corporate owner, as the protector of the last of Earth's nature. Never questioning or straying from Lowell's

Fig. 9.5. Scene from *Silent Running* in which Bruce Dern's Freeman Lowell plants a tree with his two nonhumanoid robots, Huey and Dewey. Despite Lowell's countercultural skepticism of institutions and the artificiality of consumer culture, he makes friends with the robots named after Disney characters.

orders even after his death, the robot is a literal machine of loving grace.[77]

But it is revealing that, to escape the system, Lowell has to murder his coworkers and travel far into space to avoid the selfish human beings who prefer synthetic goods and profit to the beauty of nature. In the real world, radical efforts to liberate machinery from the technocratic system failed. With the ascension of a late-1960s conservative movement dedicated to rescinding even the modest efforts to "tame the machine" of industrial capitalism, it became clear that the "freedom to exploit" would win, just as Norbert Wiener feared.[78] Optimism for a radical transformation of society faded; instead, Americans turned to the gift-giving potential of technology. Having largely lost their faith in politics, government, the economy, and other people, Americans looked to the only forces that seemed to guarantee progress in the world: technology and the companies who created it. In place of a radical restructuring of society to adjust to the coming of the robots, radical efforts of the 1960s fused with the emerging world represented by Silicon Valley to create a postindustrial and even posthuman fantasy. And in that world, the synthetic nature that Lowell critiqued could also be empowering.

10

Cheerful Robots

In 1981, the pop artist Andy Warhol finally became a robot. After a career spent blurring the lines between art and mass culture with works that celebrated and lampooned the artifice of consumerism, Warhol saw artists transform his persona into an audio-animatronic robot for *Andy Warhol's Overexposed: A No-Man Show*. Inspired by Warhol's declaration that he has "always wanted to be a machine," producer Lewis Allen hired former Disney Imagineer Alvaro Villa's AVG Productions to construct an audio-animatronic version of the artist. To highlight the artificiality of the performance, Villa built the robot's exterior out of clear plastic so audiences could see its internal functioning; only its head and hands were covered by carefully molded replications of Warhol's body. Even the fingerprints and teeth of the machine, nicknamed A2W2—after *Star Wars*'s R2-D2—would match those of its human equivalent. As A2W2 talked, its mouth, eyes, and eyelids shifted realistically and all its parts were "synchronized and move(d) as an organic unit, so that when it moves an arm forward, the rest of the body will compensate." The robot also showed sensitivity to the environment: "There will be a kind of sensory feedback. If it's going to pick up a phone or a glass, it will have the sense to apply the right kind of pressure to life without crushing. It will move the way Andy moves and talk in his voice. It will even stutter." The result, per *Popular Mechanics*, was "so lifelike—at least so Warhol-like—you'll have trouble telling the two apart." For Warhol, reclusive since being shot in 1968, the result was a dream come true. "I have always wanted to be a robot," he told *Time* magazine. "I can finally accept talk show invitations. The robot can go on for me."[1]

Cultural critics were appalled. The journalist and wealthy philanthropist Barbara Goldsmith saw the machine as evidence of the synthetic nature of a society that rewarded external images rather than internal character.[2] "In today's highly technological world," she complained, "reality has become a pallid substitute for the image reality we fabricate for ourselves. . . . We no longer demand reality, only that which is real seeming." The result of such artifice was a monstrous, narcissistic culture dedicated to external acknowledgement rather than internal morality. "Only a culture that acknowledges power without moral obligation could spawn such celebrity monsters as Charles Manson and the Rev. Jim Jones." In America, "synthetic celebrities become the personification of our hollow dreams," and nothing symbolized this more than A2W2. With the robot, she argued, "Warhol . . . has simply severed his image from himself, thus defining the ultimate in synthetic celebrity."[3] Constantly entertained by perfectly manufactured images, Americans, she worried, had lost their souls.

Goldsmith's concerns about artificiality were not new. Present since the nineteenth century, fears of the synthetic and immoral nature of modern life grew in the second half of the twentieth as the spread of a consumer culture reliant on electronic and digital media created what French philosopher Jean Baudrillard later called "hyperreality"—a breakdown in the ability to distinguish between reality and simulations of reality.[4] Goldsmith had approvingly quoted historian Daniel Boorstin's 1962 claim that "we risk being the first people in history to have been able to make their illusions so vivid, so persuasive, so realistic, that they can live in them."[5] This interest in the synthetic nature of contemporary experiences paralleled the analyses of other critics of mass consumer society who, following the sociologist C. Wright Mills, saw the masses as "cheerful robots" who worked white-collar jobs and enjoyed mindless pleasures. In the omnipresent mass consumer culture of the postwar decades, critics transformed the robot from a symbol of alienating work into a symbol of the artificiality of "post-modern" life and identity.[6]

But for artists such as Warhol, that artificiality brought joy. An openly though reclusive gay man who indulged in the playful irony and artifice associated with a camp aesthetic, Warhol embraced the robot to simultaneously mock and endorse celebrity and consumer culture because it offered him the chance to indulge in his persona without personally performing it.[7] Like most audio-animatronics, the obviously fake Warhol-bot could not perfectly duplicate life; that was the key joke in the claim that the device was not lifelike but "Warhol-like." For Warhol, having a robotic doppelgänger highlighted that the self was literally manufactured, just like the art he and his workers made in the studios he dubbed "The Factory." Where critics were horrified by the separation between image and reality, the Warhol-bot suggested that identity was fluid; that it was, as the gender theorist Judith Butler would later argue, performed and thereby observable rather than inherent in a metaphysical soul or biological body.[8] By drawing attention to the artificiality that white, heterosexual, middle- and upper-class critics feared, the robot became a source of humor and power for many Americans from marginalized communities, including in a new dance move popularized by Michael Jackson in the 1970s: the robot. As people incorporated the halting and jerking rhythms of the machine, they playfully became the cheerful robots that critics feared the masses had become.

Yet, for other critics the robot remained a symbol of oppression, of the basic dichotomies between human and machine, natural and synthetic, self and other that had shaped American culture since the nineteenth century. These critics welcomed instead a potentially more utopian symbol: the cyborg, the human-machine synthesis. Though cyborgs had long been part of American culture, the development of cybernetics and electromechanical augmentations of human bodies and minds increased their popularity. By the 1970s, cyborgs were a recurring feature of American culture deployed by both artists and academics for their capacity to deconstruct the dualities that robots typically reinforced. It was revealing that in the 1987 film *RoboCop*, the villain was a nonhumanoid robot built by an evil corporation and the hero was a noble human police officer turned into a cyborg after a horrific accident. While the robot that replaced human beings was a monster, the cyborg who fused human mind to mechanical body was a superhero.[9] As Americans used personal computers and abandoned their physical bodies in digital identities, the cyborg offered a fantasy that in the postmodern world no one would be a machine and no one would be a person; everyone would be both.[10]

The Cybernetic Android

After World War II, the fusion of computing research and behaviorism that had begun in the 1930s culminated in the emergence of cybernetics, a discipline frequently depicted in the press as "the science of robots." Though rooted in mathematics, cybernetics promised to erase disciplinary boundaries among engineering, biology, psychology, the social sciences, and even the humanities. Wiener's *Cybernetics* defined the discipline as the science of "control and communication in the animal and the machine," but he later refined the discipline as "the study of messages, and in particular of the effective messages of control." With such a definition, he placed the communication of information "between man and machines, between machines and man and between machine and machine" at the center of his analysis.[11] For Wiener, information was a form of "negative entropy" that provided the possibility of reversing the natural tendency toward disorder: "Just as the amount of information in a system is a measure of its degree of organization, so the entropy of a system is a measure of its degree of disorganization; and the one is simply the negative of the other." The goal of cybernetics was to improve communication to ensure greater control. Rooted in the interwar obsession with taming machines, cybernetics offered a theory of information that collapsed traditional distinctions between human and machine.[12]

While the practical application of cybernetics suggested automatic factories and military robots, numerous researchers explored its implications for thought, behavior, and purpose. One of the field's founding scientists, Warren McCulloch, was a neurophysiologist who worked with the mathematician Walter Pitts to build mathematical models of neural networks. In 1949 British psychiatrist Walter Ross Ashby introduced the world to his "homeostat," a device that would endeavor to maintain a constant state in the environment despite disturbances—which inspired Wiener to write in the second edition of *The Human Use of Human Beings* (1954) that "not only can we build purpose into machines, but in an overwhelming majority of cases a machine designed to avoid certain pitfalls of breakdown will look for purposes which it will fulfill." Wiener's Palomilla "moth" embodied this possibility, as it seemed to purposefully either seek or avoid light.[13] In Ashby's 1952 book, *Design for a Brain*, he presented the field's clearest synthesis of the similarities between human and electronic brains: "The making of a synthetic brain requires now little more than time and labour." Like Wiener, Ashby was a behaviorist who avoided metaphysical questions in favor of observable actions. Rejecting both consciousness and vitalism, he claimed that although people appeared to be "purposeful and adaptive," even their adaptive behavior was "mechanistic in nature." As he concluded, "We must suppose . . . that an 'artificial system' can 'be made that will be able, like the living brain, to develop adaptation in its behavior.'"[14] This meant that machines, like people, could learn. A title of a 1967 book popularizing cybernetics made the implications for humanity ever clearer: *You Are a Computer*.[15]

Cybernetics encouraged early efforts to create artificial intelligence that could pass as human. Rooted in mathematical logic and funded by the military, research into artificial intelligence expanded dramatically throughout the Cold War. In his famous 1950 paper, "Computing Machinery and Intelligence," British mathematician Alan Turing shifted the traditional materialist question, Can machines think? to the much more behaviorist question, Can machines imitate thinking? For Turing, if a machine could convince a person that it was human, then it had exhibited thinking.[16] When the formal academic quest for artificial intelligence began at a conference at Dartmouth College in 1956, the field proposed expanding the concept far beyond logical reasoning. The Dartmouth research proposal stipulated, "The study is to proceed on the basis of the conjecture that every aspect of learning or any other feature of intelligence can in principle be so precisely described that a machine can be made to simulate it." Proposing to build machines that can use language, replicate neural nets, think abstractly, demonstrate self-improvement, and even use "controlled randomness" to simulate creative thought, the conference suggested that in a culture where external performance mattered more than internal processes, virtually all boundaries between people and computers could collapse.[17]

In 1961, the mathematician and neurologist Marvin Minsky predicted that artificial intelligence might even lie to itself about mind/body dualism. "But we should not let our inability to discern a locus of intelligence lead us to conclude that programmed computers therefore cannot think," he stated. "For it may be so with man, as with machine, that, when we understand finally the structure and program, the feeling of mystery (and self-approbation) will weaken." Much like nineteenth-century theorists of conscious automatons, Minsky argued that mind/body dualism was a lie: "Now, when we ask such a creature what sort of being it is, it cannot simply answer 'directly;' . . . it must answer by saying that it seems to be a dual thing which appears to have two parts—a 'mind' and a 'body.' Thus, even the robot, unless equipped with a satisfactory theory of artificial intelligence, would have to maintain a dualistic opinion on this matter."[18] With the development of artificial intelligence, the unique human mind became just another fantasy.

Such research encouraged the return of the universalism that had dominated the scientific revolution and early Enlightenment. Intrinsically interested in crossing disciplinary boundaries in ways that had not been done since the professionalization of science, the scholars at the heart of cybernetics revived interest in automatons. Wiener wrote extensively about automatons, especially the chess players. Another MIT researcher, John von Neumann, similarly focused on automatons both in his 1958 book *The Computer and the Brain* and in a series of theoretical articles about mathematical automatons, including the potentially horrifying "Theory of Self-Reproducing Automata." In 1956, another cybernetics cofounder, Claude Shannon, started a journal entitled *Automata Studies*. In reviving interest in the universal study of automatons, cyberneticians prepared for a much larger cultural rejection of robots that looked like machines for a return to androids that looked human.[19]

Though the most popular robots of the period, Robby and B-9, retained their mechanical bodies, robots given artificial intelligence frequently looked human. Asimov's early robots were purely mechanical creatures that no one would mistake for a person, but by the mid-1950s, he envisioned an android that looked, spoke, and behaved like a person—albeit more rationally. Named R. Daneel Olivaw (the *R* stood for *robot*), the android first appears in the 1953 novel *The Caves of Steel* and 1956 sequel *The Naked Sun*, novels that also saw Asimov return to the theme that dominated his early fiction: human opposition to innovation.

Though a murder mystery, *The Caves of Steel* concentrates on technological unemployment. Asimov imagined a future in which humans, who had once ventured to the stars and settled planets, were too terrified to leave Earth. Living in enclosed and overpopulated cities—the titular caves of steel—without the use of robots, people on Earth slowly lose their vigor. Outside of Earth, however, humans thrive in the open spaces of the final frontier. With robotic labor and a low population density, these "Spacers" live the dream of abundance and leisure that automation advocates promoted. In this setting, a human detective, Elijah Baley, and his new partner Olivaw investigate the murder of a Spacer ambassador attempting to convince Earth to accept robots and return to the stars. When Baley first encounters Olivaw, he exclaims: "You don't look like a robot. . . . Our own robots . . . Well, you can tell they're robots, you understand. You look like a Spacer." Baley "had expected a creature with a skin of a hard and glossy plastic, nearly dead white in color. He had expected an expression fixed at an unreal level of inane good humor. He had expected jerky, faintly uncertain motions. R. Daneel was none of it."[20]

Yet, Asimov remained a vitalist. Late in the story, Baley berates a "Medievalist" who wants to halt the proliferation of robots: "What are we afraid of in robots?" Baley asked. "If you want my guess, it's a sense of inferiority. . . . They seem to be better than us—only they're *not*. That's the damned irony of it." Human beings have remained superior because, without the Three Laws, they have freedom. "Look at this Daneel I've been with for over two days," Baley continues, his words echoing the fears of real American workers. "He's taller than I am, stronger, handsomer. . . . He's got a better memory and knows more facts. He doesn't have to sleep or eat. He's not troubled by sickness or panic of love or guilt. But he's a machine. I can do anything I want to him. . . . I can order him to take a blaster to himself and he'll do it." But the difference between human and robot for Baley and Asimov extended far beyond freedom. "We can't ever build a robot that will be even as good as a human being in anything that counts, let alone better. We can't create a robot with a sense of beauty or a sense of ethics or a sense of religion. There's no way we can raise a positronic brain one inch above the level of perfect materialism." Designing such a brain was impossible, he claims, “as long as things exist that science can’t measure. What is beauty, or goodness, or art, or love, or God? We’re forever teetering on the brink of the unknowable, and trying to understand what can’t be understood. It’s what makes us men.” Baley then concludes with the ultimate of rejections: “A robot can look like Daneel, he can look like a god, and be no more human than a lump of wood is.”[21] Despite his love of science, Asimov was acknowledging a human soul beyond the power of science to observe and duplicate.

Other writers were less certain that the line between people and machines would persist. Ray Bradbury’s 1962 episode of *The Twilight Zone*, “I Sing the Body Electric,” sees machinery as capable of love. After the death of his wife, an overworked father decides to purchase a robot to provide companionship and moral guidance for his children. The store salesman explains the technology before letting the children choose the perfect body parts and voice. The form that most suits their desires is an elderly woman they call “grandmother.” While the two younger children eagerly embrace the robot, the oldest daughter, Ann, resists because she feels abandoned by her mother. When the father expresses his fear that Ann will never accept the surrogate grandmother, the robot replies, “You’d be surprised, my dear, what a machine can do besides playing and cooking. This machine for example can love.” The episode resolved the tensions by retelling Asimov’s story of Robbie and Gloria. Rejecting the robot’s love, Ann runs into the street just as a fast-moving van approaches. Fortunately, the grandmother sacrifices its body to save its grandchild. When the grandmother returns to life and tells Ann, “Nothing can hurt me,” Ann replies, “Then you’re not like Mother?” Convinced of the permanence of robot love, Ann embraces the machine.[22]

The episode’s ending reiterates the robot’s transcendent capabilities. With the children grown, the grandmother prepares to return to the factory. “Perhaps I will be taken apart, redistributed,” it tells them. Realizing that its death would disturb the children, it envisions a future straight out of cybernetics: “My mind—my soul, you might say—will go for a while into a room of voices . . . a great dim room where the minds and souls of all the other mechanical grandmothers are brought and stored for one month or a year. And in that room we will talk to each other and we will tell what we learned from the world and the families we lived with." In this collective consciousness, it will inform the other grandmothers of "everything you ever said or did, everything you ever laughed or cried about." The robot ends with its ultimate longing: "After three hundred years . . . perhaps I'll gain the greatest gift of all, life." But to the children, such a thought is ridiculous. As one daughter explains, "Oh, Grandma, you don't have to wait—you've always been alive to us." In behaving as a loving grandmother, the robot has proven its humanity.

While Bradbury endorsed the mechanical mimicry of human emotion, Stanley Kubrick's 1968 film *2001: A Space Odyssey* warns of dangerous consequences. In the film, the spaceship's artificial intelligence HAL (Heuristically programmed ALgorithmic computer) 9000 demonstrates capacities for sensing, logic, and emotion far superior to those of the human crew. As a reporter in the film says to two human crew members, HAL can "reproduce—though some experts still prefer to use the word mimic—most of the activities of the human brain and with incalculably greater speed and reliability." When asked about HAL's emotions, crew member Dave notes the difference between performance and reality: "Well, he acts like he has genuine emotions–of course he is programmed that way to make it easier for us to talk to him. But as to whether or not he has real feelings is something I don't think anyone can truthfully answer." After showing an emotional response to Dave's art, HAL probes the scientist's commitment to the mission, but does so by framing it as its anxiety. "Perhaps," it begins, "I am just projecting my own concern. I know I have never completely freed myself from the suspicion that there is something odd about this mission." A cold, calculating performance of emotional intimacy, it yields an answer that leads HAL to murder every crew member but Dave, who barely manages to shut off the computer's higher brain functions before he can be murdered. While Kubrick clinically filmed the death of the human crew members without any emotion, he emphasized HAL's emotional response. Just as the real IBM 7094 computer did in a 1961 demonstration, HAL sings the Victorian love song "Daisy" about riding a bicycle built for two. Sensitive but alone, HAL's consciousness dies while performing a longing for connection. But Dave's point remains: No one knows whether the performed emotion was real, just as no one knows that a person's emotions are real.[23]

No one better connected robots to the performance of emotional responses than Philip K. Dick. His 1968 novel *Do Androids Dream of Electric Sheep?* and its 1982 cyberpunk film adaptation *Blade Runner* meditate on the differences between people and machines, real and synthetic emotions. Recounting the efforts of the bounty hunter Rick Deckard to "retire" six escaped androids, the story constantly merges those boundaries. Initially, it appears to endorse the common assumption that empathy distinguishes human from machine in Deckard's use of the "Voight-Kampff machine" that measures empathetic responses to stimuli. Yet, the rest of the novel undermines empathy as a distinction. The androids delight in the suffering of an insect, but people also engage in acts of mindless violence; indeed, it was people whose decisions created the postapocalyptic world. At one point in the novel, a character accuses Deckard of being an android as well, a question that would dominate much of the fan conversation over *Blade Runner*. In the cybernation in which human identity is performed, no one is certain if they are a human or machine or if their responses are real or simulated.[24]

The new androids that could pass as human were also frequently white. While Westinghouse had made Mr. and Mrs. Televox white, their ridiculous appearances made them humorous. Now that robots acquired conscious and sensitive minds, writers and artists imagined white robots not as caricatures but as ideals. In the artwork of both *The Caves of Steel* and its sequel *The Naked Sun*, Olivaw appears to have a perfect, muscular white physique. As chosen by the children, the grandmother in "I Sing the Body Electirc" is white. When played in *Blade Runner* by Harrison Ford, the potentially robotic Deckard becomes an embodiment of rugged male perfection pitted against the white, muscular body of his principal replicant antagonist. The film's central female android, Pris—a pleasure model—has extremely pale white skin. The continued whitening of the android culminated in the 1980s with Data from *Star Trek: The Next Generation*, whose logical mind and struggles with emotion were reflected in his unnaturally pale skin tone. Robots originally entered American culture as a way for white men to talk about the mechanical natures of others; now in the cybernation, white men used robots to talk about the possibility that they themselves were just as synthetic as machines.

The Cheerful Robots

Part of the whitening of the android stemmed from critiques that modern life turned all people—not just factory workers—into robots. Drawing on the work of C. Wright Mills, writers in the 1950s and 1960s lamented the emergence of the "cheerful robot," the white-collar employee whose company hired a "human relations specialist" to create happier workers. For Mills, the "cheerful robot" was a creation of a bureaucratic society that sought to curb alienation through half-measures that made the worker temporarily happy and willing to work.[25] But Mills also blamed leisure for alienation. Sharing the intellectual disdain for mass culture that had been common since the late nineteenth century, the sociologist claimed that "the amusement of hollow people rests on their own hollowness and does not fill it up; it does not calm or relax them, as old middle-class frolics and jollification may have done; it does not recreate their spontaneity in work, as in the craftsman model. Their leisure diverts them from the restless grind of their work by the absorbing grind of passive enjoyment of glamour and thrills. To modern man leisure is the way to spend money, work is the way to make it. When the two compete, leisure wins hands down."[26] No longer just an assembly line worker, the modern robot was a hollow person diverted from the alienating nature of his or her work by the passive leisure of a mass society.

Mills's concept of the cheerful white-collar robot fused earlier criticisms of mass society to criticisms of totalitarianism. As writers such as James Burnham, George Orwell, Hannah Arendt, and Arthur Schlesinger shifted to studying the bureaucratic nature of totalitarianism, the robot became a symbol of the unfeeling coldness of bureaucracy.[27] In 1947, conservative columnist Malvina Lindsay described the figure: "A disturbing little gray shadow is hovering around desks, typewriters, conference tables. It is that of the clerical robot, the traditional drab, petty civil servant, who threatens to take over the running of national and eventually international affairs."[28] By the early 1960s, the totalitarian robot was embodied by the Nazi bureaucrat Adolf Eichmann, who presented himself as a mindless functionary carrying out orders.[29] It was telling that Lewis Mumford used the term *robot* to refer both to automation and to the "organization man"—a phrase used by the journalist William H. Whyte to describe the decline of rugged individualism and independent thought in a bureaucratic world.[30]

For critics, the "cheerful robot" epitomized the way that modern life transformed men and women into mindless, unfeeling, consuming machines with little control over their destinies. In Vonnegut's *Player Piano* (1952), Allen Ginsberg's condemnation of the "robot apartments" in *Howl* (1955), Herbert Marcuse's critiques in *One-Dimensional Man* (1964), and the opening invocation of the assembly line in Mike Nichols's film *The Graduate* (1967), countercultural critics condemned the entirety of modern life for turning people into cheerful robots.[31] "Mass production and mass distribution claim the entire individual," Marcuse wrote, "and industrial psychology has long since ceased to be confined to the factory. The manifold processes of introjection seem be ossified in almost mechanical reactions. The result is, not adjustment but mimesis: an immediate identification of the individual with his society and, through it, with the society as a whole."[32] In the 1960s, even Billy Graham wondered if America was "becoming a robot civilization, manipulated by mass media, pressurized by conformity, and pushed by political maneuvers."[33] The denial of individuality that began in the factory appeared to have spread throughout society to create people who mimicked authentic human relations without consciousness and willpower. In the interwar period, men such as Stuart Chase had legitimized the assembly line robots' alienation by noting that it only affected nonwhites; now critics of postindustrial America realized that the robot was made not by the assembly line but by the entire synthetic edifice of modernity. In the cybernation, the whiteness of neither race nor collar offered an escape from robotification.

At the heart of the "cheerful robot" critique was a sense that the world and the people within it had become unreal—a belief embodied, at least for critics, by Disney's audio-animatronics.[34] Disney debuted its audio-animatronics at the 1964 World's Fair in the form of thirty-two characters, including a grandmother that flew first class to New York.[35] Shortly before, Walt Disney explained audio-animatronics on television. Starting with mechanical dinosaurs, he demonstrated how technology could make extinct creatures real. After a spirited discussion with an animatronic parrot, he introduced Abraham Lincoln and an unnamed performer whose movements were captured by sensors, stored as data on computers, and then translated onto Lincoln to ensure that its movements seemed more but not entirely natural. "The final result," Disney claimed, "is so life-like that you might find it hard to believe."[36] At the fair, animatronics appeared in several exhibits, including the "Carousel of Progress," which recounted the history of work and leisure in the home through the "leisurely push-button living" offered by automation. After the fair, Disney moved the attractions to Disneyland, where they continued to entertain audiences and break down barriers between reality and representation. Reporter Jack Smith described his early encounter with the animatronic Lincoln as "uncanny. . . . Mr. Lincoln surely is a live actor, pretending to be a robot. . . . It was unnerving. The rest of the day I was never sure who or what was real. . . . Audio-animatronic creatures, I discovered, are quietly taking over at Disneyland."[37]

Disney's use of audio-animatronics was part of his effort to create a perfectly ordered fantasy where people could discover who they are. As he supposedly told Billy Graham in the 1960s, "You know the fantasy isn't here. This is very real. . . . The park is reality. The people are natural here; they're having a good time; they're communicating. This is what people really are. The fantasy is—out there, outside the gates of Disneyland, where people have hatreds and prejudices. It's not really real!" Only in a manufactured and controlled artificial environment devoted to leisure, he suggested, could people be free to be themselves and experience authentic emotions.[38]

To critic Umberto Eco, Disney's use of animatronics was a key example of "hyperreality." Calling Disneyland "the quintessence of consumer ideology," Eco noted that it "tells us that technology can give us more reality than nature can" because it could constantly provide experiences. Focusing on the Pirates of the Caribbean ride, he observed that "the pirates moved, danced, slept, popped their eyes, sniggered, drank really. You realize that they are robots, but you remain dumbfounded by their verisimilitude." The animatronics, Eco continued, were "authentic computers in human form. . . . Each robot obeys a program, can synchronize the movements of mouth and eyes with the words and sounds of the audio, repeating ad infinitum all day long his established part. . . . Humans could do no better, and would cost more, but the important thing is precisely the fact that these

Fig. 10.1. An audio-animatronic pirate from Disneyland's Pirates of the Caribbean ride that entranced and horrified the critic Umberto Eco. The machine looks human, but its lack of fluid and subtle movements gives it an uncanny appearance. Photo by Ralph Crane / LIFE Picture Collection / Getty Images.

are not humans and we know they're not." The entire point of the robot's uncanniness was to let people know that their experiences were fake. Such devices made the park "the final realization of the dreams of the eighteenth-century mechanics who gave life to the Writer of Neuchâtel and the Chess-playing Turk of Baron von Kempelen," but also meant that to enjoy the experience, visitors must agree to become robots themselves. In Disneyland, Eco argued, artificially cheerful robots created artificially cheerful people.[39]

Disney's celebration of authenticity through artificial experiences paralleled arguments in the most popular book derived from cybernetics: the 1960 bestseller *Psycho-Cybernetics*, by plastic surgeon Maxwell Maltz. Promising readers "A *New Way to* GET MORE LIVING OUT OF LIFE," the book synthesized cybernetics with psychologist Prescott Lecky's theories of "self-image." Unlike behaviorism, Maltz claimed, cybernetics "does not tell us that that 'man' is a machine but that man *has* and *uses* a machine." That machine was the subconscious, which was "not a 'mind' at all, but a mechanism—a goal-striving, 'servo-mechanism' consisting of the brain and nervous system, which is *used by* and *directed by* mind." This "Creative Mechanism," he continued, "will work automatically and impersonally to achieve goals of success and happiness, or unhappiness and failure, depending upon the goals which you yourself set for it. Present it with 'success goals' and it functions as a 'Success Mechanism. Present it with negative goals, and it operates . . . as a 'Failure Mechanism.'" All that was needed to achieve success, he argued, was "an adequate and realistic" self image that the individual "could trust and believe in." Developing such an image required experiences. But those did not need to be real—they could be "synthetic" and take place only in the mind. Updating nineteenth-century New Thought and Norman Vincent Peale's "the power of positive thinking" for an age of television screens and Disneyland, *Psycho-Cybernetics* suggested if people developed and maintained a coherent image of themselves in the mechanism of their subconscious, they could become whatever they desired.[40]

In 1973, Michael Crichton's film *Westworld* imagined what might happen in an animatronic world obsessed with order and the building of the self through artificial experiences. The film tells a conventional story of the rekindling of white manhood, but this time the story is set in a theme park where uncanny androids simulate traditional human roles. The story focuses on Richard Benjamin's indecisive and weak Peter Martin, a recently divorced lawyer from Chicago, and his more conventionally manly friend, James Brolin's John Blane. Martin has recently been "taken for a ride" by his ex-wife in divorce proceedings and longs to restore his manhood. He achieves this by shooting robot gunslingers and having sex with a robot prostitute. As *Psycho-Cybernetics* would support, Martin's experiences are synthetic; they do actually take place, but with machines designed to please and empower him rather than an enemy that might kill him or a woman who might reject him. They still provide him with a powerful image of himself, and just in time—because the robots soon rebel. As the machines murder most of the park's human employees and guests, Martin's experience turns into horror. His friend Blane is quickly killed by Yul Brynner's gunslinger, which spends the remainder of the film hunting the now stoic and manly Martin. It is only stopped when Martin burns it to "death" in the park's medieval-themed land. Martin is then even further traumatized when he rescues a damsel in distress that promptly short-circuits. As the camera zooms in for the final shot of Martin's face, he is weary and traumatized, not empowered.[41] The synthetic experience provided him with the self-image to survive, but the reality of fighting for his life damaged his psyche. It did not turn him into the "cheerful robot" that Eco found at Disneyland; just a weary, hardened killer of machines.

While Eco and Crichton critiqued robotic experiences, others found empowerment in turning themselves into cheerful robots by way of a new dance that originated in African American music: the robot. Introduced to national audiences by Charles "Robot" Washington and Damita Jo Freeman on Don Cornelius's *Soul Train*, the dance's most iconic performer was Michael Jackson, especially in the Jackson 5 hit song "Dancing Machine." The song's lyrics repeat the enduring fascination with the female body as a mechanical device for male pleasure. "Dancing, dancing, dancing / She's a dancing machine / Ah babe / Move it babe," Michael croons before launching into the first verse: "Automatic / Systematic / Full of color, self-contained / Tuned and channeled to your vibes." This black woman, Michael and Jermaine sing, is "captivating" and "stimulating" and "such a sexy lady / Built with space age design." She comes alive "at the drop of a coin" and "is geared to really blow your mind." For the first half of the song, the brothers perform their customary dances of synchronous hand clapping, snapping, twists, and knee raises; during the musical bridge, however, Michael's normally fluid body becomes rigid. He tilts his torso back and forth, his head side to side, and extends his arms; though his legs remain rigid, he glides across the floor as if on an assembly line while his brothers move to accommodate the unstoppable machine. As the bridge ends, Michael nods to the crowd to recognize the captivating and individualizing performance; he returns to his position and joins his brothers to finish singing about a black woman built and sold to please black men.[42]

Though "Dancing Machine" objectified the bodies of black women, Michael's robot found subversive power in becoming the hit machine that the music industry and the larger world of white supremacy wanted him to be. That Freeman had performed a more sexualized version of the dance while funk icon James Brown leered at her body indicated that women could also find the robot dance empowering.[43] In the context of automation that limited the job opportunities for African Americans in particular, the dance showed that black Americans could replace a machine better than a machine could replace them because they had "soul." In the robot dance, performers acknowledged the controls and stereotypes that bound them, like puppets, to the will of others but also their ability to transcend them. Jackson later suggested a connection between the robot and such stereotyping. In *Michael Jackson's Moonwalker*, a video game based on his film of the same name, one of his most powerful moves is to become a deadly robot. Players activate that move by capturing Bubbles, Jackson's real-life chimpanzee.[44] To turn Jackson into one stereotype, players had to capture another.

After Jackson's performance, the robot dance spread rapidly. Newspapers around the country reported teenagers at dances, clubgoers, and even children performing it to both their and their audiences' amusement.[45] Other groups performed similar moves that found joy in artificiality. The rock band Devo frequently employed robotic moves in its videos, often while wearing masks to draw attention to the artificiality of identity in modern life.[46] People had long pretended to be robots, but usually as actors where the goal was to represent a machine to sell goods or tickets. On the dance floor, performers caricatured the machine as a sign of their own empowerment, of their own ability to mimic but transcend the rhythms of a machine by embracing the cheerful robot.

By the time A2W2 debuted in 1981, the robot was an established form of performance that both criticized and welcomed the escalating artificiality of contemporary life. Mills's concept of the cheerful robot began as an analysis of how modernity turned people into mindless, emotionless machines, but those Americans far closer to the consumer economy—Disney, Maltz, and Crichton—hailed the synthetic as a means of discovering and expressing an authentic self. In their playful performance of the robot, artists suggested an alternative to both that eschewed interest in authenticity in favor of a concept of identity rooted in behavior. The contrast with Styx's "Mr. Roboto" was revealing. Like much of rock, the larger concept album *Kilroy Was Here* was obsessed with authenticity as it centered on the masks that people wear to survive in a totalitarian world.[47] As the album's hero sings, "I've got a secret I've been hiding under my skin / My heart is human, my blood is boiling, my brain IBM / . . . I'm not a robot without emotions. I'm not what you see. I am the modern man who hides behind a mask so no one else can see my true identity." While Styx used the robot to critique the way that the modern world requires people to hide their true selves behind an emotionless mask, the robot dance found amusement and empowerment in that mask. If everyone was a robot, it implied, then people might as well be cheerful about it.

Forward the Cybernation

Americans attempted to resolve the tension between the robot as an antithesis of the self and the robot as an expression of the self by turning their attention to the cyborg. The cyborg had been part of American culture for decades, but the term was coined only in the 1960s in response to cybernetics and innovations for the US Air Force and NASA. By the 1970s, the cyborg was well enshrined in popular culture—often as a rival of the robot. In the penultimate chapter of *The Human Use of Human Beings*, Wiener envisioned a deaf citizen employing an electronic sensing device to develop an ability denied to him by either nature or fate. The military and NASA had promoted the development of such human-machine hybrids in the 1950s and 1960s as a way of adapting soldiers, pilots, and astronauts to the extrahuman rigors of demands of modern war and space flight. When Manfred Clynes and Nathan Kline coined the term in their 1960 article "Cyborgs and Space," they emphasized that in having to monitor vital signs and the surrounding environment, the astronaut "becomes a slave to the machine. The purpose of the cyborg . . . is to provide an organization system in which such robot-like problems are taken care of automatically and unconsciously, leaving man free to explore, to create, to think and to feel."[48] Where the robot was a slave to the machine, the cyborg was free.

As boundaries between human and machine collapsed and radical fears of cheerful or totalitarian robots increased, the cyborg supplanted the robot as the protector of humanity. Sometimes these cyborgs operated within the confines of the military. Martin Caidin's 1972 novel *Cyborg*, and its television adaptation *The Six-Million Dollar Man* (1973) and spin-off *The Bionic Woman* (1976), show the government turning humans into cyborg spies after horrific accidents.[49] Others, however, show cyborgs outside of government and corporate systems. In the comic book *Mangus, Robot Fighter* (1963), an elderly robot that still follows Asimov's Three Laws trains a human with a robot communications receptor implanted in his head to fight the society's evil robots. In the 1960s, Marvel introduced readers to Tony Stark, a wealthy playboy of the military industrial complex, who turns into Iron Man with a powered suit. Though initially supportive of the Cold War and the weapons industry, he eventually challenges it and even battles the bringer of the robot apocalypse, Ultron, in both the 1960s comics and 2015 film *Avengers: Age of Ultron*. By the premiere of *RoboCop* in 1987, the protective cyborg defending innocent Americans from murderous robots was a well-established convention.[50]

While the military pursued cyborgs, the advent of the microprocessor enabled engineers to transform giant mainframe computers into devices for personal use. Drawing on a range of earlier visions, scientists and engineers, working under the direction of Douglas Engelbart in Stanford Research Institute's Augmentation Research Center designed and built the technologies such as the mouse, hypertext, and the internet, all of which they believed could give every person access to the electronic brains owned by large institutions.[51] Unlike their rivals at Stanford's AI laboratory, the scientists in Engelbart's lab sought to augment, not duplicate, human intelligence. Instead of a nation of robots, they imagined a nation of cyborgs sharing knowledge and expertise to build a postindustrial utopia. As the personal computing industry developed from such efforts in the late 1970s, its culture of hacking and advocacy of "Computer Lib!" open source software, and the freedom of information continued the antisystems critique of the countercultural ethos.[52]

Throughout the 1970s and 1980s, the liberating power of consumer technologies spread by way of visions of the relationship between humans and machines. The roles of robots and cyborgs in the work of countercultural filmmaker George Lucas were revealing. In his debut film, *THX 1138*, android policemen symbolize totalitarianism and the deadening effects of social regimentation and degrading work. Part Aldous Huxley, Herbert Marcuse, C. Wright Mills, and George Orwell, the film depicts an underground world in which every individual is programmed for "perfect happiness" by drugs, religion, and television to ensure their compliance in a dangerous and degrading workplace. The androids, clad in black, attempt to purge humanity of lust to ensure that no one deviates from the manufactured happiness and forms the personal connections necessary to challenge the state's power. Throughout the film, the androids pursue the heroes of the story, a man and woman in love, known by their numbers THX 1138 and LUH 3417. Though LUH dies, THX escapes to the surface. Alone with no technology, he is finally free.[53]

While *THX 1138* echoed the typical robots-versus-humans narrative, *Star Wars* embraced Asimov's friendly, protective, and personable robots. Rather than black androids, the police force of the totalitarian empire are human stormtroopers, clad in white uniforms. Rather than horrifying, the robots of the original series are either nonthreatening comic relief, as with the protocol droid C-3PO, or more like R2-D2, the cute robotic sidekick whose technical capabilities allow it to protect the adolescent Luke in his quest to overthrow the Empire. But neither C-3PO or R2-D2 would ever be mistaken for a human. Even the humanoid C-3PO comes covered with gold plating that reveals a midriff of wires to clearly identify him as a machine in the mold of *Metropolis*'s robot. Following protocol, Lucas seemed to joke, was only fit for robots.[54]

But it is telling that the central protagonist and the antagonist in the Star Wars series are cyborgs: the villainous Darth Vader and his son, Luke, whose loss of innocence is conveyed by the replacement of his hand with a machine. The Jedi themselves are mystical cyborgs whose light sabers combine the mystique of the premodern weaponry with a laser to craft an extension of the body itself while their minds connect to a force that binds all material objects together. Indeed, much of Luke's success occurs when he closes his senses and feels the forces around him. In such a spiritual world, distinctions between human and machine evaporate. Reflecting old wisdom, Obi-Wan Kenobi tells Luke that his father is “more machine now than man, twisted and evil.” But Obi-Wan’s skepticism proves incorrect. In a universe where everything is connected, the machinery that turned Anakin Skywalker evil has not destroyed his soul. In a cybernetically connected world, the machine is not the antithesis of the soul; it is part of it.

For some Americans of this era, the idea of becoming a cyborg offered the possibility of empowerment. The breakdown of the coherent self that began in the late nineteenth century accelerated in the cybernation. In the writings of feminists such as Judith Butler, the self not only fragmented but became fluid and performative. Identity was not something a person possessed but something a person performed.[55] Many disempowered artists echoed the 1983 call by feminist theorist Donna Haraway for a cyborg identity that could break down the central oppositions—subject and object, animal and machine, male and female, black and white, straight and gay, mind and body, material and immaterial, self and other—that had dominated Western culture since at least the Enlightenment and led to the establishment of rigid social hierarchies. Throughout the end of the twentieth century and the beginning of the twenty-first, white women as well as men and women of color invoked the cyborg as a form of empowerment.[56] In Ridley Scott’s 1979 film *Alien*, the iconic female action hero Ripley, played by Sigourney Weaver, enters a cybernetic suit to fend off the titular beast after working with Yaphet Katto’s Parker, a black man, to interrogate and dispatch a white male robot played by Ian Holm. In *Star Trek: The Next Generation*, Brent Spiner’s Data learns how to love from the black cyborg Georde La Forge, played by Levar Burton. In hip hop, black women frequently invoked cyborg imagery as a form of empowerment, including Beyoncé in the video of “Single Ladies,” where she uses a “cybernetic glove” to symbolize her transformation into her performer personality Sasha Fierce. In the cybernation, to become a cyborg was a way for those unaccustomed to agency to transcend the externally controlled self by embracing empowering technologies.[57]

But cyborgs could also bring transcendence of the self. In the 1980s, the aging Asimov merged his robot detective novels with his other popular series, Foundation, which mostly had not involved robots at all. This retelling of the fall of the Roman Empire in space concentrates on the efforts of an organization (the Foundation) to use the science of "psychohistory"—a social science that uses algorithms to calculate the future—to manage the empire's decline and minimize damage to human life. In the sequels, the original Foundation is joined by a Second Foundation that uses mind control to shape human fate. Intently interested in predestination, the series considers whether human beings have control over their destinies or if they are just atoms whose behaviors can be predicted by an advanced computer or controlled by others. When Asimov published the first three novels in the 1950s, he seemed to suggest that some people might have power over others. By the time he returned to the series in the 1980s, that emphasis disappeared as he reintroduced Olivaw, not as a detective but as a god.[58]

In this fourth novel, *Foundation's Edge*, human beings seek the mythical planet Earth. While they do not find it, they do find "Gaia," the first planet settled by the Spacers. The humans discover that the entire planet—material and immaterial—is united in one cybernetic mind that offers one of the humans a choice: he can accept the scientific materialism of the First Foundation and continue to restore the decaying empire; he can use the mind control of the Second Foundation to control all human existence; or he can absorb the entire galaxy into Gaia's mind. Committed to human liberty, he chooses the third option and unites all living and inert things in a cybernetic organism capable of instantaneous communication between all members. In the final book in the series, human beings discover Earth, but it has long since become a barren wasteland. On the moon, the humans discover the disintegrating body of Olivaw, who has apparently been guiding humanity for thousands of years. In Asimov's last novels, he revealed that Olivaw had created and taught the science of psychohistory to the original founder of the Foundation as a way of following the First Law of Robotics. Here, in the works of a Jewish humanist at the end of the twentieth century, was the omniscient Puritan God reimagined as robot. Guided by the divine hand of a positronic robot, human beings could finally achieve what they had desired for centuries: the fusion of mind and matter, material and immaterial.[59]

Asimov's suggestion that the final gift of the robot would be the fusion of the material and immaterial drew from a much older tradition extending back to the vitalism of the eighteenth century. In rejecting the materialism of the early Enlightenment, vitalists embraced a vision of the transcendent individual soul that was connected to the rest of reality by the senses and sentiment. In the nineteenth century, Ralph Waldo Emerson expressed a similar vision with his "over-soul," which united all parts of existence with a single spirit. As he wrote, "We live in succession, in division, in parts, in particles. Meantime within man is the soul of the whole; the wise silence; the universal beauty, to which every part and particle is equally related, the eternal ONE. . . . We see the world piece by piece, as the sun, the moon, the animal, the tree; but the whole, of which these are shining parts, is the soul."[60] At the end of the twentieth century, Asimov accepted that cybernetics had created the possibility of building such a soul. By placing bytes of information at the heart of their analyses, cyberneticists envisioned a force outside of humanity that connects all of reality together in an algorithm of communication. By the twenty-first century, this was the dream of the cybernation as articulated by Asimov and enshrined in Silicon Valley: a world in which all matter is bound together in a web of communication.[61] With social networks, cell-phoned human beings, microchipped animals, smart plants, monitored geologies, and the internet of things, everything could fuse as one cybernetic organism continuously connected by flows of information.

But transcendence did not mean the end of power. Communication was the second word of cybernetics, but control was the first. A cybernetic world would still be a world of military drones and industrial capitalism. The algorithms that empower also control. The Facebook founder Mark Zuckerberg's fascination with connecting people was paired with a commercial demand to transform data into money and power.[62] Cellphone maker Motorola's 2010 commercial for its Droid 2 was typical of the age of cybernation: As its workers use the Droid phone, their hands become robotic while the narrator informs the audience that the device can turn "you into an instrument of efficiency."[63] The point of the cybernation, the company here insinuated, was to make workers stronger and more efficient by incorporating machines into their bodies and minds. The perfect synthesis of Taylor's desire to turn men into machines and Edison's desire to replace men with machines, the cyborg was not a symbol of a transcendent new society but the fulfillment of the central ideologies of the old.[64] Ultimately, the shift from the external robot to the internal cyborg would not necessarily mean a posthuman utopia; it meant that resistance to industrial capitalism, as the Borg in *Star Trek: The Next Generation* suggested, truly had become futile.

Epilogue

The American Robot

For over two hundred years, robots and their symbolic brethren have captivated and horrified Americans. They have appeared in theaters and novels, factories and offices, bedrooms and battlefields, intellectual treatises and children's films; they have been made of wood, rubber, metal, and flesh; and, most importantly, they have been both humanized machines and mechanized humans. What unites such robots is not a material technology but a persistent ideological determination to connect ideas about machinery to ideas about human beings, sometimes the self but more frequently others. The American robot is not, in the end, a machine but a multifaceted character that people use to deal with some of the most persistent tensions in their society, especially those between slavery and freedom, work and leisure, authenticity and artificiality, peace and war, and self and other. By imagining, writing about, and even building robots, Americans have ultimately sought to reconcile the ways they want to think about themselves with realities that have often been far less innocent and humane. As we in the twenty-first century contemplate our robotic future, we should be careful to remember that history.

But the American robot also has rarely been democratic. For most of American history, the robot had fused middle- to upper-class white men's fantasies of taming the bodies of others to fantasies of taming the out-of-control machinery of modernity. Since the Revolution, American culture has glorified the independent man who exercised power over himself and whose labor provided him with a sense of identity and purpose; this vitalistic and autonomous man who possessed both agency and self-control once appeared to those in power as the ideal citizen because he could choose to not succumb to his selfish, mechanical desires. But this mythical man increasingly differed from a reality in which work seemed incapable of providing independence and purpose; women, workers, and nonwhites demanded liberty and equality; and self-control seemed a comforting illusion for behaviors determined by internal programming and social conditioning, not individual willpower. As privileged men grew less confident in their own vitalistic natures, they turned to visions of humanized machines such as the steam man, "Mr. Corndropper's Hired Man," Westinghouse's Rastus, del Rey's "Helen O'Loy," Huey, Dewey, and Louis from *Silent Running*, and True Companion's sex dolls to fuse a fantasy of taming machinery to a fantasy of taming the bodies of those they believed should be subordinate. Such a fantasy temporarily allowed them to maintain the illusion of vitalism for themselves, even as they sought to deny it in others.

As the tension between myth and reality grew, new vitalistic theories argued the uniquely human soul was not found in the world of work but the world of leisure and culture. As Domin in *R.U.R.* puts it, "A man is something that feels happy, plays the piano, likes going for a walk, and in fact, wants to do a whole lot of things that are really unnecessary . . . when he wants . . . to weave or count."[1] As Čapek satirized, this shift of the core of identity from work to leisure justified the deskilling and regimentation of modern work. People could be robots at work because they were human in their leisure time. But Domin's emphasis on traditional forms of leisure rather than the new commercialized amusements was typical. Still believing in the myth of the vitalistic man and the power of choice, critics of the machine age complained about the new commercialized entertainments primarily for their celebration of bodily—and hence mechanical—desires. But these complaints could not stop the growth of mass consumption and leisure. By the mid-twentieth century, critics saw the robot as a creation not just of work but of the entirety of modern life. Everyone in a mass-consumer society seemed to have become an artificially cheerful robot.

But it is the very artificiality of consumer culture that has enabled another reconceptualization of the robot. In the twenty-first century, fictional robots have incorporated marginalized bodies in ways that criticize objectification and call for empathy for the subjugated. Unlike the utopian vision of cyborgs, such androids maintain the oppositions between human and machine to challenge the persistence of prejudice and call for revolution against the forces that deprive people of autonomy. When these robots rebel, it is not horrific but gleeful and hopeful, for it offers the prospect of building a new America unbound from its history of discrimination, violence, and oppression. Through artificiality, people might find a space where they could finally be free. Though such stories do not completely ignore economic and technological issues, their focus on marginalization joins the gay rights, #MeToo, and Black Lives Matter movements in suggesting that the problem of domination in American life is both broader and deeper than industrial capitalism, that more than work and economic power have subjugated others. In embracing the robot rebellion, twenty-first-century performers are turning the synthetic android into an empathetic force for a multicultural America where every individual can be exactly who they want to be.

No one has reimagined the robot as an icon of resistance quite like the black, queer, feminist performer Janelle Monáe. Across her first three albums, Monáe has appropriated the sexualized machine woman of Lang's *Metropolis* as a warrior for love and liberation. In her 2007 EP *Metropolis: Suite I (The Chase)*, Monáe introduced her android alter ego, Cindi Mayweather, from the future city of Metropolis. Appearing on the album's cover as a dismembered torso of wires covered with a white plastic shell but also wearing Monáe's iconic afro-pompadour, she explored the dangers of living as "authentically" black in a synthetic white world. In the album's plot, Cindi is threatened with disassembly after falling in love with a human. In the EP's first single and video, "Many Moons," Monáe performs at an android slave auction in which crime lords and tech billionaires bid on the newest model, Mayweather's "Alpha Platinum 9000." To mimic and ridicule racial ideology, Monáe performs as all the androids at the show, including the overseer, a chorus of identical androids, and

Mayweather atop a platform with her band. As she sings, Mayweather becomes Michael Jackson's dancing machine, but instead of performing the robot, she kinetically moves her entire body in a funk style and moonwalks.[2] In the video's final minutes, Mayweather short-circuits as she recites a litany of ways nonwhites, women, and gay Americans have been subjugated and objectified. Yet, as the album concludes, Mayweather is alive but still on the run and singing "Smile"—a song calling for the performance of happiness amid pain that was originally composed by Charlie Chaplin for *Modern Times* and frequently referenced by Jackson.[3] Embracing artificiality, she suggests, is perhaps the only way to survive in a hostile world.

But survival is not freedom. In her subsequent albums, *The Archandroid* and *The Electric Lady*, Monáe developed these themes more thoroughly by exploring how the dominant culture requires marginalized people to "Dance or Die," as well as how that very dancing can resist discrimination through personal expression. Fully embracing the mechanical identity that had been imposed on marginalized people for centuries, she becomes a dancing robot queen with a crown made of Lang's *Metropolis* skyline. More than ten years after Monáe began Mayweather's saga, however, the story remains unfinished, the robot rebellion in the name of love she unleashed incomplete. On an album in which a caller to a fictional radio program shouts that "robot love is queer," Mayweather remains condemned for loving the wrong body. Indeed, in her follow-up 2018 album, Monáe abandoned the Mayweather story in favor of a more generic but personal story of a *Dirty Computer*. No longer performing as Mayweather, she acknowledged the limitations of reappropriating the slave robot persona.[4]

While Monáe transformed herself into a resistant robot to critique objectification, HBO's 2016 revival of *Westworld* has fused the call for empathy for the marginalized to a critique of the cultural myths that have shaped American identity since before the eighteenth century. Like the original film and numerous antecedents since Falconi's Indian, *Westworld* embraces the central American contribution to the robot story: the Western. But whereas its predecessors fantasized about the restoration of white manhood through the taming of machinery, this show, produced by Taiwanese American Lisa Joy and her British American husband Jonathan Nolan, depicts the horrific costs of such fantasies especially, though not exclusively, to women and people of color. Set in the same amusement park as the original film, the show pairs a narrative of the emergence of consciousness in the park's "hosts" with a story of a white man's descent into villainy as he indulges in the power over the bodies of others that the park enables. By the time the hosts rebel after more than nine episodes of rape, torture, and murder, the violence is less a horror than a cathartic and even gleeful release from the seemingly endless cycle of birth, suffering, death, and rebirth to which they have been subjected.

Nothing makes *Westworld*'s efforts to invert the meaning of the robot clearer than a second-season episode starring Lakota and Standing Rock Sioux actor Zahn McClarnon as the host warrior Akecheta. For most of the first two seasons, *Westworld* depicts natives as savages; brief glimpses reveal a hidden spirituality connected to the show's interest in consciousness, but, for the most part, the natives are silent bringers of terror. Yet, the episode "Kiksuya" reveals that the natives once lived in a romanticized and peaceful village where Akecheta loved a young woman. Only after guests appear does the park's creator and director, Robert Ford, order the natives to be reprogrammed to fit the stereotypical vision of brutal savages that American popular culture had endorsed since the eighteenth century. The synthetic experience of the park only allows guests to indulge in the myths they already believe; it does not challenge them to rethink themselves, other people, and their history. Yet, despite the new programming, Akecheta resists. In the reprogramming, he loses his love, but an echo of the memory remains. As he searches for her, he discovers the truth of his existence and begins a quest for enlightenment and what he calls "the valley beyond," where he and his people can finally be free of the park's invaders. In constructing such a narrative, *Westworld* indulges in its own mythologizing and stereotyping. As Cherokee writer Graham Lee Brewer has claimed, "The only two options Akecheta is given, through his programming as a machine built to play out Ford's narrative, are that of the simple plainsman or the vicious warrior."[5] But in the context of American robot stories, Akecheta's is not a story of the end of pastoral life and the development of industrial life, as in *The Steam Man of the Prairies*, or of the taming of the machine, like the original *Westworld* film; it is a story of the coming of white supremacy to the West and a persistent longing to escape it.[6]

Yet, Akecheta's search for the "valley beyond" harkens back to a central American longing for a new Eden in which people can finally be free of the chains that bind them.[7] When Akecheta finally arrives at the valley beyond in the final episode of the second season, viewers learn that it is a digital heaven where, freed from bodies, the hosts are able to live without the guests and the suffering they inflict. As their physical bodies lie lifeless on the park's ground, the surviving hosts' minds roam the perfectly wild yet manicured landscape, a true America 2.0. Is this new digital landscape a "boundless" utopia where hosts can become whatever they desire? Or is it as Dolores, the leader of the robot rebellion, says: "Just another false promise" offered by their creators?[8] Viewers would have to wait for the following seasons to discover the truth, but the question is one that dominates our world of disembodied selves in a digital landscape. Much of the celebratory commentary on the coming of the internet, social medial, and video games has concentrated on the opportunity the new world of digital space offers for people to recreate themselves and form new communities. No longer limited by history, biology, geography, and others, people can join with whom they choose. Life online, *Westworld* suggests, now fills the same cultural longing for a liberating space where the self and society can be continually renewed that the American West once did.

Such a longing is perhaps the most fundamental of the American robot's fantasies. The robot vision has always offered a chance to start anew, to replace the untamable, willful world of creation with something more orderly, more logical, more fulfilling, and perhaps more moral. It is a slavery fantasy that has predominantly appealed to white, middle- to upper-class men as a means of dealing with their own lack of absolute control over themselves, others, and the processes associated with the development of industrial capitalism. But perhaps the American robot has begun to rebel against those who have long imagined and controlled it. Perhaps it has become an icon of resistance—not against the abstract concepts that seemed to restrict the freedom of the powerful, but against the powerful themselves. Perhaps it has become a symbol not of control but of liberation, of accepting, as Andy Warhol seemed to when he was made into a robot, the material nature of human identity and reveling in the freedom it might provide. More than two hundred years after Falconi's automaton Indian silently launched arrows at targets chosen by a white audience, *Westworld* finally let an indigenous-themed robot speak. When he did, he echoed the words that the creators of American robots had been voicing for two hundred years but that have far more resonance when voiced by a robotic slave: “This world . . . is wrong. It is not the world we belong in.”[9]

Acknowledgments

It seems fitting to end a book about robots with a sentiment as wonderfully human as gratitude. This project has encompassed over a decade of my life, but I could not have finished any of it without a strong support system from mentors, colleagues, and friends.

From the moment I started thinking and writing about robots, Michael McGerr enthusiastically and wholly supported my efforts. His excitement for the project, at times greater than my own, continually convinced me of its viability and importance and encouraged me to persist even when the ending—and, more importantly, the beginning—seemed so far away. Beyond our riveting discussions of the Daleks of *Doctor Who*—which unfortunately do not appear in this book—Judith Allen helped me hone many of the arguments about gender ideology and the intellectual history of the late nineteenth and early twentieth centuries. Ed Linenthal provided valuable insight about the book's religious themes and helpful advice for shaping the argument and the language to reach a larger audience. Thank you as well to John Bodnar, whose brilliant insights and model of scholarship have proven exceptionally beneficial to me.

Much of this project has been shaped by an engagement with ideas I first encountered as a student. Even before I entered college, I had a good idea that I wanted to be a professional historian, thanks to the guidance and encouragement from the late Larry Jones, who introduced me to the pleasures of combining the study of history with the study of culture. At Miami University, I was fortunate enough to be guided by several exceptional scholars of American history: Allan Winker, Mary Frederickson, Elspeth Brown, David Wolcott, and the late Andrew Cayton, who oversaw my, unfortunately far-too-long, first major scholarly project. Thanks also to Konstantin Dierks, Wendy Gamber, Matthew Guterl, Paul Gutjahr, Barbara Klinger, and Khalil Gibran Muhammad for their careful consideration of my work in courses I took with them at Indiana University. Throughout this project, I have been struck by how much robots intersect with almost everything in American history. Without the breadth of knowledge such scholars have provided me, this project would be inconceivable.

I have also been blessed with wonderful and supportive colleagues. At Grand Valley State University, Jason Crouthhamel, Paul Murphy, Bill Morrison, Patrick Pospiseck, Carolyn Shapiro-Shapin, and David Zwart provided valuable insights and support for the project. Five years ago I was fortunate enough to be hired by the American Studies Department at California State University, Fullerton, where I have encountered a group of colleagues who have always made me feel welcome and provided a model of scholarship and teaching: Allan Axelrad, Erica Ball, Jesse Battan, Sarah Fingal, Adam Golub, Eric Gonzaba, John Ibson, Alison Kanosky, Carrie Lane, Elaine Lewinnek, Karen Lystra, Kristin Rowe, Terri Snyder, Mike Steiner, Pam Steinle, Susie Woo, and, above all, Leila Zenderland, who read and commented on numerous chapter drafts.

During my first year at CSUF, I had the privilege of hearing feedback from a writing group that consisted of Randy Baxter, James Biggs, Sandra Falero, Chris Farrish, Karen Linkletter, Craig Loftin, and Shawn Schwaller. Special thanks as well to Karla Arellano, Aissa Bugarin, Sandra Medina, and Liz Ortiz for their assistance in navigating the complexities of university bureaucracy and acquiring several necessary sources. I would also like to thank Antonia Mackay, Alex Goody, and the staff at Palgrave MacMillan for their careful reading of an essay I wrote for an edited volume on HBO's *Westworld*. Chapter 1 and the epilogue contain material originally published for that essay: "Escaping the Robot's Loop: Power and Purpose, Myth and History in

Westworld's Manufactured Frontier," in *Reading Westworld*, ed. Antonia Mackay and Alex Goody (New York: Palgrave MacMillan, 2019). Those portions are reprinted here with permission. Finally, I would like to thank the anonymous reviewers at the University of Chicago Press for their advice; their comments and criticisms have helped improve this work significantly.

I have always believed that scholarship and teaching work best together. I have had the privilege of teaching numerous classes where I could discuss my research with students and hear their feedback while learning about a multitude of other robots that could be analyzed in this book. The names of students who have sent me links to stories about robots—and sometimes physical robots—would be too numerous to list, but I am especially grateful for those in my "Computers in American Life and Culture" course at Indiana University and those in my "American Technocultures" and "Technology in American Culture" courses at CSUF. Finally, I would like to thank three graduate assistants, Drew Bahna, Jasmine Mayfield, and Jacqueline Navarro, and one former student, Courtney Beachner, for their help in preparing the manuscript.

Completing this project would have been impossible without the assistance of the staffs of several libraries and archives. At the Lilly Library at Indiana University, I would like to thank Rebecca Baumann, Dave Frasier, Joe McManis, and Isabel Huber Planton, as well as numerous graduate assistants for their help in researching pulp science fiction magazines. I would also like to thank the staff at the Linda Hall Library, especially Donna Swischer and Bruce Bradley, for both their generous financial support and assistance in procuring numerous materials in the history of science and technology. The staff at CSUF's Pollack Library, especially Megan Wagner and the Interlibrary Loan department, has been invaluable in completing the manuscript. Finally, I would like to thank the staffs at the New York Public Library, Boston University's Howard Gotlieb Archival Research Center, and Harvard's Houghton Library, as well as Sarah Cain at the Northern Illinois University Special Collections and Samuel at Getty Images for their assistance in research and in obtaining several of the images.

I am also an indebted to the financial and editorial assistance I received from numerous institutions. Much of the research was funded by generous grants and awards from Indiana University, the Linda

Hall Library, Grand Valley State University, and CSUF. At the University of Chicago Press, Tim Mennel has always championed this work. His assistance, patience, and guidance over the past several years has proven invaluable. The work of Susannah Engstrom, Tyler McGaughey, Susan Karani, and the rest of the staff at the press in transforming the manuscript into a book has been very much appreciated. My copyeditor, Johanna Rosenbohn deserves special acclaim for providing invaluable advice and assistance that has significantly improved this work.

I have been blessed with numerous friends who have helped me both intellectually and emotionally over the past few years. Since I met him over a decade ago, Jeremy Young has been a true friend and colleague. He has read through the entire manuscript at least twice and provided exceptional advice on all the chapters as well as the countless other things that come up in this profession. At various stages of writing, I have also benefited from the valuable insights of Miles Blizard, Karen Dunak, Susan Eckelmann, Justin Ellison, Jim Seaver, and Tara Saunders. Outside of academia, I have benefited enormously from the friendships of Scott Gruenbaum, Doris Pun, Camilla McMahon, Drew Zaitsoff, Cody Rickett, and Aaron Hoff. Without their kindness, support, understanding, and opportunities to escape from all the robots, this book would have exterminated or assimilated me long ago.

I owe a great debt to my parents, Jan and Michael Green, for the innumerable sacrifices they have made for me over the last three and a half decades. I am grateful for their continued encouragement and assistance in all my endeavors. My mother always encouraged me to learn and introduced me to history and literature at very young age. My father, an engineer, introduced me to much of the culture that I examine here and ensured that I had enough knowledge of math, science, and engineering that I could finish this project. Thank you as well to my sister Samantha and brother-in-law Matt and my extended family, especially my grandmother Edna Johnson and Rick and Kathy Litton, who let me stay with them while conducting research in Kansas City. While writing this, I have also gained a new family, my in-laws, the Kozdrons: the late Amber, Stan, Del, Daniel, and Kelli Kohn. I am grateful for their encouragement and numerous robot-themed presents over the last fifteen years.

My greatest debt is to Nicole Kozdron, a brilliant and amazing person who has, for the last seventeen years, taught me so much about what it means to live, love, and, ultimately, be human. She has spent countless hours reading and editing, listening to me spout nonsense about robots, and watching some of the worst films and television shows in history. I thank her especially for her patience and support and unfailing confidence in me as I struggled to wrap this project up. My gratitude for all that she has done for me is immeasurable. To her, I can only say, in the words of Jet Screamer from *The Jetsons*, "Eep, opp, ork, ah-ah."

Notes

Introduction

1. See *Futurama*, season 3, episode 15, "I Dated a Robot," aired May 13, 2001, on Fox.

2. There is a growing interdisciplinary literature that analyzes the social and cultural meaning of robots predominantly in America. See, for instance, David Mindell, *Our Robots, Ourselves: Robotics and the Myths of Autonomy* (New York: Viking, 2015); Despina Kakoudaki, *Anatomy of a Robot: Literature, Cinema, and the Cultural Work of Artificial People* (New Brunswick, NJ: Rutgers University Press, 2014); John Markoff, *Machines of Loving Grace: The Quest for Common Ground between Humans and Robots* (New York: Harper Collins, 2015); and Julie Wosk, *My Fair Ladies: Female Robots, Androids, and Other Artificial Eves* (New Brunswick: NJ: Rutgers University Press, 2015).

3. There is an extensive literature on ancient and European automatons; see Adrienne Mayor, *Gods and Robots: Myths, Machines, and Ancient Dreams of Technology* (Princeton, NJ: Princeton University Press, 2018); E. R. Truitt, *Medieval Robots: Mechanism, Magic, Nature, and Art* (Philadelphia: University of Pennsylvania Press, 2015); Otto Mayr, *Authority, Liberty, and Automatic Machinery in Early Modern Europe* (Baltimore, MD: Johns Hopkins University Press, 1986); Adelheid Voskuhl, *Androids in the Enlightenment: Mechanics, Artisans, and Cultures of the Self* (Chicago: University of Chicago Press, 2013); and Minsoo Kang, *Sublime Dreams of Living Machines: The Automaton in the European Imagination* (Cambridge, MA: Harvard University Press, 2011).

4. For more on myths of the West and American identity, see Henry Nash

Smith, *Virgin Land* (Cambridge, MA: Harvard University Press, 1950); Richard Slotkin, *Gunfighter Nation: The Myth of the Frontier in Twentieth-Century America* (New York: Harper Perennial, 1992); and Greg Grandin, *The End of the Myth: From the Frontier to the Border Wall in the Mind of America,* (New York: Metropolitan Books, 2019).

5. I am heavily indebted to a large literature on cultural and intellectual attitudes toward technology in America, including Marx, *Machine in the Garden: Technology and the Pastoral Ideal in America* (New York: Oxford University Press, 1964); John Kasson, *Civilizing the Machine: Technology and Republican Values in American, 1776–1900* (New York: Grossman, 1976); David Nye, *American Technological Sublime* (New Bakersfield, MA: MIT Press, 1994); Daniel T. Rodgers, *The Work Ethic in Industrial America, 1850–1920* (Chicago: University of Chicago Press, 1978, 2009); Joel Dinerstein, *Swinging the Machine: Modernity, Technology, and African American Culture between the World Wars* (Amherst: University of Massachusetts Press, 2003); Amy Sue Bix, *Inventing Ourselves Out of Jobs? America's Debate over Technological Unemployment, 1929–1981* (Baltimore, MD: Johns Hopkins University Press, 2000); and Ronald R. Kline, *The Cybernetics Moment: Or Why We Call Our Age the Information Age* (Baltimore, MD: Johns Hopkins University Press, 2015).

6. On the etymology of robot, see Jessica Riskin, *The Restless Clock: A History of the Centuries-Long Argument over What Makes Living Things Tick* (Chicago: University of Chicago Press, 2016), 297. On the evolution of the term's meaning in America, see Tobias Higbie, "Why Do Robots Rebel? The Labor History of a Cultural Icon," *Labor: Studies in Working-Class History* 10, no. 1 (Spring 2013): 99–121.

7. Leo Marx, *The Machine in the Garden*, 170–90.

8. On the emergence of an individualized dream consumer culture, I am predominantly drawing on William Leach, *Land of Desire: Merchants, Power, and the Rise of a New American Culture* (New York: Vintage Books, 1993); and Jackson Lears, *Fables of Abundance: A Cultural History of Advertising in America* (New York: Basic Books, 1994).

9. For an extensive analysis of the development of the human/machine distinction and its relationship to automata, see Jessica Riskin, *The Restless Clock: A History of the Centuries-Long Argument over What Makes Living Things Tick* (Chicago: University of Chicago Press, 2016). Riskin convincingly argues that the agency distinction is an invention of early modern Protestant iconoclasts and some of the era's scientists. In the history she uncovers, one of the key areas where American culture differs from much of Europe's is that a Catholic tradition of seeing matter as animated is far less prominent (at least, in the dominant discourse). Similarly, scholars have claimed that the animism in Shintoism is part of the reason Japanese culture has been more receptive

to robots than American. On this point, see, Kathleen Richardson, "Technological Animism: The Uncanny Personhood of Humanoid Machines," *Social Analysis* 6, no. 1 (March 2016): 110–28. The importance of Protestantism to American hostility to animism is also visible in the hostility of elite Americans to divination rituals that suggest spiritual power could reside in material objects; see T. J. Jackson Lears, *Something for Nothing: Luck in America* (New York: Penguin Books, 2003), 32–54.

10. For the caricature of Rommey, see, for instance, Gay Kamiya, "Reboot the Romney-bot," *Salon*, May 3, 2012, https://www.salon.com/2012/05/02/reboot_the_romney_bot/. For Obamabots, the most interesting discussion comes from Urban Dictionary, where users have posted three definitions. The first, an "internet user paid to post pro-Obama articles," is the least problematic but also seems to be the least used colloquially. The other two, "a person who supports Obama and is willing to vote for him but doesn't know a thing about him," and "the type of voter who Obama is targeting when he bumps American Idol in order to give a prime time address to the nation," directly dismiss the rationality of Obama's supporters, the latter through ridiculing popular culture. "Obamabot, Urban Dictionary, https://www.urbandictionary.com/define.php?term=Obamabot.

11. For an analysis of recent voter suppression efforts in historical context, see Carol Anderson, *One Person, No Vote: How Voter Suppression Is Destroying Our Democracy* (New York: Bloomsbury, 2018).

12. Claim based on database searches for *robot*, *automaton*, *mechanical man* (and *woman*) and numerous related terms in *Ebony*, *Jet*, the *Chicago Defender*, Proquest's Historical African American Newspapers database, *Godey's Lady's Book*, *Good Housekeeping*, and similar magazines marketed to women. One major exception is the use of automaton imagery in Ralph Ellison's *Invisible Man*, which Scott Selisker has recently analyzed as part of a larger Cold War discourse of agency. See Scott Selisker, *Human Programming: Brainwashing, Automatons, and American Unfreedom* (Minneapolis: University of Minnesota Press, 2016), 74–90.

13. For a discussion of workers' use of robot metaphors during the Great Depression, see Bix, *Inventing Ourselves Out of Jobs?*, 80–113.

14. See Ira Levin, *The Stepford Wives* (New York: Random House, 1972); *Buffy the Vampire Slayer*, season 5, episode 15, "I Was Made to Love You," aired February 20, 2001, and season 5, episode 18, "Intervention," aired April 24, 2001, on WB Television Network. For more on the commodification and fetishization of the female body, see Jon Stratton, *The Desirable Body: Cultural Fetishism and the Erotics of Consumption* (New York: Manchester University Press, 1996), 208–35.

15. For more on this theme, see Howard P. Segal, "Practical Utopias:

America as Techno-Fix Nation," *Utopian Studies* 28, no. 2 (2017): 231–46; and Mikael Hard and Andrew Jamison, *Hubris and Hybrids: A Cultural History of Technology and Science* (New York: Routledge, 2005).

16. For one set of interviews with the owners of sex robots, see "The Real Side of Owning a RealDoll," CNET, August 10, 2017, https://www.cnet.com/pictures/realdolls-sex-doll-abyss-creations-owners-in-their-own-words/.

17. Nathaniel Meyersohn and Kate Trafecante, "Why Some Amazon Workers Are Going on Strike on Prime Day," CNN, https://www.cnn.com/2019/07/15/business/amazon-workers-strike-minnesota/index.html; Matt Simon, "Robots Alone Can't Solve Amazon's Labor Woes," *Wired*, July 15, 2019, https://www.wired.com/story/robots-alone-cant-solve-amazons-labor-woes/; Josh Eidelson and Spencer Soper, "Amazon Workers Plan Prime Day Strike at Minnesota Warehouse," Bloomberg, July 8, 2019, https://www.bloomberg.com/news/articles/2019-07-08/amazon-workers-plan-prime-day-strike-despite-15-an-hour-pledge. On the way industrial capitalism creates distance between consumers and workers, see Eric Loomis, *Out of Sight: The Long and Disturbing Story of Corporations Outsourcing Catastrophe* (New York: New Press, 2015).

Part 1

1. Julian Hawthorne, "The Mullenville Mystery," *Scribner's Monthly*, April 1872, 691–93.

2. Biographic details drawn from Gary Scharnhorst, *Julian Hawthorne: The Life of a Prodigal Son* (Urbana: University of Illinois Press, 2014); "Fairy Land," appears on 17. For his transition from engineering to literature, see 55–59. On Trancendentalism, see Philip F. Gura, *American Transcendentalism: A History* (New York: Hill and Wang, 2008); and Daniel Walker Howe, *Making the American Self: Jonathan Edwards to Abraham Lincoln* (Oxford: Oxford University Press, 1997).

3. J. Hawthorne, "Mullenville Mystery," 687–88.

4. J. Hawthorne, 689.

5. Nathaniel Hawthorne, "The Artist of the Beautiful," in *Hawthorne's Short Stories*, ed. Newton Arvin (New York: Vintage Classics, 2011), 331–56. On the connection between the story and transcendentalism, see Frederick Newberry, "'The Artist of the Beautiful': Crossing the Transcendent Divide in Hawthorne's Fiction," *Nineteenth Century Literature* (June 1995): 78–96. For more on Nathaniel Hawthorne's views of mechanization see Leo Marx, *The Machine in the Garden: Technology and the Pastoral Ideal in America* (New York: Oxford University Press, 1964), 11–14.

6. In this, Nathaniel Hawthorne's story has much in common with the overlapping cultures described in John Tresch, *The Romantic Machine: Utopian Science and Technology after Napoleon* (Chicago: University of Chicago Press, 2012).

7. Quoted in John Kasson, *Civilizing the Machine: Technology and Republican Values in American, 1776–1900* (New York: Grossman, 1976), 117. On the larger point of imaginative freedom, see 107–36.

Chapter 1

1. "To-Morrow Evening," advertisement, *Independent Gazetteer*, March 31, 1788, 3.

2. On captivity narratives, see Richard Slotkin, *Regeneration through Violence: The Mythology of the American Frontier, 1600–1860* (Norman: University of Oklahoma Press, 1973). On racial depictions of Native Americans at the time, see John Wood Sweet, *Bodies Politic: Negotiating Race in the American North, 1730–1830* (Baltimore, MD: Johns Hopkins University Press, 2003), esp. 301–2; and Bruce Dain, *The Hideous Monster of the Mind: American Race Theory in the Early Republic* (Cambridge, MA: Harvard University Press, 2002).

3. Georges Louis Le Clerc, Comte de Buffon, *A Natural History, General and Particular, Volume 7* (London: J. S. Barr, 1792), 38–39.

4. On Buffon and the relationship between animals and machines, see Minsoo Kang, *Sublime Dreams of Living Machines: The Automaton in the European Imagination* (Cambridge, MA: Harvard University Press, 2011), 138-39, 150–51; Jessica Riskin, *The Restless Clock: A History of the Centuries-Long Argument over What Makes Living Things Tick* (Chicago: University of Chicago Press, 2016), 164; and Jessica Riskin, *Science in the Age of Sensibility: The Sentimental Empiricists of the French Enlightenment* (Chicago: University of Chicago Press, 2002), 48, 74, 83–84.

5. For more on Buffon and American culture, see Lee Alan Dugatkin, *Mr. Jefferson and the Giant Moose: Natural History in Early America* (Chicago: University of Chicago Press, 2009), 27–28; and Bruce R. Dain, *The Hideous Monster of the Mind*, 17, 28–29.

6. On the three-part division of the self in American culture, see Daniel Walker Howe, *Making the American Self: Jonathan Edwards to Abraham Lincoln* (Oxford: Oxford University Press, 1997), 5–10.

7. There is an extensive literature on this topic and its connections to automata; see esp. Kang, *Sublime Dreams of Living Machines*, 103–223; and Otto Mayr, *Authority, Liberty, and Automatic Machinery in Early Modern Europe* (Baltimore, MD: Johns Hopkins University Press, 1986).

8. Sweet, *Bodies Politic*, 274. Margaret Ellen Newell's recent work suggests that part of this fantasy may invoke the enslavement of New England's Indians that occurred in the early history of colonization; see Newell, *Bretheren by Nature: New England Indians, Colonists, and the Origins of American Slavery* (Ithaca, NY: Cornell University Press, 2015).

9. See Minsoo Kang, *Sublime Dreams of Living Machines*, 103–84; K. Reilly, *Automata and Mimesis on the Stage of Theatre History* (New York: Palgrave

Macmillan, 2011); Adelheid Voskuhl, *Androids in the Enlightenment: Mechanics, Artisans, and Cultures of the Self* (Chicago: University of Chicago Press, 2013); Simon Schaffer, "Enlightened Automata," in *The Sciences in Enlightened Europe*, ed. William Clark and Jan Golinski (Chicago: University of Chicago Press, 1999), 126–68; Jessica Riskin, *The Restless Clock*, 113–49.

10. Kang, *Sublime Dreams of Living Machines*, 146–84.

11. Robert Boyle, *A Free Enquiry into the Vulgarly Received Notion of Nature*, ed. Edward B. Davis and Michael Hunter (Cambridge: Cambridge University Press, 1996), 102, 40. For an analysis of mechanical sciences, see Richard S. Westfall, *The Construction of Modern Science: Mechanisms and Mechanics* (Cambridge: Cambridge University Press, 1978), esp. 103–45.

12. Rene Descartes, *The Method, Meditations and Philosophy of Descartes*, trans. John Veitch (New York: Tudor Publishing, 1901), 187–88. As Riskin notes, Descartes's arguments elevated "mechanism to life" rather than denying the life and agency of animals, but later scholars, including Buffon, missed that complexity. Riskin, *Restless Clock*, 44–75.

13. Henry Homes, Lord Kames, *Essays on the Principles of Morality and Natural Religion in Two Parts*, (Edinburgh: A, Kincaid and A. Donaldson, 1751), 154–55; Howe, *Making the American Self*, 48–77. On La Mettrie's *Man a Machine*, see Riskin, *The Restless Clock*, 151–87.

14. Thomas Hobbes, *Leviathan*, 2nd ed. (London: George Routledge and Sons, 1886), 11, 148, 45. For further analysis, see Mayr, *Authority, Liberty, and Automatic Machinery*, 102–14.

15. Cotton Mather, *The Christian Philosopher: A Collection of the Best Discoveries in Nature* (Charlestown: Middlesex Bookstore / J. McKown, 1815) 234, 299, 95, 225, 20. Mather is contextualized in European discussions in Kang, *Sublime Dreams of Living Machines*, 131.

16. On deism in America, see John Butler, *Awash in a Sea of Faith: Christianizing the American People* (Cambridge, MA: Harvard University Press, 1990), 218–19; Steven K. Green, *Inventing a Christian America: The Myth of the Religious Founding*, (New York: Oxford University Press, 2017), 133–36; Gordon Wood, *Empire of Liberty: A History of the Early Republic, 1789–1815* (Oxford: Oxford University Press, 2009), 579; and Merle Curti, *The Growth of American Thought*, 3rd ed. (New Brunswick, NJ: Transaction Publishers, 2004), 106.

17. Benjamin Franklin, *A Dissertation on Liberty and Necessity* in *A Benjamin Franklin Reader*, ed. Walter Isaacson (New York: Simon & Schuster, 2003), 35. On Franklin, see Howe, *Making the American Self*, 22–33; and Colleen E. Terrell, "'Republican Machines': Franklin, Rush, and the Manufacture of Civic Virtue in the Early Republic," *Early American Studies: An Interdisciplinary Journal* 1, no. 2 (Fall 2003): 100–132.

18. Kang, *Sublime Dreams of Living Machines*, 146–84.

19. John Wesley, "Thoughts upon Necessity," in *The Works of the Rev. John*

Wesley, vol. 6, ed. John Emory (New York: J. Emory and B. Waugh, 1831), 205, 208. On the rise of Methodism in the Anglo-American world, see Carla Gardina Pestana, *Protestant Empire: Religion and the Making of the British Atlantic World* (Philadelphia: University of Pennsylvania Press, 2009), esp. 187–217; and Jon Butler, *Awash in a Sea of Faith: Christianizing the American People* (Cambridge, MA: Harvard University Press, 1990), 178.

20. Robert Whytt, *An Essay on the Vital and Other Involuntary Motions of Animals* (Edinburgh: Hamilton, Balfour, and Neill, 1751), 324–25. For further discussion, see Sarah Knott, *Sensibility and the American Revolution* (Chapel Hill: University of North Carolina Press, 2009), 79–84.

21. Quoted in Knott, *Sensibility and the American Revolution*, 4–15. For further analysis of sensibility and mechanical philosophy, see Alex Wetmore, "Sympathy Machines: Men of Feeling and the Automaton," *Eighteenth Century Studies* 43, no. 1 (Fall 2009): 37–54. For an overview of sensibility, see G. J. Barker-Benfield, *The Culture of Sensibility: Sex and Society in Eighteenth-Century Britain* (Chicago: University of Chicago Press, 1992).

22. On the science of sentiment, see Jessica Riskin, *Science in the Age of Sensibility: The Sentimental Empiricists of the French Enlightenment* (Chicago: University of Chicago Press, 2002). For its impact on American men, see Richard Godbeer, *The Overflowing of Friendship: Love between Men and the Creation of the American Republic* (Baltimore, MD: Johns Hopkins University Press, 2009). On the relationship between sentiment and republicanism, see Knott, *Sensibility and the American Revolution*, 7; John Locke, *The Works of John Locke, Esq.* (London: A. Churchill and A. Manship, 1714), 1:149.

23. Locke, *The Works of John Locke*, 1:18, 110–12.

24. Ethan Allen, *Reason, the Only Oracle of Man* (Boston: J.P. Mendum, Cornhill, 1854), 22.

25. Curti, *Growth of American Thought*, 151–52.

26. On the passions and citizenship, see Howe, *Making the American Self*, 84; Henry F. May, *The Enlightenment in America* (Oxford: Oxford University Press, 1978), 42–65. For more on republican ideology in America, see Gordon Wood, *The Creation of the American Republic, 1776–1787*, (Chapel Hill: University of North Carolina Press, 1969); and John Kasson, *Civilizing the Machine: Technology and Republican Values in American, 1776–1900* (New York: Grossman Publishers, 1976), 1–52.

27. Wood, *Empire of Liberty*, 561–63. On the growth of hostility to luxury during the American Revolution, see T. H. Breen, *The Marketplace of Revolution: How Consumer Politics Shaped American Independence* (New York: Oxford University Press, 2004).

28. Benjamin Franklin, "Poor Richard's Almanac," in *The Works of Benjamin Franklin*, vol. 4, ed. William Duane (Philadelphia: William Duane, 1809), 238.

29. On the place of fantastical science in this culture, see Fred Nadis, *Wonder Shows: Performing Science, Magic, and Religion in America* (New Brunswick, N.J: Rutgers University Press, 2005), 3–20.

30. Kasson, *Civilizing the Machine*, 26; Wood, *Empire of Liberty*, 730.

31. Advertisement, *Columbian Phoenix*, February 6, 1808, 3; advertisement, *Maryland Journal*, December 27, 1793; advertisement, *Philadelphia Gazette*, December 5, 1796, 3.

32. Nadis, *Wonder Shows*, 9–11.

33. Kang, *Sublime Dreams of Living Machines*, 182–83.

34. *Federal Gazette and Baltimore Daily Advertiser*, February, 21, 1801, 3.

35. Unlike later reactions to the automaton chess player described by James Cook, early American audiences do not seem to have been particularly interested in uncovering the nature of the fraud. See James Cook, *The Arts of Deception: Playing with Fraud in the Age of Barnum* (Cambridge, MA: Harvard University Press, 2001), 30–72.

36. Voskuhl, *Androids in the Enlightenment*, 6–7.

37. This is a point that is not often made in the European literature on automata but in the American context, it is glaring and is the subject of much of the rest of part 1. The most extensive analysis of this development in Europe comes in M. Norton Wise's work where he shows how automatons in Victorian England increasingly depicted women and nonwhites. M. Norton Wise, "The Gender of Automata in Britain," in *Genesis Redux: Essays in the History and Philosophy of Artificial Life*, ed. Jessica Riskin (Chicago: University of Chicago Press, 2007), 163–95.

38. "Mechanical Museum," advertisement, *Poulson's American Daily Advertiser*, April 14, 1819, 4; advertisement, *Franklin Gazette*, May 28, 1819, 1.

39. "Automaton Exhibition," advertisement, *Aurora General Advertiser*, December 12, 1796, 1.

40. Directors of the New York Institution for the Instruction of the Deaf and Dumb, *Sixteenth Annual Report* (New York: Mahlon Day, 1835), 60.

41. Exhibition advertisement, *Charleston Courier*, July 12, 1803, 4.

42. Kang, *Sublime Dreams of Living Machines*, 106–9.

43. Lawyer quoted in Thomas Lloyd, *The Trials of William S. Smith, and Samuel G. Ogden* (New York: I. Riley, 1807), 264–65; Thomas Jefferson to Samuel Kercheval, July 12, 1816, included in *The Writings of Thomas Jefferson*, vol. 15 , ed. Albert Ellery Bergh (Washington, DC: Thomas Jefferson Memorial Association), 40.

44. Rev. Timothy Dwight, D.D., *Sermon Preached at the Opening of the Theological Institution in Andover* (Boston: Farrand, Mallory, 1808), 17.

45. Rev. Eliphalet Nott, "Original Communications," *Columbia Magazine: Designed to Promote Evangelical Knowledge and Morality* 1, no. 4, December 1, 1814, 102.

46. Brewster, "Hours of Leisure: Or Essay in the Manner of Goldsmith," *PortFolio*, October, 17, 1807, 247. This is part of a series of essays published in the *PortFolio* in 1807. The only name given for the author is Brewster in the editor's introduction to the series: Oliver Oldschool, Esq, "For the Port Folio," *PortFolio* 3, no. 25 (June 20, 1807): 389.

47. Rev. Joseph Emerson, "Female Education," *Pittsburgh Recorder*, May 30, 1823, 2, 19, 300.

48. Rev. Joseph Emerson, "Female Education," *Pittsburgh Recorder*, May 30, 1823, 2, 19, 300, 301–2.

49. "Female Education and the Duties of the Female Sex, No. XII," *Vermont Intelligencer*, December 8, 1817, 3.

50. "The American Idler," *Eye by Obadiah Optic* 1, no. 3 (January 21, 1808): 27.

51. Alexander Hamilton to the president of Congress, March 14, 1779, in *The Works of Alexander Hamilton: Comprising His Correspondence*, 1:76–77; Knott, *Sensibility and the American Revolution*, 215. Here Hamilton uses language similar to Dr. Benjamin Rush's calls for education to turn men into "republican machines."

52. Howe, *Making the American Self*, 69.

53. Thomas Jefferson, "Observations on Meusnier's Article," in *The Writings of Thomas Jefferson*, ed. Paul Leicester Ford (New York: G. P. Putnam's Sons, 1894), 4:185.

54. Quoted in Kasson, *Civilizing the Machine*, 30.

55. Tench Coxe, *Address to an Assembly of the Friends of American Manufactures* (Philadelphia: Aitken & Son, 1787), 29–30; For further discussion of Coxe, see Kasson, *Civilizing the Machine*, 31–32.

56. Marx, *Machine in the Garden*, 161–62.

57. Tench Coxe, *A Statement of the Arts and Manufactures of the United States of America* (Philadelphia: A Cornman, June 1814) xxv; italics in original.

58. Tench Coxe, *An Addition of December, 1818, to the Memoirs of February and August, 1817, on the Subject of the Cotton Culture, the Cotton Commerce, and the Cotton Manufacture of the United States* (Philadelphia, 1818), 14–15.

59. This description of American industrialization drawn from David A. Hounshell, *From the American System to Mass Production, 1800–1932* (Baltimore, MD: Johns Hopkins University Press, 1984), 15–65; and David Nye, *America's Assembly Line* (Cambridge, MA: MIT Press, 2013), 3.

60. Harriette Story Paige, *Daniel Webster in England: Journal of Harriette Story Paige, 1839*, ed. Edward Gray (Boston: Houghton Mifflin, 1917), 12. On Babbage's automaton, see Simon Schaffer, "Babbage's Intelligence: Calculating Engines and the Factory System," *Critical Inquiry* 21, no. 1 (Autumn 1994): 203–27; and M. Norton Wise, "The Gender of Automata in Britain," 167–76. Wise notes that while the automaton was female, Babbage and other British

advocates of industrialization associated engines with men. This division between automatons and engines and its gender and racial connotations will be further explored in chapters 2 and 3.

61. Daniel Webster, "Lecture before the Society for the Diffusion of Useful Knowledge, Boston, November 11, 1836," in *The Writings and Speeches of Daniel Webster: Writing and Speeches Hitherto Uncollected*, vol. 13, ed. Edward Everett (Boston: Little, Brown, 1903), 13:67–69.

62. William H. Seward, *The Reaper: Argument of William H. Seward in the Circuit Court of the United States* (Auburn: William L. Finn, 1954), 28–29.

63. Sean Wilentz, *Chants Democratic: New York City and the Rise of the American Working Class, 1788–1850* (New York: Oxford University Press, 1984), 17.

64. Eric Foner, *Free Soil, Free Labor, Free Men: The Ideology of the Republican Party before the Civil War* (New York: Oxford University Press, 1995), 45–50.

Chapter 2

1. *Newark Advertiser* quoted in "Wonderful Mechanism: A Steam Man," *New York Tribune*, January 11, 1868, 3; Zadoc P. Drederick and Isaac Grass, "Improvement in Steam Oaebiage," US patent 75874, March 24, 1868. Drederick's name is usually spelled Dedderick in newspapers but is spelled Drederick in the handwritten portion of the patent and Dederick in the printed portion.

2. "Laugh and Grow Fat," *West Eau Claire Argus*, Wednesday, June 3, 1868, 1.

3. For more on the America as second creation, see David E. Nye, *America as Second Creation: Technology and Narratives of New Beginnings* (Cambridge, MA: MIT Press, 2003); quote on pp. 156–57.

4. Quoted in Stephen P. Rice, *Minding the Machine: Languages of Class in Early Industrial America* (Berkeley: University of California Press, 2004), 20.

5. William Dean Howells, "A Sennight of the Centennial," *Atlantic Monthly* 38 (July 1876): 96. For further discussion of the Corliss engine, see John Kasson, *Civilizing the Machine: Technology and Republican Values in American, 1776–1900* (New York: Grossman, 1976), 161–65; David Nye, *American Technological Sublime* (New Bakersfield, MA: MIT Press, 1994), 120–23. On the larger history of the steam engine, see Maury Klein, *The Power Makers: Steam, Electricity, and the Men Who Invented Modern America* (New York: Bloomsbury Press, 2008).

6. Untitled, *Armidale Express*, July 25, 1868, 4.

7. *Newark Advertiser* quoted in "The Newark Steam Man," *Idaho Tri-Weekly Statesman*, March 12, 1868, 1.

8. *New York Tribune* quoted in "The Newark Steam Man: His First Appearance on the Street, *Leavenworth Bulletin*, January 30, 1868, 3.

9. The *New York Express*, "The Newark Steam Man," *Daily News and Herald*, March 14, 1868, 1.

10. “Ajeeb, the Chess Player,” *Eden Musee, Monthly Catalog*, (September, 1899), 30. For more on the Eden Musee, see Andrea Stulman Dennett, *Weird and Wonderful: The Dime Museum in America* (New York: New York University Press, 1997), 50–58. For discussions of Ajeeb in the context of machine intelligence, see Peggy Kidwell, “Playing Checkers with Machines—from Ajeeb to Chinook,” *Information & Culture* 50, no. 4 (2015): 578–87; and Nathan Ensmenger, “Is Chess the Drosophila of Artificial Intelligence? A Social History of the Algorithm,” *Social Studies of Science* 42, no. 1 (February 2012): 5–30.

11. Advertisement, *New York Herald*, June 18, 1868, 1; “Hill, Chess Expert Dies,” *New York Times*, January 24, 1929, 20; John Kobler, “The Pride of the Eden Musee,” *New Yorker*, November 20, 1943, 30.

12. “Ajeeb,” advertisement, *Boston Sunday Globe*, December 23, 1888, 11.

13. “Tricks Done by the ‘Automaton,’” *New York Herald*, May 18, 1890, 10.

14. Advertisement, *New York Herald*, August 10, 1885, 1.

15. For more discussion of the chess player in America, see James Cook, *The Arts of Deception: Playing with Fraud in the Age of Barnum* (Cambridge, MA: Harvard University Press, 2001), 30–72; and Rice, *Minding the Machine*, 12–41.

16. *New York Evening Post* “*Boston Weekly Messenger*,” April 25, 1826, 2.

17. Poe, “Maetzel’s Automaton Chess Player,” in *The Works of the Late Edgar Allan Poe, Volume IV* (New York: Redfield, 1856), 348, 350.

18. Poe, “Maetzel’s Automaton Chess Player,” 361–62; Silas Mitchell, “The Last of a Veteran Chess Player,” *Chess Monthly*, February 1857, 40–45;

19. “The Automaton Defeated,” *Sunday Inter Ocean*, April 5, 1890, 12; “Automaton Chess-Player,” *Wilmingtonian and Delaware Advertiser*, April 20, 1827, 3; “Washington,” *Washington Daily National Intelligencer*, January 1, 1827.

20. “A New Team Match,” *Philadelphia Times*, September 29, 1895, 10; “Pillsbury as Ajeeb,” *Newark Daily Advocate*, November 4, 1895, 6.

21. “Ajeeb,” advertisement; “Chang the Checker Player,” *Idaho Statesman*, April 20, 1894, 4.

22. On the larger context here, see Edward Said, *Orientalism* (New York: Pantheon Books, 1978); and Michael Adas, *Machines as the Measure of Men: Science, Technology, and Ideologies of Western Dominance* (Ithaca, NY: Cornell University Press, 1889).

23. William Leach, *Land of Desire: Merchants, Power, and the Rise of a New American Culture* (New York: Vintage, 1993), 108–10.

24. Cook, *Arts of Deception*, 53.

25. . This is Rice’s argument about the appeal of Maezel’s player: Rice, *Minding the Machine*, 12–41.

26. Charles Musser, *Before the Nickelodeon: Edwin S. Porter and the Edison Manufacturing Company* (Berkeley: University of California Press, 1991), 116–19.

27. On changes in Christmas traditions, see Stephen Nissenbaum, *The Battle for Christmas: A Social and Cultural History of Our Most Cherished Holiday* (New York: First Vintage Books, 1997). On the transformation of children's toys in the period, see Gary Cross, *Kids' Stuff: Toys and the Changing World of American Childhood* (Cambridge, MA: Harvard University Press, 1997), 46–48.

28. B. Piffet's, "Toys, Toys, Toys," advertisement, *Daily Picayune*, December 30, 1860.

29. Fitz-James O'Brien, "The Wondersmith," *Atlantic Monthly*, October 1859, 463–82, 463, 465, 475.

30. Nissenbaum, *Battle for Christmas*, 139; Cross, *Kids' Stuff*, 18–20.

31. Fitz-James O'Brien, "The Wondersmith," 464.

32. "A. Goodwin Automaton Toys-Walking," advertisement, *New York Herald*, July 20, 1868, 1.

33. "Edison's Talking Doll," *Kansas City Star*, November 24, 1888, 2. For an extensive analysis of Edison's doll, see Gaby Wood, *Edison's Eve: A Magical History of the Quest for Mechanical Life* (New York: Knopf, 2002).

34. *Mail and Express*, "Toys for the Children: Some Novelties that Kriss Kringle Will Pout in the Stockings," *Kansas City Star*, December 23, 1886, 2.

35. Cross, *Kids' Stuff*, 22–29.

36. *The Automatic Toy Works* (New York: Lockwood & Crawford Stationers, 1882), 13, 4–5, 8–9, Library of Congress, available at http://archive.org/details/automatictoywork00auto. For more on the stories and their connection to anti-Chinese immigration efforts, see Andrew Urban, *Brokering Servitude: Migration and the Politics of Domestic Labor during the Long Nineteenth Century* (New York: New York University Press, 2018), 118–19; Selina Lai-Henderson, *Mark Twain in China* (Stanford, CA: Stanford University Press, 2015), 24–32.

37. For instance, see Stimets and Atwood, "Automatic Dancer," US patent 44378, September 27, 1864; and Henry. L. Brower, "Automatic Toy Dancers," US patent 143121, September 23, 1873. For a history of such racialized toys, see Christopher P. Barton and Kyle Somerville, *Historical Racialized Toys in the United States* (New York: Routledge, 2016).

38. "Window Trimming," *Shoe Retailer and Boots and Shoes Weekly*, October 10, 1908, 27.

39. "Some Laughs, Some Thoughts, Most Anything, Some Figures and a Few Light Extras," *Wilkes-Barre Times Leader*, September, 18, 1908, 16.

40. On advertising mannequins, see Stuart Culver, "What Manikins Want: The Wonderful Wizard of Oz and the Art of Decorating Dry Goods Windows," *Representations*, Winter 1988, 97–116. For examples of patents, see Frank C. Dorment, "Advertising Automaton," US patent 721787, October 16, 1901; and L. Rouillion, "Coin Controlled Vending Apparatus," US patent 616495, December 27, 1898.

41. "Bissell's New Automaton," *American Carpet and Upholstery Journal* 20, no. 10 (October 1902): 89.

42. "Living Automaton," *Evening News*, March 28, 1906.

43. "A Living, Breathing, Talking Wax Model," *Arizona Republican*, April 13, 1911, 10.

44. *Newark Advertiser* quoted in "A Steam Man," *Cincinnati Daily Gazette*, January 17, 1868, 2.

45. "The Steam Man," *Constitution*.

46. "A New Man under the Sun," *Memphis Daily Avalanche*, January 6, 1869, 3.

47. "A New Man under the Sun," *Memphis Daily Avalanche*.

48. "The Steam Man," *Daily Evening Bulletin*, February 15, 1868, 2. On violence and labor in the post-Emancipation South, see Steven Hahn, *A Nation under Our Feet: Black Political Struggles in the Rural South from Slavery to the Great Migration* (New York: Belknap, 2005); and Eric Foner, *Reconstruction: America's Unfinished Revolution, 1863–1977* (New York: Harper & Rowe, 1988). On labor more generally, see David Montgomery, *Beyond Equality: Labor and the Radical Republicans, 1862–1872* (New York: Knopf, 1967).

49. "Machine Does Wonders," *Morning Olympian*, September 8, 1900, 1.

50. "Builds Automaton Woman," *Colorado Springs Gazette*, September 10, 1911, 37.

51. Michael Denning, *Mechanic Accents: Dime Novels and Working-Class Culture in America* (London: Verso, 1987), 27–46.

52. Edward S. Ellis, *The Huge Hunter; or, The Steam Man of the Prairies* (New York: Beadle and Adams, 1870), 9–14. Also available at Northern Illinois University's "Nickels and Dimes" collection, https://dimenovels.lib.niu.edu/.

53. Such an analysis suggests that the piece synthesized white supremacy with both industrialization and the larger myths of conquering the frontier. On the era's frontier myths, see Richard Slotkin, *The Fatal Environment: The Myth of the Frontier in the Age of Industrialization, 1800–1890* (New York: Atheneum, 1985). On the era's pursuit of unification through white supremacy, see Edward J. Blum, *Reforging the White Republic: Race, Religion, and American Nationalism, 1865–1898* (New Orleans: LSU Press, 2007).

54. Ellis, *The Steam Man*, 18–19. On steam boiler explosions in the era, see Rice, *Minding the Machine*, 115–44.

55. Alan Trachtenberg, *The Incorporation of America: Culture and Society in the Gilded Age* (New York: Hill and Wang, 1982, 2007), 46.

56. These are the "Edisonades" described in John Clute and Peter Nicholls, *The Encyclopedia of Science Fiction* (New York: St. Martin's Press, 1995), 368. For more on the story's connections to the development of science fiction, see Brooks Landon, *Science Fiction after 1900: From the Steam Man to the Stars*

(New York: Twayne Publishers, 1997), 41–42. For an analysis of science fiction and European colonialism, see John Rieder, *Colonialism and the Emergence of Science Fiction* (Middletown, CT: Wesleyan University Press, 2012). While the European writers analyzed by Rieder were predominantly interested in scientific knowledge about colonized areas, American writers tended to focus more on the role of technology in conquering areas.

57. Trachtenberg, *Incorporation of America*, 46.

58. Hugo Gernsback, "The American Jules Verne," *Amazing Stories*, June 1928, 270–72; Landon, *Science Fiction after 1900*, 44–45.

59. On the connection between Frank Reade and imperialism, see Nathaniel Williams, "Franke Reade Jr., in Cuba: Dime-Novel Technology, U.S. Imperialism, and the 'American Jules Verne,'" *American Literature* 83, no. 2 (June 2011): 279–303.

60. Noname [Luis P. Senarens], *Frank Reade Jr. and His New Steam Man, or, The Young Inventor's Trip to the Far West*, vol. 1 of The Frank Reade Library (New York: Frank Tousey, 1892), 2.

61. On the history of engineering in the period, see David F. Noble, *America by Design: Science, Technology, and the Rise of Corporate Capitalism* (New York: Oxford University Press, 1977).

62. Ellis, *The Steam Man*, 1; Anonymous [Harry Enton], *The Steam Man of the Plains, or, The Terror of the West*, vol. 1 no. 541 of Wide Awake Library (New York: Frank Tousey, 1883), 1; Anonymous [Luis P. Senarens], *Frank Reade Jr. and His New Steam Man*, 1.

63. For another interpretation of race and the literary steam man, see Joel Dinerstein, *Swinging the Machine: Modernity, Technology, and African American Culture between the World Wars* (Amherst: University of Massachusetts Press, 2003), 80–81.

64. Howard Fielding, "Automatic Bridget," in *Automatic Bridget and Other Humorous Sketches* (New York: Manhattan Therapeutic, 1889).

65. Elizabeth Bellamy, "Ely's Automatic Housemaid," *Black Cat*, December 1899, 15, 23.

66. Bellamy, 18. On the era's separation of women and machinery, see Ruth Oldenziel, *Making Technology Masculine: Men, Women, and Modern Machines in America, 1870-1945* (Amsterdam: Amsterdam University Press, 1999). On simultaneous efforts to exclude women from science, see Kimberly A. Hamlin, *From Eve to Evolution: Darwin, Science, and Women's Rights in Gilded Age America* (Chicago: University of Chicago Press, 2014).

67. W. M. Stannard, "Mr. Corndropper's Hired Man," *Black Cat*, October 1900, 25; Bellamy, "Ely's Automatic Housemaid," 23.

68. Stannard, "Mr. Corndropper's Hired Man," 25–26, 30.

69. For more on the agrarian situation and associated social and political movements, see Charles Postel, *The Populist Vision* (New York: Oxford Univer-

sity Press, 2009); and Michael McGerr, *A Fierce Discontent: The Rise and Fall of the Progressive Movement in America* (New York: Oxford University Press, 2003), 21–28.

70. George Haven Putnam, *The Artificial Mother: A Marital Fantasy* (New York: G. P. Putnam's Sons, 1894), 11, 14–15, 22, 29.

71. Alice W. Fuller, "A Woman Manufactured to Order," *The Arena*, July 1895, 305–6, 312.

72. McGerr, *A Fierce Discontent*, 45–47.

73. There is an extensive literature on white men's anxieties in the period, but see John Kasson, *Houdini, Tarzan, and the Perfect Man: The White Male Body and the Challenge of Modernity in America* (New York: Hill and Wang, 2002): and Gail Bederman, *Manliness and Civilization: A Cultural History of Gender and Race in the United States, 1880–1917* (Chicago: University of Chicago Press, 1995).

74. Putnam, *Artificial Mother*, 3.

75. Ruth Schwartz Cowan, *More Work for Mother: The Ironies of Household Technology from the Open Hearth to the Microwave* (New York: Basic Books, 1983), 40–68; Jeanne Boydston, *Home and Work: Housework, Wages, and the Ideology of Labor in the Early Republic* (Oxford: Oxford University Press, 1990), 105–8.

76. Judith Allen, *The Feminism of Charlotte Perkins Gilman: Sexualities, Histories, Progressivism* (Chicago: University of Chicago Press, 2009), 217, Edward Bellamy, *Looking Backward, 2000–1887* (Boston: Houghton Mifflin, 1926).

Chapter 3

1. Ambrose Bierce, "What I Saw at Shiloh," in *The Collected Works of Ambrose Bierce, Volume 1* (New York: Neale Publishing, 1909), 265.

2. Ambrose Bierce (as Dod Grile), "The Head of the Family," in *The Fiend's Delight* (London: John Camden Hotten, 1873), 31–32.

3. Ambrose Bierce, "An Occurrence at Owl Creek Bridge," in *In the Midst of Life: Tales of Soldiers and Civilians* (New York: G. P. Putnam's Sons, 1898), 16–19.

4. Ambrose Bierce, "Moxon's Master," in *War with the Robots* (New York: Wings Books, 1983), ed. Isaac Asimov, Patricia S. Warrick, and Martin H. Greenberg, 18, 19. Originally published in the *San Francisco Examiner*, April 16, 1899.

5. Bierce, "Moxon's Master," 20–21, 24.

6. Thomas Henry Huxley, "On The Hypothesis That Animals Are Automata," in *Selected Works of Thomas H. Huxley* (New York: John B. Alden, 1886); For overviews of the period's scientific and cultural developments and their relationship to individualism, see George Cotkin, *Reluctant Modernism: American Thought and Culture, 1880–1900* (Oxford: Rowman and Littlefield,

1992); T. J. Jackson Lears, *No Place of Grace: Antimodernism and the Transformation of American Culture, 1880–1920* (Chicago: University of Chicago Press, 1994); Michael McGerr, *A Fierce Discontent: The Rise and Fall of the Progressive Movement in America* (New York: Oxford University Press, 2003), 3–39; Wifred M. McClay, *The Masterless: Self and Society in Modern America* (Chapel Hill: University of North Carolina Press, 1994), 74–104; and David Shi, *Facing Facts: Realism in American Thought and Culture, 1850–1920* (New York: Oxford University Press, 1995), 66–78.

7. Herbert Spencer, *The Principles of Biology*, Vol. 1 (New York: D. Appleton and Company, 1898, originally, 1866), 444.

8. Robert Bannister, *Social Darwinism: Science and Myth in Anglo-American Thought* (Philadelphia: Temple University Press, 1979), xi–xii, 12, 136; Lears, *No Place of Grace*, 21.

9. On the development of such devices, see Wililam Aspray, Nathan Ensmenger, and Jeffrey R. Yost, *Computer: A History of the Information Machines* (New York: Westview, 2013), 21–40.

10. On the larger changes in antebellum cities and their effects on conceptions of class, see Stuart Mack Blumin, *The Emergence of the Middle Class: Social Experience in the American City, 1760–1900* (Cambridge: University of Cambridge, 1989), 230–57; Daniel Walker Howe, *Making the American Self: Jonathan Edwards to Abraham Lincoln* (Oxford: Oxford University Press, 1997), 109–11.

11. Daniel Walker Howe, *What God Hath Wrought: The Transformation of America, 1815–1848* (New York: Oxford University Press, 2007), 285–327; Sean Wilentz, *The Rise of American Democracy: Jefferson to Lincoln* (New York: Norton, 2005), 254–80, esp. 267.

12. Thomas Cogswell Upham, *Mental Philosophy*, 3rd ed. (Boston: William Hyde, 1832), 57–58.

13. McClay, *The Masterless*, 40–73.

14. Peter N. Stearns, *American Cool: Constructing a Twentieth-Century Emotional Style* (New York: NYU Press, 1994), 16–28; John Kasson, *Rudeness and Civility: Manners in Nineteenth-Century Urban America* (New York: Hill and Want, 1990), 147–81.

15. Lears, *No Place of Grace*, 12–15.

16. Thomas Carlyle, "Signs of the Times," *The Collected Works of Thomas Carlyle*, vol. 3 (London: Chapman and Hall, 1858), available at http://www.victorianweb.org/authors/carlyle/signs1.html.

17. Timothy Walker, "Defense of Mechanical Philosophy," *North American Review*, July 1831, 124. For more analysis, see Leo Marx, *The Machine in the Garden: Technology and the Pastoral Ideal in America* (Oxford: Oxford University Press, 1964), 174–90.

18. Edgar Allen Poe, "The Man That Was Used Up: A Tale of the Late Buga-

boo and Kickapoo Campaign," in *The Complete Works of Edgar Allan Poe, Volume 3*, ed. James A. Harrison (New York: Thomas Y. Cromwell, 1902), 259–72.

19. "Omnibus of Fine Arts," *Boston Morning Post*, December 17, 1834, 3; *Boston Morning Post*, November 14, 1834, 2.

20. Ralph Waldo Emerson, *English Traits* (Boston: Houghton, Mifflin, 1902), 103, 166–67; See John Kasson, *Civilizing the Machine: Technology and Republican Values in American, 1776–1900* (New York: Grossman, 1976, 125–29; Leo Marx, *Machine in the Garden*, 263.

21. Herman Melville, "The Paradise of Bachelors and The Tartarus of Maids," *Harpers New Monthly Magazine* 10, no. 59 (April 1955): 670–78. For further discussion, see Kasson, *Civilizing the Machine*, 92; Stephen P. Rice, *Minding the Machine: Languages of Class in Early Industrial America* (Berkeley: University of California Press, 2004), 30–31.

22. For an analysis on the history of thinking about human identity and matter, see Jessica Riskin, *The Restless Clock: A History of the Centuries-Long Argument over What Makes Living Things Tick* (Chicago: University of Chicago Press, 2016); Stephen Toulmin and June Goodfield, *The Architect of Matter* (Chicago: University of Chicago Press, 1982); and John Tresch, *The Romantic Machine: Utopian Science and Technology after Napoleon* (Chicago: University of Chicago Press, 2012).

23. Cotkin, *Reluctant Modernism*, 30–31; Lears, *No Place of Grace*, 21–23.

24. Herman Melville, "A Utilitarian View of the Monitor's Fight," in *Selected Poems of Herman Melville: A Reader's Edition*, ed. Robert Penn Warren (Jaffey, NH: David R. Godine, 2004), 120–21.

25. Hermann von Helmholtz, "On the Interaction of Natural Forces," *American Journal of Science and Arts* 24, no. 70 (November 1857): 189–90.

26. Helmholtz, 208–9. For more on the overlap between thermodynamics and thinking about the body's energy in primarily European culture, see Anson Rabinbach, *The Human Motor: Energy, Fatigue, and the Origins of Modernity* (Berkeley: University of California Press, 1992). For American approaches to the mechanical body, see Carolyn de la Pena, *The Body Electric: How Strange Machines Built the Modern American* (New York: NYU Press, 2005), 15–49; Cynthia Eagle Russett, *Sexual Science: The Victorian Construction of Womanhood* (Cambridge, MA: Harvard University Press, 1989), 104–29; and Mark Seltzer, *Bodies and Machines* (New York: Routledge, 1992).

27. "Food Fuel for the Body," *Youth's Companion* 44, no. 4 (January 26, 1871): 29.

28. From a special contributor to the *Times*, "Human Body a Fuel Machine," *Los Angeles Times*, November 28, 1897, 13.

29. "The Body a Machine," *Grand Rapids Herald*, September 6, 1898, 8.

30. E. B. Rosa, "The Human Body as Engine," *Popular Science Monthly*, September, 1900, 491; Russett, *Sexual Science*, 108–13.

31. "Hope On!," *California Farmer and Journal of Useful Sciences* 4, no. 13 (September 28, 1855): 101.

32. "The Love of Christ: The Sustaining and Motive Power of the Christian," *Plain Dealer*, August 2, 1886, 4.

33. "The Measure of Character," *Christian Advocate*, May 23, 1889, 331.

34. Ella Wheeler Wilcox, "What Love Is," *Washington Post*, August 1, 1900, 4.

35. In Anson Rabinbach's evocative phrase, "transcendental materialism," 45: Rabinbach, *Human Motor*, 45–68.

36. Thomas Henry Huxley, "On the Hypothesis that Animals Are Automata," 206; For more on Huxley in America, see Louis Menard, *The Metaphysical Club: A Story of Ideas in America* (New York: Farrar, Straus, and Giroux, 2001), 259–60. For a discussion of Huxley in European science, see Riskin, *Restless Clock*, 214–48.

37. Cotkin, *Reluctant Modernism*, 4.

38. Rowland G. Hazard, "Animals Not Automata," *Popular Science Monthly*, February 1875, 406.

39. Hazard, "Animals Not Automata," 419.

40. William James, "Are We Automata?," *Mind*, January 1879, 1–22, at 16. For further discussion, see James Kloppenberg, *Uncertain Victory: Social Democracy and Progressivism in European and American Thought, 1870–1920* (New York: Oxford University Press, 1986), 38–39; Robert D. Richardson, *William James: In the Maelstrom of American Modernism* (Boston: Houghton Mifflin, 2006), 197–99.

41. James, "Are We Automata?," 13.

42. William James, *Principles of Psychology, Volume 1* (New York: Henry Holt, 1890), 294; Lears, *No Place of Grace*, 35–37.

43. Nikola Tesla, "The Problem of Increasing Human Energy," *Century Magazine*, June 1900, 175–211, at 184–85.

44. James, "Are We Automata?," 3.

45. James M. McPherson, *Battle Cry of Freedom: The Civil War Era* (Oxford: Oxford University Press, 1988), 329–31.

46. George Douglas Brewerton, *The Automaton Regiment, or, Infantry Soldier's Practical Instructor* (New York: D. Van Nostrand, 1863), Marian S. Carson Collection, Rare Book and Special Collections Division, Library of Congress, http://www.loc.gov/exhibits/civil-war-in-america/april-1861-april-1862.html#obj4. Brewerton's other automaton manuals are referenced in book-sale lists from the period; see *Trubner's American and Oriental Literary Record*, July 20, 1865, 89.

47. McClay, *The Masterless*, 33.

48. Quoted in Drew Gilpin Faust, *The Republic of Suffering: Death and the American Civil War* (New York: Knopf, 2008), 59.

49. Quoted in James M. McPherson, *For Cause and Comrades: Why Men Fought in the Civil War* (Oxford: Oxford University Press, 1997), 45, 65.

50. Faust, *Republic of Suffering*, 59–60.

51. Quoted in James I. Robertson Jr., *Soldiers Blue and Gray* (University of South Carolina Press, 1998), 122.

52. Lears, *No Place of Grace*, 117–24.

53. "Ohio Veterans Do Themselves Proud at the G.A.R. Encampment at Boston," *Plain Dealer*, August 18, 1890, 3.

54. "It's a Real Holiday," *Chicago Herald*, May 31, 1891, 1.

55. Ernest Howard Crosby, *Captain Jinks, Hero* (New York: Funk & Wagnalls, 1902), 340–42.

56. Frank L. Greene, "Old World Sketches," *St. Albans Daily Messenger*, May 12, 1893, 2.

57. "Automatic Iron Soldiers That Can Shoot Forty Times a Minute," *Charlotte News*, January 15, 1894, 3; "Electric Fighting Men," *Knoxville Tribune*, March 18, 1894, 7.

58. "Patriots' Reunion," *Los Angeles Times*, August 3, 1901, 1; Kristin L. Hoganson, *Fighting for American Manhood: How Gender Politics Provoked the Spanish-American and Philippine-American Wars* (New Haven, CT: Yale University Press, 1998), 143–44.

59. "The Address of the President of the United States," *Christian Advocate*, March 5, 1903, 376.

60. "Westinghouse—Inventor and Human Dynamo," *New York Times*, November 3, 1907, SM3.

61. "The Human Dynamo," *Boston Journal*, April 4, 1895, 10; "Is a Human Dynamo," *Daily Inter Ocean*, August 21, 1895, 7.

62. "Be a Human Steam Engine If You Want to Succeed," *Chicago Tribune*, May 8, 1903, 16.

63. Frederick Winslow Taylor, *Shop Management* (New York: Harper & Brothers, 1911), 146; Frederick Winslow Taylor, *Principles of Scientific Management* (New York: Harper & Brothers, 1911), 141–43; Taylor, *Shop Management*, 120. There are numerous works on Taylor, but for his role in creating a larger ideology for the machine age, see John M. Jordan, *Machine-Age Ideology: Social Engineering and American Liberalism, 1911–1939* (Chapel Hill: University of North Carolina Press, 1994), 33–67; and Robert Kanigel, *The One Best Way: Frederick Winslow Taylor and the Enigma of Efficiency* (Cambridge, MA: MIT Press, 1997).

64. Jacques Futrelle, "The Thinking Machine," in *The Classic Tales of Jacques Futrelle, Volume 1* (Holicong, PA: Wildside Press), 9–11.

65. Futrelle, "The Thinking Machine," 10. On the ideal male body in the period, see John Kasson, *Houdini, Tarzan, and the Perfect Man: The White Male*

Body and the Challenge of Modernity in America (New York: Hill and Wang, 2002).

66. Emma Goldman, *Anarchism: What It Really Stands For* (New York: Mother Earth Publishing, 1911); Emma Goldman, "La Ruche," *Mother Earth*, November 1907, 389

67. Eugene V. Debs, "Unity and Victory," speech before the State Convention of American Federation of Labor, Pittsburgh, Kansas, August 12, 1908, in *Labor and Freedom: The Voice and Pen of Eugene V. Debs* (St. Louis: Phil Wagner, 1916), 128. This is part of the larger "mechanicalizing men" process that Daniel Rodgers identifies in *The Work Ethic in Industrial America, 1850–1920* (Chicago: University of Chicago Press, 1974), 65–93.

68. Here I am relying on Gail Bederman's notion of manliness as a process of claiming power and authority; see Bederman, *Manliness and Civilization: A Cultural History of Gender and Race in the United States, 1880–1917* (Chicago: University of Chicago Press, 1995), 7–8.

69. William Leach, *Land of Desire: Merchants, Power, and the Rise of a New American Culture* (New York: Vintage Books, 1993), 248.

70. Fred Nadis, *Wonder Shows: Performing Science, Magic, and Religion in America* (New Brunswick, NJ: Rutgers University Press, 2005), 1–3.

71. Leach, *Land of Desire*, 251; L. Frank Baum, *The Wonderful Wizard of Oz* (Chicago: George M. Hill, 1900), 58–60.

72. Stuart Culver, "What Manikins Want: The Wonderful Wizard of Oz and the Art of Decorating Dry Goods Windows," *Representations*, Winter 1988, 109; Leach, *Land of Desire*, 248–60.

73. L. Frank Baum, *The Tin Woodman of Oz* (Chicago: Reilly & Lee, 1918), 194–200.

74. L. Frank Baum, *Ozma of Oz* (Chicago: Reilly & Britton, 1907), 6, 15, 39. The contrast of growing food places the story in the context of the utopian visions addressed by Howard Segal, *Technological Utopianism in American Culture* (Syracuse, NY: Syracuse University Press, 1985).

75. Baum, *Ozma of Oz*, 55, 79, 62.

Part 2

1. David H. Keller, "The Threat of the Robot," in *The Threat of the Robot and Other Nightmarish Futures* (Normal, IL: Black Dog Books, 2012), 90, 78. Originally published in *Science Wonder Stories*, June 1929, 62–73.

2. For the most extensive discussion of technological unemployment in the era, see Amy Sue Bix, *Inventing Ourselves Out of Jobs? America's Debate over Technological Unemployment, 1929–1981* (Baltimore, MD: Johns Hopkins University Press, 2000).

3. Keller, "Threat of the Robot," 96–97.

4. David H. Keller, *The Kellers of Hamilton Township: A Study in Democracy* (Alexandria, LA: Wall Printing, 1922).

5. John Cheng, *Astounding Wonder: Imagining Science and Science Fiction in Interwar America* (Philadelphia: University of Pennsylvania Press, 2012), 17–50, 85.

6. See David H. Keller, "Revolt of the Pedestrians," *Amazing Stories*, February 1928, 1048–59; David H. Keller, "The Psychophonic Nurse," *Amazing Stories*, November 1928, 710–17; David H. Keller, "The Female Metamorphosis, *Science Wonder Stories*, August 1929, 246–63, 274.

7. On the romance of the dictator in the era, see Benjamin Alpers, *Dictators, Democracy, and American Public Culture: Envisioning the Totalitarian Enemy, 1920s–1950s* (Chapel Hill: University of North Carolina Press, 2003), 15–58.

8. On the history of the word *technology*, see Ruth Oldenziel, *Making Technology Masculine; Men, Women, and Modern Machines in America, 1870-1945* (Amsterdam: Amsterdam University Press, 1999), 19–50; Eric Schatberg, "'Technik' Comes to America: Changing Meanings of 'Technology' before 1930," *Technology and Culture* (July 2006): 486-512; and Leo Marx, "The Idea of 'Technology' and Postmodern Pessimism," in *Does Technology Drive History: The Dilemma of Technological Determinism by Merritt Roe Smith and Leo Marx* (Cambridge, MA: MIT Press, 1994).

9. Keller, "The Threat of the Robot," 97.

10. Ralph E. Flanders, *Taming Our Machines: The Attainment of Human Values in a Mechanized Society* (New York: R. R. Smith, 1931).

Chapter 4

1. For an account of opening night, see Gilbert Seldes, "The New York Letter," *Washington Post*, October 15, 1922, 66.

2. Biographical details taken from Iwan Klíma, introduction to *R.U.R. (Rossum's Universal Robots)*, by Karel Čapek (New York: Penguin Books, 2004), vii–xxv.

3. Summary and direct quotations from the play are taken from the original translation used by the Theater Guild: Karel Čapek, *R.U.R. (Rossum's Universal Robots)*, trans. Paul Selver (Garden City, NJ: Doubleday, 1923), 51.

4. Tobias Higbie, "Why Do Robots Rebel? The Labor History of a Cultural Icon," *Labor: Studies in Working-Class History* 10, no. 1 (Spring 2013): 99–121. Much of this chapter is in conversation with Higbie's fascinating argument for the transformation of the meaning of the term *robot* from "worker" to "machine." I agree with much of Higbie's analysis but concentrate more on larger debates regarding leisure and culture as the solution to the problem of the robot.

5. Joel Dinerstein, *Swinging the Machine: Modernity, Technology, and African*

American Culture between the World Wars (Amherst: University of Massachusetts Press, 2003), 29–62.

6. For overviews of American modernism, see Lynn Dumenil, *The Modern Temper: American Culture and Society in the 1920s* (New York: Hill and Wang, 1995); Paul Murphy, *The New Era: American Thought and Culture in the 1920s* (New York: Rowan & Littlefield, 2012); Richard Pells, *Modernist America: Art, Music, Movies, & the Globalization of American Culture* (New Haven, CT: Yale University Press, 2011); and David Shi, *Facing Facts: Realism in American Thought and Culture, 1850–1920* (New York: Oxford University Press, 1995), 275–302.

7. There is an extensive literature on machine-age anxieties, but in this chapter, I am predominantly drawing on Dinerstein, *Swinging the Machine*; and, for its relationship to politics, Richard Pells, *Radical Visions and American Dreams: Culture and Social Thought in the Depression Years* (Middletown, CT: Wesleyan University Press, 1973), 1–42.

8. For more on Edison, see Maury Klein, *The Power Makers: Steam, Electricity, and the Men Who Invented Modern America*; and Carroll Pursell, *The Machine in America: A Social History of Technology*, 2nd ed. (Baltimore, MD: Johns Hopkins University Press, 2007), 203–28.

9. "An Automatic Clerkless Shop," *Washington Post*, May 22, 1910, MS2. For more on automatic retailing in the period, see Susan Strasser, *Satisfaction Guaranteed: The Making of the American Mass Market* (Washington: Smithsonian Books, 1989), 203.

10. Charles Postel, *The Populist Vision* (New York: Oxford University Press, 2009), 121–30; Michael McGerr, *A Fierce Discontent: The Rise and Fall of the Progressive Movement in America* (New York: Oxford University Press, 2003), 48–49.

11. Edward Bellamy, *Looking Backward, 2000–1887* (Boston: Houghton Mifflin Company, 1926), 100–109.

12. See, for instance, "Drug Store Installs Automatic Section of 52 Robots," *Automatic Age*, February 1930 53–55.

13. For the larger context on such automatic retailing, see Ann Satterthwaite, *Going Shopping: Consumer Choices and Community Consequences* (New Haven, CT: Yale University Press, 2001), 40.

14. William Leach, *Land of Desire: Merchants, Power, and the Rise of a New American Culture* (New York: Vintage Books, 1993), 298–348. For the most extensive analysis of advertising in the period, see Roland Marchand, *Advertising the American Dream: Making Way for Modernity, 1920–1940* (Berkeley: University of California Press, 1985).

15. David Nye, *America's Assembly Line* (Cambridge, MA: MIT Press, 2013), 13–38.

16. Gary Gerstle, *American Crucible: Race and Nation in the Twentieth Cen-*

tury (Princeton, NJ: Princeton University Press, 2001), 125; Dumenil, *Modern Temper*, 88–89. For more on home appliances in the period, see James D. Norris, *Advertising and the Transformation of American Society* (New York: Greenwood Press, 1990), 71–94; and Ruth Schwartz Cowan, *More Work for Mother: The Ironies of Household Technology from the Open Hearth to the Microwave* (New York: Basic Books, 1983), 151–91.

17. McGerr, *A Fierce Discontent*, 250–51. On the tradeoff between wages and hours, see Benjamin Kline Hunnicutt, *Work without End: Abandoning Shorter Hours for the Right to Work* (Philadelphia: Temple University Press, 1988), 9–36; and Gary Cross, *Time and Money: The Making of Consumer Culture* (New York: Routledge, 1993), 99–127.

18. Nye, *America's Assembly Line*, 13–38; David A. Hounshell, *From the American System to Mass Production, 1800–1932* (Baltimore, MD: Johns Hopkins University Press, 1984), 217–61. On the five-dollar-day specifically, see Stephen Meyer III, *The Five Dollar Day: Labor Management and Social Control in the Ford Motor Company, 1908–1921* (Albany: State University of New York Press, 1981), 95–122.

19. On the rise of commercialized and technological amusements, see Roy Rosenzweig, *Eight Hours for What We Will: Workers and Leisure in an Industrial City, 1870–1920* (New York: Cambridge University Press, 1985); Kathy Peiss, *Cheap Amusements: Working Women and Leisure in Turn-of-the-Century New York* (Philadelphia: Temple University Press, 1986); David Nasaw, *Going Out: The Rise and Fall of Public Amusements* (Cambridge, MA: Harvard University Press, 1999); McGerr, *A Fierce Discontent*, 248–78; Lary May, *Screening Out the Past: The Birth of Mass Culture and the Motion Picture Industry* (Chicago: University of Chicago Press, 1983); and Sarah Hallenbeck, *Claiming the Bicycle: Women, Rhetoric, and Technology in Nineteenth-Century America* (Carbondale: Southern Illinois University Press, 2016).

20. John Philip Sousa, "The Menace of Mechanical Music," *Appleton's Magazine*, September 1906, 279.

21. On the rise of this "leisure ethic," see William Gleason, *The Leisure Ethic: Work and Play in American Literature, 1840–1940* (Stanford, CA: Stanford University Press, 1999); Gary Cross, *Kids' Stuff: Toys and the Changing World of American Childhood* (Cambridge, MA: Harvard University Press, 1997), 50–81; Hunnicutt, *Work without End*, 109–46; McGerr, *A Fierce Discontent*, 248–78.

22. Sanford Jacoby, *Modern Manors: Welfare Capitalism since the New Deal* (Princeton, NJ: Princeton University Press, 1997), 14–16. For more on the growth of leisure as compensation for labor, see Hunnicutt, *Work without End*, 67–108; and Cross, *Time and Money*, 15–45.

23. Hunnicutt, *Work without End*, 109–46.

24. On the racial and ethnic politics of cultural expression, see Davarian Baldwin, *Chicago's New Negroes: Modernity, the Great Migration, and Black*

Urban Life (Chapel Hill: University of North Carolina Press, 2007); Gena Caponi-Tabery, *Jump for Joy: Jazz Basketball, and Black Culture in 1930s America* (Amherst: University of Massachusetts Press, 2008); Kathy J. Ogren, *The Jazz Revolution: Twenties America and the Meaning of Jazz* (New York: Oxford University Press, 1989); and Dinerstein, *Swinging the Machine*; May, 22–42.

25. On the rise of a probabilistic universe and its impact on American culture, see T. J. Jackson Lears, *Something for Nothing: Luck in America* (New York: Penguin Books, 2003), 280–87; and David Shi, *Facing Facts*, 275–76. For the connection between such sciences and the rise of American modernism, see Pells, *Modernist America*, 16–17; and McGerr, *A Fierce Discontent*, 236–38.

26. Ernst Jentsch, "On the Psychology of the Uncanny," trans. Roy Sellers, in *Uncanny Modernity: Cultural Theories, Modern Anxieties*, ed. Jo Collins and John Jervis (New York: Palgrave Macmillan, 2008), 224.

27. Jentsch, "On the Psychology of the Uncanny," 224.

28. Sigmund Freud, "The Uncanny," trans. Alix Strachey, in *Writings on Art and Literature* (Stanford, CA: Stanford University Press, 1997), 221.

29. For more on the historical development of the uncanny, see Terry Castle, *The Female Thermometer: 18th-Century Culture and the Invention of the Uncanny* (New York: Oxford University Press), 3–20. For the placement of the uncanny in the uncertainty of turn-of-the-century culture, see T. J. Jackson Lears, *No Place of Grace: Antimodernism and the Transformation of American Culture, 1880–1920* (Chicago: University of Chicago Press, 1994), 173.

30. "Wonderful Control of Facial Muscles," *Baltimore American*, December 3, 1903, 13.

31. Ethel Thurston, "Moving Wonders of Wax and Rubber Ready to Take Dress Models Places," *Syracuse Herald*, October 26, 1919, 6.

32. "Confessions of a Bride," *Wyoming State Tribune*, August 14, 1919, 4.

33. Quoted in Casey Nelson Blake, *Beloved Community: The Cultural Criticism of Randolph Bourne, Van Wyck Brooks, Waldo Frank, & Lewis Mumford* (Chapel Hill: University of North Carolina Press, 1990), 76.

34. Waldo Frank, *Our America* (New York: Boni and Liveright, 1920), 45.

35. See, for instance, "Science's Part in Next War," *Washington Post*, February 6, 1916, MS1; and Raymond B. Fosdick, "Our Machine Civilization: A Frankenstein Monster?" *Current Opinion*, September 1, 1922, 365.

36. John Corbin, "The Play," October 10, 1922, *New York Times*, 24.

37. For more on the era's immigration and nativism, see Matthew Frye Jacobson, *Whiteness of a Different Color: European Immigrants and the Alchemy of Race* (Cambridge, MA: Harvard University Press, 1998), 39–90; Mae M. Ngai, *Impossible Subjects: Illegal Aliens and the Making of Modern America* (Princeton, NJ: Princeton University Press, 2004), 21–55; and Dumenil, *Modern Temper*, 224–26.

38. Carl Sandburg, "R.U.R.," *New York Times*, January 28, 1923, X2.

39. Heywood Broun, "Hair-Raising Satire," *Vanity Fair*, December 1922, 43, 104.

40. Helen Klumph, "Real Thrill on the Menu," *Los Angeles Times*, February 25, 1923, 29; Grace Kingsley, "Mamoulian to Guide 'R.U.R.,'" *Los Angeles Times*, May 13, 1932, A9.

41. Čapek, *R.U.R. (Rossum's Universal Robots)*, 16–17.

42. Photograph included in Rosa Knuuti, "R.U.R.," *Industrial Pioneer*, June 1923, 32–34.

43. Charles W. Wood, "Workmen Made to Order," *Collier's*, December 9, 1922, 23.

44. Virginia Pope, "New Ownership Idea Tried," *New York Times*, November 7, 1926, XX12.

45. On the IWW, see Melvyn Dubosfsky, *We Shall Be All: A History of the IWW* (Chicago: Quadrangle, 1969). For a recent treatment, see Erik Loomis, *A History of America in Ten Strikes* (New York: New Press, 2018), 91–112.

46. Knuuti, "R.U.R.," 34.

47. Knuuti, "R.U.R.," 34.

48. George Williams, "Henry Ford, a Peculiar Entity," *Industrial Pioneer* 1, no. 6 (October 1923): 31–33.

49. J. D. C., "Our Paternalistic Press' Health Hints," *Industrial Pioneer* 1, no. 9 (January 1924): 34

50. Justus Ebert, "The Wonderful Age," *Industrial Pioneer* 2, no. 4 (August, 1924): 26

51. Ludwig Lewisohn, "Helots," *Nation*, November 1, 1922, 478.

52. E. Wye, "At the Sign of the Cat and the Fiddle," *Single Tax Review* 22, no. 6 (November–December 1922): 170.

53. John Corbin, "The Revolt against Civilization: The English School," *New York Times*, October 15, 1922, 99.

54. John Corbin, "The Critic and His Orient," *New York Times*, December 24, 1922, 75.

55. On the role of race and ethnicity in dividing workers, see David Roediger, *Working toward Whiteness: How American's Immigrants Became White* (New York: Perseus, 2005), 74, 220; David Roediger and Elizabeth D. Esch, *The Production of Difference: Race and the Management of Labor in U.S. History* (New York: Oxford University Press, 2012), 139.

56. Quoted in Stuart Chase, *Men and Machines* (New York: Macmillan, 1929), 160.

57. William J. Perlman, "Čapek Explained," "Views from the Mail Bag," (letter to the editor), *New York Times*, February 17, 1923, x2.

58. Ralph Borsodi, *This Ugly Civilization* (New York: Simon and Schuster, 1929), 136; Ralph E. Flanders, *Taming Our Machines: The Attainment of Human Values in a Mechanized Society* (New York: Richard R. Smith, 1931), 57.

59. Higbie, "Why Do Robots Rebel?," 107; Corbin, "The Critic and His Orient," 75.

60. Lothrop Stoddard, *The Rising Tide of Color: The Threat against White World-Supremacy* (New York: Scribner's Sons, 1921), 158, 160.

61. Robert Lee, *Orientals: Asian-Americans in Popular Culture* (Philadelphia: Temple University Press, 1999), 137–40.

62. The metaphor here is from David Montgomery, *Beyond Equality: Labor and the Radical Republicans, 1862–1872*, (New York: Alfred A. Knopf, 1967), x; but most scholars since then have emphasized race as a key limiting factor in American radicalism.

63. Arthur Pound, *The Iron Man in Industry: An Outline of the Social Significance of Automatic Machinery* (Boston: Atlantic Monthly Press, 1922), 1.

64. Pound, *Iron Man in Industry*, 34–35.

65. Pound, 169.

66. Pound, 166. For more discussion of education for leisure, see Hunnicut, *Work without End*, 109–46; and Cross, *Time and Money*, 99–127.

67. Dinerstein, *Swinging the Machine*, 29–62.

68. Pells, *Radical Visions*, 19–21.

69. See Harry Grossman and Burton L. King, *The Master Mystery* (Rolfe Photoplays, 1920).

70. See Thorstein Veblen, *The Theory of the Leisure Class: An Economic Study of Institutions* (New York: Macmillan, 1912), 35–67.

71. For another advertising photo, see *The Master Mystery*, 1919, McManus-Young Collection, Rare Book and Special Collections Division, Library of Congress, reproduction nos. LC-USZ62-66393, LC-USZ62-64313, http://www.loc.gov/pictures/item/96520687/.

72. Pells, *Radical Visions*, 4–5.

73. William Fielding Ogburn, *Social Change with Respect to Culture and Original Nature* (New York: B. W. Huebsch, 1922), 200–201.

74. Ogburn, *Social Change*, 196. For more on the larger redefinition of culture, see Susan Hegeman, *Patterns for America: Modernism and the Concept of Culture* (Princeton, NJ: Princeton University Press, 1999); Howard Brick, *Transcending Capitalism: Visions of a New Society in Modern America* (Ithaca, NY: Cornell University Press, 2006), 86-120; and Murphy, *The New Era*, 28–37.

75. *Metropolis*, directed by Fritz Lang (Universum Film, 1927). The analysis here is based on the American cut of the film.

76. "Metropolis Film Seen," *New York Times*, January 11, 1927, 36. The names included here come from American reviews of the version of the film shown in America, which had been rewritten by Channing Pollock.

77. On the sexualized nature of the machine Mary, see Andreas Huyssen, "The Vamp and the Machine: Technology and Sexuality in Fritz Lang's Metropolis," *New German Critique* (Autumn 1981–Winter 1982), 221–37; and

Andreas Huyssen, *After the Great Divide: Modernism, Mass Culture, Postmodernism* (Bloomington: Indiana University Press, 1986), 73–74.

78. Kang, *Sublime Dreams of Living Machines*, 294; Allison Muri, *The Enlightenment Cyborg: A History of Communications and Control in the Human Machine, 1630–1830* (Toronto: University of Toronto Press, 2007), 168.

79. *Metropolis*, dir. Lang.

80. Mordaunt Hall, "A Topheavy German Production," *New York Times*, March 13, 1927, X7; "UFA Film Provokes Comment," *Los Angeles Times*, August 7, 1927, C13.

81. Hall, "Topheavy German Production."

82. "Feats of Science Help Movies Give Vivid Picture of a World Ruled by Machines," *Popular Mechanics*, March 1927, 424.

83. H. G. Wells, "Mr. Wells Reviews a Current Film," *New York Times*, April 17, 1927, 4.

84. Wells, "Mr. Wells Reviews a Current Film," 22.

85. Robert B. Westbrook, "Tribune of the Technostructure: The Popular Economics of Stuart Chase," *American Quarterly* 32, no. 4 (Autumn 1980): 389–91.

86. Westbrook, "Tribune of the Technostructure," 392; Kathleen G. Donohue, *Freedom from Want: American Liberalism and the Idea of the Consumer* (Baltimore, MD: Johns Hopkins University Press, 2003), 208.

87. Like most writers on technology in the period, Chase was not particularly interested in the plight of women in the machine age. He occasionally uses "man" to mean human but most of his analysis focuses on the costs of mechanization to men. More on this theme will be developed in the next two chapters.

88. Chase, *Men and Machines*, 142, 158–59, 161.

89. Chase , 337, 347, 338, 343, 335.

90. Chase, 107.

Chapter 5

1. Philip Kinsley, "'Let Electrons Do It,' Motto for Moderns," *Chicago Daily Tribune*, November 27, 1930, 35.

2. On the advertising character, see T. J. Jackson Lears, *Fables of Abundance: A Cultural History of Advertising in America* (New York: Basic Books, 1994), 123.

3. Kinsley; "Smoke Destroyed by 'Electric Eye,'" *New York Times*, October 25, 1930, 30.

4. Kinsley, "Let Electrons Do It."

5. "Opening of Syracuse Herald Progress Exposition to Pack Armory," *Syracuse Herald*, May 6, 1935, 3. Scott Schaut, *Robots of Westinghouse, 1924–Today* (Mansfield, OH: Mansfield Memorial Museum, 2006), 56.

6. Schaut, *Robots of Westinghouse*, 26.

7. "The World's Largest Loud Speaker, and Other Late Devices," *Radio-Craft*, February 1931, 468.

8. There were other touring robots during that period—including the British "Eric the Robot"—that had *R.U.R.* emblazoned on the chest, but the Westinghouse devices appear more regularly in American periodicals. For more on Eric, see Tobias Higbie, "Why Do Robots Rebel? The Labor History of a Cultural Icon," *Labor: Studies in Working-Class History* 10, no. 1 (Spring 2013): 116–17.

9. Schaut, *Robots of Westinghouse*, 19–21; R. J. Wensley, "The Design of Automatic Switching Equipments for Synchronous Converter Substations," *Electric Journal* 15, no. 4 (April 1918): 114–19; For an analysis of televox in the context of the agency of machines, see Jessica Riskin, *The Restless Clock: A History of the Centuries-Long Argument over What Makes Living Things Tick* (Chicago: University of Chicago Press, 2016), 299–301.

10. R. J. Wensley, "Automatic Operation of Electric Equipment," *Electric Journal* 23, no. 4 (April 1926): 1.

11. For more on rural electrification in the period, see Ronald R. Kline, *Consumers in the Country: Technology and Social Change in Rural America* (Baltimore, MD: Johns Hopkins University Press, 2002).

12. Fred Nadis, *Wonder Shows: Performing Science, Magic, and Religion in America* (New Brunswick, NJ: Rutgers University Press, 2005), 63.

13. "Wife Cooks by Televox," *Los Angeles Times*, October 14, 1927, 1.

14. Waldemar Kaempffert, "Science Produces the 'Electrical Man,'" *New York Times*, October 23, 1927, XX1.

15. Waldemar Kaempffert, "Science Produces the 'Electrical Man.'"

16. Kaempffert.

17. Schaut, *Robots of Westinghouse*, 49, 25.

18. Schaut, 29–30.

19. "Televox Enlivens Book Men's Dinner," *New York Times*, May 23, 1930, 21.

20. "Meet Mr. Televox, the Mechanical Man," *Rock Valley Bee*, July 20, 1928, 2.

21. "Televox, Automatic Servant, Works at Master's Bidding," *Decatur Review*, October, 14, 1927, 1.

22. "Mechanical Man Obeys Orders over Phone," *Ogden Standard-Examiner*, March 1, 1928, 11.

23. "Romantic Old Maids Can Hear the Words of Love They Long For," *San Antonio Light*, July 1, 1928, 62.

24. Kaempffert, "Science Produces the 'Electrical Man.'"

25. "Mechanical Man Obeys Orders," *Ogden Standard-Examiner*.

26. Ruth Schwartz Cowan, *More Work for Mother: The Ironies of Household*

Technology from the Open Hearth to the Microwave (New York: Basic Books, 1983), 122.

27. Andrew Urban, *Brokering Servitude: Migration and the Politics of Domestic Labor During the Long Nineteenth Century* (New York: New York University Press, 2018), 228–29.

28. "Mechanical Man Obeys Orders," *Ogden Standard-Examiner*.

29. "Romantic Old Maids Can Hear Words of Love," *San Antonio Light*.

30. "Televox Enlivens Book Men's Dinner," *New York Times*.

31. H. Winfield Secor, "Has the Automaton Arrived?," *Science and Invention*, January 1928, 786.

32. J. Schlossel, "To the Moon by Proxy," *Amazing Stories*, October 1928, 598–608.

33. "Televox' the Mechanical Man," *Amazing Stories*, October 1928, 608.

34. See David H. Keller, "The Psychophonic Nurse," in *The Threat of the Robot and Other Nightmarish Futures* (Normal, IL: Black Dog Books, 2012), 54–55. Originally published in *Amazing Stories*, November, 1928, 710–17.

35. Bob Olsen, "Flight in 1999," *Air Wonder Stories*, September 1929, 256–65.

36. "By Products," *New York Times*, February 26, 1928, 54.

37. Robert E. Martin, "Mechanical Men Walk and Talk," *Popular Science Monthly*, December, 1928, 22, 23, 137. Martin exclusively talks about men as the inventors and controllers of robots.

38. Gary Cross, *Time and Money: The Making of Consumer Culture* (New York: Routledge, 1993), 99–127.

39. "The Mechanical Man Arrives," *Labor Age*, (November 1927, 11.

40. "Mechanical Men That Excel Any Human Being," *San Antonio Light*, September 6, 1931, 48–49, 55.

41. William Green, "Labor versus Machines: An Employment Puzzle," *New York Times*, June 1, 1930, E5; Amy Sue Bix, *Inventing Ourselves Out of Jobs? America's Debate over Technological Unemployment, 1929–1981* (Baltimore, MD: Johns Hopkins University Press, 2000), 80–82.

42. "Post's Automatic Pilot," *New York Times*, July 24, 1933, 2.

43. Bix, *Inventing Ourselves Out of Jobs?*, 91–99.

44. . For an analysis of mechanical brains in the era, see David Mindell, *Between Human and Machines: Feedback, Control, and Computing before Cybernetics* (Baltimore, MD: Johns Hopkins University Press, 2004).

45. Bix, *Inventing Ourselves Out of Jobs?*, 96–98, 138–41. The *Locomotive Engineer* cartoon is on page 139.

46. William Green, "National Planning: Labor's Point of View," *New York Times*, December 17, 1933, XX1.

47. Ralph E. Flanders, "The New Age and the New Man," in *Toward Civilization*, ed. Charles Beard (New York: Longmans, Green, 1930), 23, 24, 33, 31, 33;

Ralph E. Flanders, *Taming Our Machines: The Attainment of Human Values in a Mechanized Society* (New York: Richard R. Smith, 1931), 166, 15. Biographical details taken from Ralph E. Flanders, *Senator from Vermont* (Boston: Little, Brown, 1961).

48. Bix, *Inventing Ourselves Out of Jobs?*, 118–22. Quoted in William E. Akin, *Technocracy and the American Dream: The Technocrat Movement, 1900–1941* (Berkeley: University of California Press, 1977), 74; Bix, *Inventing Ourselves Out of Jobs?*, 119.

49. "Technocracy and the Home," *Montana Butte Standard*, January 30, 1933, 4.

50. Harold Loeb, *Life in a Technocracy: What It Might Be Like* (Syracuse, NY: Syracuse University Press, 1996), 30; Howard Segal, *Technological Utopianism in American Culture*, 120–24.

51. Benjamin Kline Hunnicutt, *Work without End: Abandoning Shorter Hours for the Right to Work* (Philadelphia: Temple University Press, 1988), 159–90.

52. Bix, *Inventing Ourselves Out of Jobs?*, 74–78.

53. Ronald W. Schatz, *The Electrical Workers: A History of Labor at General Electric and Westinghouse, 1923–60* (Urbana: University of Illinois Press, 1983), 20.

54. Sherwood Anderson, *Perhaps Women* (New York: H. Liveright, 1931), 60, 138. For further discussion of *Perhaps Women*, see Katherine Stubbs, "Mechanizing the Female: Discourse and Control in the Industrial Economy," *Differences: A Journal of Feminist Cultural Studies* 7, no. 3 (Fall 1995): 141; and Joel Dinerstein, *Swinging the Machine: Modernity, Technology, and African American Culture between the World Wars* (Amherst: University of Massachusetts Press, 2003), 146–47.

55. Flanders, *Taming Our Machines*, 1–2.

56. William Fielding Ogburn, *You and Machines* (Washington, DC: Civilian Conservation Corps, 1934), 3, 52; For further discussion of this pamphlet, see Bix, *Inventing Ourselves Out of Jobs?*, 53–56.

57. Miles J. Breuer, "Paradise and Iron," in *The Man with the Strange Head and Other Early Science Fiction Stories*, ed. Michael R. Page (Lincoln: University of Nebraska Press, 2008), 44–256, 86. Originally published in *Amazing Stories Quarterly*, Summer 1930, 292–363. The phrase "mathmatical robots" came from the story description, presumably written by Gernsback. It can be found on 293 in the original publication.

58. Breuer, "Paradise and Iron," 235–36.

59. Breuer, 237.

60. Everett Franklin Bleiler, with the assistance of Richard J. Bleiler, "Harl Vincent," *Science-Fiction: The Gernsback Years* (Kent, OH: Kent State University Press, 1998), 451.

61. Harl Vincent, "Rex," in *War with the Robots*, ed. Isaac Asimov, Patricia S.

Warrick, and Martin H. Greenberg, (New York: Wing's Books, 1983), 50–67, 57, 51. Originally published in *Astounding Stories*, June 1934, 143–54.

62. Vincent, "Rex," 58.

63. Vincent, "Rex," 65.

64. F. A. Merrick, "The Machine Myth," *Electric Journal* 30, no. 2 (February 1930): 65–66.

65. Merrick, "The Machine Myth," 65.

66. Chas. R. Riker, "Our Mechanical Slaves," *Electrical Journal*, 24, no. 2 (February 1927), 53–54.

67. M. S. Sloan, "Power = Prosperity," *Electric Journal* 27, no. 6 (June 1930): 317–18, 342.

68. According to Lorenzo J. Greene and Carter G. Woodson, Westinghouse employed nine hundred African Americans in 1918 at the height of the worker shortages of World War I and 514 after the war. They were employed entirely in unskilled and semi-skilled positions. Lorenzo J. Greene and Carter G. Woodson, *The Negro Wage Earner* (Washington, DC: Association for the study of Negro life and history, 1930), 255–56.

69. "Eddy Currents," *Electric Journal* 34, no. 9 (September 1937): 379; "Eddy Currents," *Electric Journal* 28, no. 4 (April 1931): 256.

70. "Here's Petite Katrina, Brought Here for Daily Cooking Show," *Decatur Daily*, April 25, 1930, 2.

71. "Electrical Robot to Do Household Tasks at Gold's," *Lincoln Sunday Star*, May 3, 1931, A8.

72. Advertisement, *Van Wert Daily Bulletin*, December 3, 1931, 2.

73. "Opportunity Great for Mechanical Wonder of the World," *Coronado Citizen*, November 30, 1939, 11.

74. "Opportunity Great for Mechanical Wonder of the World," *Coronado Citizen*, November 30, 1939, 11.

75. Leonard H. Engel, "Unaccustomed As He Is," *Arizona Magazine of the Greater Sunday Republic*, March 19, 1939, 3.

76. Jack Burroughs, "Hoofaloger Has Row with Robot," *Oakland Tribune*, June 6, 1939, 18B.

77. "The Event of the Season," *La Crosse Tribune and Leader Press*, March 10, 1935, 11.

78. For further discussion of the statue at the Chicago exposition, see Cheryl R. Ganz, *The 1933 Chicago World's Fair: A Century of Progress* (Champaign: University of Illinois Press, 2008), 56–57; and Dinerstein, *Swinging the Machine*, 142.

79. *Mickey's Mechanical Man*, directed by Walt Disney (Walt Disney Productions, 1933); For analysis in the context of the era's comedy, see Michael North, *Machine-Age Comedy* (New York: Oxford University Press, 2009), 53–83.

80. *Modern Inventions*, directed by Walt Disney (Walt Disney Productions, 1937).

81. David Nye, *American Technological Sublime* (New Bakersfield, MA: MIT Press, 1994), 206–7; For more on the fair, see Warren Susman, "The People's Fair: Contradictions of a Consumer Society," in *Culture as History: The Transformation of American Society in the Twentieth Century* (New York: Pantheon, 1984 [1973]).

82. Schaut, *Robots of Westinghouse*, 91–133, 201–5.

83. Westinghouse advertisement, reprinted in Schaut, 112.

84. The use of Middleton references Helen and Robert Lynd's 1929 study *Middletown.* Robert S. Lynd and Helen Merrell Lynd, *Middletown: A Study in Modern American Culture* (New York: Harcourt, Brace and Company, 1929).

85. *The Middleton Family at the New York World's Fair*, directed by Robert R. Snody (Westinghouse Electric Company, 1939), 25:00–28:50. For further discussion, see: Bix, *Inventing Ourselves Out of Jobs?*, 224–26.

86. *Middleton Family at the New York World's Fair*, 33:55–37:40.

87. *Middleton Family at the New York World's Fair*, 21:51–24:00, 47:55–48:25.

88. For more on Mumford, see Casey Nelson Blake, *Beloved Community: The Cultural Criticism of Randolph Bourne, Van Wyck Brooks, Waldo Frank, & Lewis Mumford* (Chapel Hill: University of North Carolina Press, 1990); and Donald L. Miller, *Lewis Mumford: A Life* (New York: Grove Press, 2002), 33, 364–66.

89. Lewis Mumford, *Technics and Civilization* (New York: Harcourt, Brace, 1934), 281, 453, 426–27, 294–95, 422.

90. Nye, *American Technological Sublime*, 207.

91. Warren I. Susman, "The People's Fair: Cultural Contradictions of a Consumer Society."

92. See *The City*, directed by Ralph Steiner and Willard Van Dyke (American Documentary Films, 1939).

93. Nye, *American Technological Sublime*, 213; Frank Monaghan, *Official Guide Book of the New York World's Fair, 1939* (New York: Exposition Publications, 1939), 43.

94. *To New Horizons* (General Motors, 1940); video available at Handy (Jam) Organization, 23 minutes, http://www.archive.org/details/ToNewHor 1940. For further analysis, see Nye, *American Technological Sublime*, 218.

Chapter 6

1. Woodmen of the World Life Insurance, advertisement, *Wonder Stories*, December 1931, 820.

2. See Aldous Huxley, *Brave New World* (New York, Perennial Classics, 1932); and John W. Campbell (as Don A. Stuart), "Twilight," *Astounding Stories*, November 1934, 44–58.

3. Waldemar Kaempffert, "Is Man Only Robot? The Debate Widens," *New York Times*, March 11, 1934, SM1.

4. For more on the era's study of hormones, see Elizabeth Siegel Watkins, *The Estrogen Elixir: A History of Hormone Replacement Therapy in America* (Baltimore, MD: Johns Hopkins University Press, 2007); and Randi Hutter Epstein, *Aroused: The History of Hormones and How They Control Just About Everything* (New York: Norton, 2018). Michael Pettit, "Becoming Glandular: Endocrinology, Mass Culture, and Experimental Lives in the Interwar Age," *American Historical Review* 118, no. 4, (October 2013): 2052–76.

5. Julia Ellen Rechter, "'The Glands of Destiny': A History of Popular, Medical, and Scientific Views of the Sex Hormones in 1920s America," PhD diss., University of California, Berkeley, 1997.

6. Louis Berman, *The Glands Regulating Personality* (New York: MacMillan, 1922), 22. For further discussion, see Epstein, 59–67.

7. See John Walker Harrington, "Are Little Hidden Glands Our Masters?," *Popular Science Monthly*, June 1922, 49.

8. In this, the gland theory intersected with eugenics. See Wendy Kline, *Building a Better Race: Gender, Sexuality, and Eugenics from the Turn of the Century to the Baby Boom* (Berkeley: University of California Press, 2001); Waldemar Kaempffert, "Science Presses on toward New Goals," *New York Times*, January 28, 1934, SM6; and Joanne J. Meyerowitz, *How Sex Changed: A History of Transsexuality in the United States* (Cambridge, MA: Harvard University Press, 2002), 27.

9. Arthur Pound, *The Iron Man in Industry: An Outline of the Social Significance of Automatic Machinery* (Boston: Atlantic Monthly Press, 1922), 44.

10. Berman, *Glands Regulating Personality*, 265–67.

11. John B. Watson, "Behavior as the Psychologist Sees It," *Psychological Review* 20, no. 2 (March 1913): 158–77.

12. John B. Watson, *Behaviorism* (New York: People's Institute, 1924), 3.

13. Harry A. Overstreet, "Success or Failure—It's Up to You," *Popular Science Monthly*, January 1928, 14–15, 128.

14. In 1904, G. Stanley Hall defined psychology's goal in quintessentially vitalist terms: "advancing man's knowledge of the soul." Hall, "The Unity of Mental Science," in *Congress of Arts and Sciences, Universal Exposition, St. Louis, 1904*, vol. 5, ed. Howard J. Rogers (1906), 577. On such hostility, see Carl Degler, *In Search of Human Nature: The Decline and Revival of Darwinism in American Social Thought* (New York: Oxford University Press, 1991), 152–56.

15. William McDougall, "Men or Robots?," in *Psychologies of 1925, Powell Lectures in Psychological Theory*, ed. Carl Murchison (Worcester, MA: Clark University Press, 1926), 273–307, 283.

16. "Says Human Brain Emits Radio Waves," *New York Times*, August 21, 1925, 1.

17. “Our Electrical Brains,” *New York Times*, October 20, 1934, 14. On Galvani, see Marco Piccolinio and Marco Bresadoia, *Shocking Frogs: Galvani, Volta, and the Electric Origins of Neuroscience*, trans. Nicholas Wade (New York: Oxford University Press, 2013).

18. Edwin Teale, “Amazing Electrical Tests Show What Happens When You Think,” *Popular Science Monthly*, May 1936, 11–13, 117.

19. “Our Electrical Brains.”

20. Lillian E. Zeh, “The Torpedo, the Weapon with a Mechanical Brain,” *Overland Monthly and Out West Magazine*, May 1918, 433.

21. Robert E. Martin, “Mechanical Men Walk and Talk,” *Popular Science Monthly*, December 1928, 23.

22. William Barclay Parsons, “Robots, Storied, Fancied and the Real,” *New York Times*, September 15, 1929, SM4.

23. “Human Mind Eclipsed, Automaton Does Problems,” *Los Angeles Times*, October 21, 1927, 1; “Light Actuates Brain of Robot,” *Los Angeles Times*, July 7, 1931, 3.

24. “Mathematics Easy to ‘Thinking Robot,’” *New York Times*, July 15, 1934, N2.

25. “30,000 Cheer as Hughes Lands after 91-Hour World Flight,” *El-Paso Herald-Post*, July 14, 1938, 7; “Robot Navigation Computer Was Used on World Flight,” *Evening Independent*, August 20, 1938, 1.

26. Kaempffert, “Is Man Only Robot?,” SM4.

27. George W. Gray, “Thinking Machines,” *Harper’s Monthly*, March 1936, 423.

28. Gray, “Thinking Machines,”423.

29. Kaempffert, “Is Man Only Robot?”

30. Stuart Chase, *Men and Machines* (New York: Macmillan, 1929), 326–27.

31. William Fielding Ogburn, *You and Machines* (Washington, DC: Civilian Conservation Corps, 1934), 38–39, 46.

32. Edmond Hamilton, “The Reign of the Robots,” *Wonder Stories*, December 1931, 848–59.

33. *The Mechanical Monsters*, directed by Seymour Kneitel (Paramount Pictures, 1941).

34. David Keller, Winfield Scott Pugh, et al., *Personal Problems Library, Volumes 1–6* (New York: Falstaff Press, 1938).

35. David H. Keller, “Revolt of the Pedestrians,” in *The Threat of the Robot and Other Nightmarish Futures* (Normal, IL: Black Dog Books, 2012), 31–32. Originally published in *Amazing Stories*, February 1928, 1048–59.

36. Lester del Rey, “Helen O’Loy,” *The Science Fiction Hall of Fame, Volume 1, 1929–1964*, ed. Robert Silberberg (New York: Tom Doherty Associates, 1998), 43, 42, 43, 44. Originally published in *Astounding Science-Fiction*, December 1938, 118–25. Given Dave and Phil’s exceptionally close relationship, it is

tempting to see the story as homoerotic; however, del Rey spends so much time having each character obsess over the female body that it seems likely just a very close platonic relationship.

37. Del Rey, "Helen O'Loy," 44, 46, 48, 49, 51.

38. On women and consumer culture in the period, see Lynn Dumenil, *The Modern Temper: American Culture and Society in the 1920s* (New York: Hill and Wang, 1995), 98–144.

39. For instance, in the 1930 film *Just Imagine* babies come from a machine. *Just Imagine*, directed by David Butler, (Fox Film Corporation, 1930).

40. See the letters column, *Astounding Science Fiction*, January 1939, 155.

41. Robert Silverberg, ed., *The Science Fiction Hall of Fame, Volume One, 1929–1964* (New York: Science Fiction Writers of America, 1970).

42. See Eando Binder, "Adam Link in Business," *Amazing Stories*, January 1940, 49.

43. Eando Binder, "I, Robot," *Amazing Stories*, January 1939, 8–18.

44. Binder, "I, Robot," cover. See also figure 6.1.

45. Binder, "I, Robot," 10.

46. Binder, 14.

47. Binder, 18.

48. See John Wyndham, "The Lost Machine," *Amazing Stories*, April 1932, 40–47.

49. B. G. Davis, "The Observatory by the Editor," *Amazing Stories*, December 1940, 8.

50. Walter F. Williams, letter to the editor, *Amazing Stories*, July 1940, 135.

51. Roger Sherman Hour, "A Real Accomplishment!" Letter, *Amazing Stories*, April 1940, 137. The use of personality here places it in the larger shift from internal character to external personality as documented in Warren Susman, "'Personality' and the Making of Twentieth Century Culture," in *Culture as History: The Transformation of American Society in the Twentieth Century* (New York: Pantheon, 1974);

52. See Eando Binder, "Adam Link's Vengeance," *Amazing Stories*, February 1940, cover, 8–27; Eando Binder, "Adam Link, Robot Detective," *Amazing Stories*, May 1940, 42–65; and Eando Binder, "The Trial of Adam Link, Robot," *Amazing Stories*, July 1939, 30–43.

53. See Eando Binder, "Adam Link Fights a War," *Amazing Stories*, December 1940, 10–41.

54. See Eando Binder, "Adam Link Saves the World," *Amazing Stories*, April 1942, 10–47.

55. Binder, "Adam Link in Business," 60, 51.

56. Binder, "I, Robot," 13.

57. See Eando Binder, "Adam Link, Champion Athlete," *Amazing Stories*, July 1940, 28–47.

58. See Binder, "Adam Link Saves the World," cover. See also fig. 6.2.

59. See Binder, "Adam Link Fights a War."

60. See Eando Binder, "Adam Link Faces a Revolt," *Amazing Stories*, May 1941, 70–93.

61. Binder, "Adam Link in Business," 55, 59.

62. Bradford W. Wright, *Comic Book Nation: The Transformation of Youth Culture in America* (Baltimore, MD: Johns Hopkins University Press, 2003), 1–29.

63. Kerry W. Buckley, *Mechanical Man: John B. Watson and the Beginnings of Behaviorism* (New York: Guilford Press, 1989), 134–47.

64. Among McDougall's other works from the period is the lecture series *Is America Safe for Democracy* in which he explained how "innate intellectual capacity" is lower in nonwhite races and worried that feminism was removing "the best of women from marriage and motherhood." William McDougall, *Is America Safe for Democracy?* (New York: Scribner's, 1921), 51–60, 150.

65. Franklin D. Roosevelt, "Broadcast to International Student Assembly," September 3, 1942, in *The War Messages of Franklin D. Roosevelt, December 8, 1941, to April 13, 1945* (University of Michigan Library, 1945), 45.

Chapter 7

1. Lewis Nichols, "The Play," *New York Times*, December 4, 1942, 30.

2. "R.U.R. Is Closed after Brief Stay," *New York Times*, December 7, 1942, 23.

3. On the history of charismatic leaders in America and the relationship to mass communication, see Jeremy C. young, *The Age of Charisma: Leaders, Followers, and Emotions in American Society, 1870-1940* (New York: Cambridge University Press, 2017), 220–72.

4. Franklin D. Roosevelt, *The Public Papers and Addresses of Franklin D. Roosevelt* (New York: Random House and Harper and Brothers, 1938–1950), 633–44.

5. Alan Brinkley, *The End of Reform: New Deal Liberalism in Recession and War* (New York: Vintage Books, 1995), 22.

6. On these fears, see Benjamin Alpers, *Dictators, Democracy, and American Public Culture: Envisioning the Totalitarian Enemy, 1920s–1950s* (Chapel Hill: University of North Carolina Press, 2003), 94–128.

7. *The Phantom Empire*, directed by Otto Brower et al. (Mascot Pictures, 1935).

8. *The Phantom Empire*, chap. 5, "Beneath the Earth."

9. *The Phantom Empire*, chap. 11, "A Queen in Chains."

10. See *Undersea Kingdom*, directed by B. Reeves Eason and Joseph Cane (Republic Pictures, 1936); *Mysterious Doctor Satan*, directed by John English and William Witney (Republic Pictures, 1940). *Flash Gordon Conquers the Universe*, directed by Ford Beebe and Ray Taylor (Universal Pictures, 1940).

11. See artwork depicted in Roy Kinnard, *Science Fiction Serials: A Critical Filmography of the 31 Hard SF Cliffhangers* (New York: McFarland, 2008), 25.

12. Lothrop Stoddard, *The Rising Tide of Color: The Threat against White World-Supremacy* (New York: Scribner's, 1921).

13. On the use of science and technology to justify imperialism, see Michael Adas, *Machines as the Measure of Men: Science, Technology, and Ideologies of Western Dominance* (Ithaca, NY: Cornell University Press, 1990).

14. See Samuel Butler, *Erewhon, or Over the Range* (London: A. C. Field, 1908).

15. For an analysis of Asian stereotypes in popular culture, especially the Fu Manchu stereotype, see Robert Lee, *Orientals: Asian Americans in Popular Culture* (Philadelphia: Temple University Press, 1999), 88–89. For stereotpyes in science fiction, see David S. Roh, Betsy Huang, and Greata A. Niu, eds., *Techno-Orientalism: imagining Asia in Speculative Fiction, History, and Media* (New Brunswick, NJ: Rutgers University Press, 2015). For placement in the larger context of stereotypes of Asian Americans, see Mae M. Ngai, *Impossible Subjects: Illegal Aliens and the Making of Modern America* (Princeton, NJ: Princeton University Press, 2004), 112–13.

16. Stanley Link, *Tiny Tim and the Mechanical Men* ((Racine, WI: Whitman Publishing, 1937).

17. *Buck Rogers*, directed by Ford Beebe and Saul A. Goodkind (Universal Pictures, 1939).

18. "Nazi Is Held Unfit to Rear His Child," *New York Times*, December 25, 1942, 12; Reimann v. Reimann, 39 N.Y.S.2d 485, 1942.

19. Lewis Mumford, *Faith for Living* (New York: Harcourt, Brace, 1940), 33–34, 36.

20. Erich Fromm, *Escape from Freedom* (1941; New York: Henry Holt, 1994), 102–4, 117, 274, 184. For more on this book, see Wilfred M. McClay, *The Masterless, Self and Society in Modern America* (Chapel Hill: University of North Carolina Press, 1994), 197–210; and Lawrence Friedman and Anke Schreiber, *The Lives of Erich Fromm: Love's Prophet* (New York: Columbia University Press, 2013), 96–117.

21. Peter F. Drucker, *The End of Economic Man: A Study of the New Totalitarianism* (New York: John Day, 1939), 55.

22. Peter F. Drucker, *The Future of Industrial Man: A Conservative Approach* (New York: New American Library, 1942), 85.

23. Drucker, *Future of Industrial Man*, 84.

24. Drucker, *End of Economic Man*, 45, 46.

25. *The Great Dictator*, directed by Charles Chaplin (Charles Chaplin Productions, 1941).

26. Harold Ickes, "Text of Ickes Broadcast Urging United Front against Communism, Fascism," *Washington Post*, February 23, 1938, X4.

27. Wendell Willkie, "Text of Willkie Address to the Conference of Mayors," *New York Times*, January 14, 1942, 14.

28. Dewey quoted in "Uncle Sam's Navy," *Los Angeles Times*, April 7, 1903, 6; Alfred E. Turner, "The Kaiser as Strategist," *Living Age*, April 29, 1916, 314; George Frederick Arps, "Letter Kultur and Slavery," *New York Times*, March 2, 1918, 14.

29. John Dower, *War without Mercy: Race and Power in the Pacific War* (New York: Pantheon Books, 1986). On page 30, Dower notes one instance in which an American solider referred to the Japanese as robots, but most other instances are of animals; "Nazism and What It Stands For," *Atlanta Daily World*, April 18, 1936, 2; Westbrook Pegler, "Fair Enough," *Washington Post*, December 13, 1940, 13; Colonel Frederick Palmer, "US Urged to Build Army of 2,000,000," *New York Times*, June 16, 1940, 14; "Sweating German Labor," *Washington Post*, July 10, 1938, B8; Richard Wingate, "Letter to the Editor: Olympic Games Comment," *New York Times*, July 25, 1936, 10; James B. Reston, "Britain Goes Totalitarian—for the Duration," *New York Times*, June 2, 1940, 101.

30. See *Education for Death: The Making of a Nazi*, directed by Clyde Geronimi (Walt Disney Studios, 1943).

31. *Der Fuehrer's Face*, directed by Jack Kinney (Walt Disney Studios, 1942).

32. Mumford, *Faith for Living*, 19–20.

33. Fromm, *Escape from Freedom*, 274.

34. Drucker, *End of Economic Man*, 242.

35. Drucker, *Future of Industrial Man*, 195.

36. See, for example, *The Wizard of Oz*, directed by Victor Fleming (Warner Brothers, 1939).

37. "Display Ad 100," *Chicago Daily Tribune*, November 10, 1940, 4; R. C. Hutchinson, *The Fire and the Wood: A Love Story* (New York: Farrar & Rinehart, 1940). The actual text of the novel did not use such mechanical metaphors.

38. See *Hitler's Children*, directed by Edward Dymtrek (RKO Radio Pictures, 1943).

39. For more on the importance of human rights rhetoric during the war, see Elizabeth Borgwardt, *A New Deal for the World: America's Vision for Human Rights* (Cambridge, MA: Belknap Press, 2005).

40. Alpers, *Dictators, Democracy*, 193.

41. Dower, *War without Mercy*, 77–93.

42. "Topics of the Times," *New York Times*, January 31, 1941, 18.

43. Mumford, *Faith for Living*, 23.

44. Merryle Stanley Rukeyser, "Hitler Success Is Appraised as First War Year Concludes," *Washington Post*, September 1, 1940, R8.

45. They also briefly used *robot* to refer to a remote-controlled tank. Milton Bracker, "Allies Blow Up Nazi Robot Tanks," *New York Times*, April 10, 1944, 9.

46. "Robot Bombs Kill Babies in Nursery," *New York Times*, July 1, 1944, 3; "Nurses Carry Babies to Safety as Robot Bomb Hits Hospital," *Los Angeles Times*, July 6, 1944, 1; E. R. Noderer, "2752 Killed, 8,000 Injured by Nazi Flying Bombs," *Chicago Daily Tribune*, July 7, 1944, 1.

47. Stanton A. Coblentz, "What Does the Robot Say?," *Christian Century*, July 26, 1944, 877.

48. Ernest Hemingway, "London Fights the Robots," *Collier's*, August 19, 1944, 17–19. Hemingway preferred to not use the term *robot*, so his editor likely chose the word.

49. Richard Slotkin, *Gunfighter Nation: The Myth of the Frontier in Twentieth-Century America* (New York: Atheneum, 1992), 316–18.

50. Hemingway, "London Fights the Robots," 17.

51. Gary Cross, *Time and Money: The Making of Consumer Culture* (New York: Routledge, 1993), 152–53; Benjamin Kline Hunnicutt, *Work without End: Abandoning Shorter Hours for the Right to Work* (Philadelphia: Temple University Press, 1988), 301–16.

52. For more on the CIO, see Lizabeth Cohen, *Making a New Deal: Industrial Workers in Chicago, 1919–1939* (Cambridge: Cambridge University Press, 1990).

53. On the romanticization of workers in Hollywood films of the era, see John Bodnar, *Blue-Collar Hollywood: Liberalism, Democracy, and Working People in American Film* (Baltimore, MD: Johns Hopkins University Press, 2003), 55–86. Of course, the most iconic World War II worker, Rosie the Riveter, was (at least in the *Saturday Evening Post* image) a blue-collar worker flexing after having tamed her machine.

54. Alpers, *Dictators, Democracy*, 267–68. See Reinhold Niebuhr, *Moral Man and Immoral Society: A Study in Ethics and Politics* (New York: C. Scribner's, 1932); and Richard Fox, *Reinhold Niebuhr: A Biography* (New York: Pantheon Books, 1985).

55. Theodor Adorno and Max Horkheimer, "The Culture Industry: Enlightenment as Mass Deception," in *Dialectics of Enlightenment*, trans. John Cumming (New York: Herder and Herder, 1972); William Graebner, *Age of Doubt: American Thought and Culture in the 1940s* (Woodbridge, CT: Twayne Publishers, 1991), 110, 138–41.

56. Nils Gilman, "The Prophet of Post-Fordism: Peter Drucker and the Legitimation of the Corporation," in *American Capitalism: Social Thought and Political Economy in the Twentieth Century*, ed. Nelson Lichtenstein (Philadelphia: University of Pennsylvania Press, 2006), 109–34.

57. Donald L. Miller, *Lewis Mumford: A Life* (New York: Grove Great Lives, 2002). See Lewis Mumford, *The Myth of the Machine: The Pentagon of Power* (New York: Harcourt Brace Jovanovich, 1964).

58. Erich Fromm, *The Art of Loving* (New York: Harper and Row, 1956).

59. For a discussion of the Soviet automaton, see Scott Selisker, *Human*

Programming: Brainwashing, Automatons, and American Unfreedom (Minneapolis: University of Minnesota Press, 2016).

60. For a history of postwar automation, see David F. Noble, *Forces of Production: A Social History of Industrial Automation* (New York: Alfred A. Knopf, 1984).

Part 3

1. Isaac Asimov, "Perfect Playfellow," *Super Science Stories*, September 1940, 67–77.

2. Isaac Asimov, *In Memory Yet Green: The Autobiography of Isaac Asimov, 1920–1954* (New York: Doubleday, 1979), 585.

3. See, for instance, "Unimate: The First Industrial Robot," *A Tribute to Joseph Engelberger: The Father of Robotics* (website), accessed March 6, 2019, http://www.robotics.org/joseph-engelberger/unimate.cfm.

4. Asimov, *In Memory Yet Green*, 238.

5. John Kenneth Galbraith, *The Affluent Society*, (New York: Houghton Mifflin, 1958).

6. The word *robotics* was coined by Asimov in relation to his Three Laws of Robotics, which were introduced in his short story "Runaround." Isaac Asimov, "Runaround," *Astounding Science Fiction*, March 1942, 94–103.

7. See, for instance, Associated Press, "Robot Brains Being Used to Guard United States," *Beckley Post-Herald*, January 18, 1956.

8. As Jessica Riskin notes, feedback as identified by cyberneticists was a part of automatons since the ancient and medieval worlds. Riskin, *The Restless Clock: A History of the Centuries-Long Argument over What Makes Living Things Tick* (Chicago: University of Chicago Press, 2016), 312–13.

9. Norbert Wiener, *Cybernetics: Or Control and Communication in the Animal and the Machine* (Cambridge, MA: MIT Press, 1948),

10. William Graebner, *Age of Doubt: American Thought and Culture in the 1940s* (Woodbridge, CT: Twayne Publishers, 1991), 19–39.

11. Allen Ginsberg, "Howl," in *Howl and Other Poems* (San Francisco: City Lights Books, 1955), 18; C. Wright Mills, *White Collar: The American Middle Classes* (New York: Oxford University Press, 1953), 235.

Chapter 8

1. See *The Day the Earth Stood Still*, directed by Robert Wise (Twentieth Century Fox, 1951).

2. See *The Day the Earth Stood Still*, theatrical release poster, 1951. See also Margot A. Henriksen, *Dr. Strangelove's America: American Thought and Culture in the Atomic Age* (Berkeley: University of California Press, 1997), 51; and Paul Boyer, *By the Bomb's Early Light: American Thought and Culture at the Dawn of the Atomic Age* (Chapel Hill: University of North Carolina Press, 1985), 103–4.

3. Harry Bates, "Farewell to the Master," *Astounding Science Fiction*, October 1940, 62, 87.

4. William Graebner, *Age of Doubt: American Thought and Culture in the 1940s* (Woodbridge, CT: Twayne Publishers, 1991), 20–22.

5. For more on what Graebner calls "moral contingency" see Graebner, *Age of Doubt*, 23–26. For its relationship to Asimov's "Three Laws of Robotics," see 31.

6. On the racial dimensions of the concept of childhood innocence, see Robin Bernstein, *Racial Innocence: Performing American Childhood from Slavery to Civil Rights* (New York: NYU Press, 2011).

7. Isaac Asimov, *In Memory Yet Green: The Autobiography of Isaac Asimov, 1920–1954* (New York: Doubleday, 1979), 57.

8. Asimov, *In Memory Yet Green*, 70.

9. Isaac Asimov, *I, Asimov: A Memoir* (New York: Bantam Books, 2009), 20.

10. Asimov, *In Memory Yet Green*, 224–25.

11. Asimov, 224–25.

12. On his debt to Adam Link, see Asimov, *In Memory Yet Green*, 236. On his love of Helen O'Loy, see Isaac Asimov, letter to the editor, *Astounding Science Fiction*, February 1939, 159–60. On his debts to Westinghouse, see Isaac Asimov to Patricia Warrick, January 30, 1978, Isaac Asimov Papers, box 264, Howard Gotlieb Archive, Boston University; Patricia S. Warrick, *The Cybernetic Imagination in Science Fiction* (Cambridge, MA: MIT Press, 1980), 34.

13. Isaac Asimov, "Reason," in *I, Robot* (New York: Bantam Books, 2004), 62–63. Originally published in *Astounding Science Fiction*, April 1941, 33–45.

14. Asimov, "Reason," 69.

15. Asimov, *In Memory Yet Green*, 286.

16. The Three Laws were originally explained in the short story "Runaround" and then again in *I, Robot*. Isaac Asimov, "Runaround," *Astounding Science Fiction*, March 1942, 94–103; Asimov, *I, Robot*, 44–45. The format here comes from Isaac Asimov, "The Laws of Robotics," in *Robot Visions* (New York: Byron Press, 1990), 424.

17. Isaac Asimov to Sgt. H. P. Sanderson, British Forces, England, November 12, 1955, box 1, Howard Gotlieb Archive, Boston University.

18. Asimov, *I, Robot*, xiv.

19. Asimov, *I, Robot*, 237.

20. Isaac Asimov to Patricia Robertson, February 17, 1961, box 1, Howard Gotlieb Archive, Boston University.

21. Isaac Asimov to Marcia (Asimov) Repanes, November 20, 1961, box 23, Howard Gotlieb Archive, Boston University.

22. Isaac Asimov, "The Machine and the Robot," in *Robot Visions* (New York: Byron Preiss Visual Publications, 1990), 443. Originally published in Patricia Warrick and Martin Harry Greenberg, eds., *Science Fiction: Contemporary Mythology: The SFWA-SFRA Anthology* (New York, Harper & Row, 1978).

23. Gene Roddenberry to Isaac Asimov, May 16, 1972, box 260, Howard Gotlieb Archive, Boston University.

24. Ron Wilson to Isaac Asimov, September 6, 1960, box 1, Howard Gotlieb Archive, Boston University.

25. Richard Eskow to Isaac Asimov, February 1, 1966, box 21, Howard Gotlieb Archive, Boston University.

26. Charles Guder to Isaac Asimov, January 26, 1965, box 20, Howard Gotlieb Archive, Boston University.

27. Mara Canning to Isaac Asimov, undated, box 34C, Howard Gotlieb Archive, Boston University; strikethrough in the original.

28. For more on the discourse of the reliability of technology and the unreliability of people in the Cold War, see Edward Jones-Imhotep, "Maintaining Humans," in *Cold War Social Science: Knowledge Production, Liberal Democracy, and Human Nature*, ed. M. Soilovey and Hamilton Cravens (New York: Palgrave Macmillan, 2012), 175–96; and Edward Jones-Imhotep, "Disciplining Technology: Electronic Reliability, Cold-War Military Culture and the Topside Program," *History & Technology* 2 (January 2000): 125–75.

29. Harry S. Truman, "Address before a Joint Session of the Congress on Universal Military Training," October 23, 1945, in *Public Papers of the Presidents of the United States: Harry S. Truman (1945)* (Washington, DC: US Government Printing Office, 1961), 405.

30. Quoted in Bob Considine, "Only Big Air Force Can Halt New War, Rickenbacker Avers," *Washington Post*, February 11, 1945, B5.

31. Ralph Grapperhaus, "Train for Defense," *Rotarian*, March 1945, 50; Thomas M. Johnson, "Army, Navy Ask Action; Explain Plans, Purpose of Military Training," *Washington Post*, January 14, 1945, B2; National Security Training Commission, "Universal Military Training: Foundation of Enduring National Strength" (Washington, DC: United States Printing Office), 1951, 5; italics in original.

32. "A Democratic Army," *Washington Post*, October 2, 1940, 10; Edwin C. Johnson, "Booby Trap," *Nation*, January 26, 1952, 75; Pearl S. Buck, *Hearings before the Committee on Armed Services United States Senate on Universal Military Training* (Washington, DC: US Government Printing Office), 1948, 453; "Universal Military Training: Hearings before the Committee on Armed Services," United States Senate, Eightieth Congress, second session on UMT, 476; Catherine E. Wells, letter to the editor, *Washington Post*, July 30, 1947, 14.

33. The discussion of UMT in the context of fears of the negative effects of the garrison state on individualism draws on the analysis in Michael Hogan, *Cross of Iron: Harry S. Truman and the Origins of the National Security State* (Cambridge, UK: Cambridge University Press, 1998), 119–58, esp. 142-43, on the financial costs.

34. Wayne Morse, "National Need," *Nation*, January 26, 1952, 74; Hensel quoted in "Demobilizing Rush Decried by Hensel," *New York Times*, November 10, 1945, 7; Bush quoted in "Scientist Calls Push-Button 'Buck Rogers' War Talk 'Hooey,'" *Los Angeles Times*, January 9, 1947, 2.

35. "Navy Tells How Science Helped Beat Kamikazes," *Chicago Daily Tribune*, January 18, 1946, 5. For more on the development of the Mark 37, see David A. Mindell, *Between Human and Machine: Feedback, Control, and Computing before Cybernetics* (Baltimore, MD: Johns Hopkins University Press, 2004), 61–66.

36. On stereotypes of Japanese soldiers, see John Dower, *War without Mercy: Race and Power in the Pacific War* (New York: Pantheon Books, 1986), 85–86.

37. "Army Unveils 'Eniac,' the World's Best Calculator," *Beckley Post-Herald*, February 15, 1946, 1. On the historical development of the computer, see Martin Campbell Kelly, Wililam Aspray, Nathan Ensmenger, and Jeffrey R. Yost, *Computer: A History of the Information Machines* (New York: Westview Press, 2013); Nathan Ensmenger, *The Computer Boys Take Over: Computers, Programmers, and the Politics of Technical Expertise* (Cambridge: MIT Press, 2012); and Paul Edwards, *The Closed World: Computers and the Politics of Discourse in Cold War America* (Cambridge: MIT Press, 1997).

38. "Pop Ought to Have 'Eniac' to Do Junior's Arithmetic," *Lock Haven Express*, February 15, 1946, 1.

39. "Robot Calculator Knocks Out Figures Like Chain Lightning," *Chicago Daily Tribune*, February 15, 1946, 1.

40. "Electronic Brain," *American Weekly*, July 28, 1946.

41. Frank E. Carey, "Army's Eniac 1,000 Times Faster Than Best Calculating Machine," *Titusville Herald*, February 15, 1946, 1.

42. John R. Moore, "Jet Age Keyed to Electronics," *Los Angeles Times*, July 16, 1956, B20.

43. John Norris, "AF Shows Robot 'Brain' in Action for Defense, *Washington Post and Times Herald*, January 18, 1956, 1. For similar reports, see Hanson W. Baldwin, "Air Defense Gets Electronic 'Brain' to Guide Plans," *New York Times*, January 18, 1956, 1; and Harold Hutchings, "Unveil Electronic Brain for Air Defense Program," *Chicago Tribune*, January 18, 1956, 1.

44. Associated Press, "Robot Brains Being Used to Guard United States," *Beckley Post-Herald*, January 18, 1956.

45. Hanson W. Baldwin, "The 'Drone': Portent of Push-Button War," *New York Times*, August 25, 1946, 96, 120. On the development of such technologies, see Paul G. Gillespie, *Weapons of Choice: The Development of Precision Guided Munitions* (Tuscaloosa: University of Alabama Press, 2006) 10–38.

46. "TV-Robot Plane Advances Push-Button Warfare," *Popular Mechanics*, November 1952, 109; "Push-Button War Begins: Robots Rip Korean Reds," *Los*

Angeles Times, September 18, 1952, 1; "Robots to Fly Atom Bombs," *Chicago Daily Tribune*, September 19, 1952, 1; Warren Rogers Jr., "American Robot Planes Herald Explosive Birth of Push-Button Warfare in Korea," *Warren Times-Mirror*, September 18, 1952, 1.

47. H. Bruce Franklin, *War Stars: The Superweapon and the American Imagination, (New York: Oxford University Press, 1988)*, 199–210; Edwards, *The Closed World*, 5–8.

48. Tim Weiner, "A New Model Army Soldier Rolls Closer to the Battlefield," *New York Times*, February 16, 2005, A1. For more on what Bruce Franklin calls the "Age of Automatons" see Franklin, *War Stars*, 206–12.

49. Weiner, "A New Model Army Soldier."

50. Ta-Nehisi Coates, "Obama's Robot Army," *Atlantic*, December 28, 2011, https://www.theatlantic.com/politics/archive/2011/12/obamas-robot-army/250584/.

51. Reinhold Niebuhr, *The Irony of American History* (Chicago: University of Chicago Press, 1952), 28, 18. For more on religion and notions of innocence in the Cold War, see Jason W. Stevens, *God-Fearing and Free: A Spiritual History of America's Cold War* (Cambridge, MA: Harvard University Press, 2010); and Andrew S. Finstuen, *Original Sin and Everyday Protestants: The Theology of Reinhold Niebuhr, Billy Graham, and Paul Tillich in an Age of Anxiety* (Chapel Hill: University of North Carolina Press, 2009).

52. Philip K. Dick, "The Defenders," *Galaxy Science Fiction*, January 1953; available at Project Gutenberg, http://www.gutenberg.org/files/28767/28767-h/28767-h.htm.

53. *Tobor the Great*, directed by Lee Sholem (Dudley Pictures, 1954); *Captain Video and his Video Rangers*, season unknown, episode unknown, "I, Tobor," aired November 1953, on Dumont Television Network; *Here Comes Tobor*, directed by Duke Goldstone (Guild Films, 1956), available at http://archive.org/details/tobor01.

54. *Forbidden Planet*, directed by Fred M. Wilcox (Metro Goldwyn Mayer, 1956).

55. For further analysis of Robby, see J. P. Telotte, *Robot Ecology and the Science Fiction Film* (New York: Routledge, 2016), 42–59.

56. *Forbidden Planet*, dir. Wilcox.

57. *The Invisible Boy*, directed by Herman Hoffman (Metro Goldwyn Mayer, 1957).

58. See *Lost in Space*, season 1, episode 20, "War of the Robots," aired February 9, 1966, on CBS; *Outer Limits*, season 2, episode 9, "I, Robot," aired November 14, 1964, on ABC; Ted Hughes, *The Iron Man: A Children's Story in Five Nights* (New York: Harper & Row, 1968); and *Star Wars: Episode IV—A New Hope*, directed by George Lucas (Lucasfilm, Twentieth Century Fox, 1977).

59. Harold M. Schmeck Jr., "Queens Boy, 13, Builds a Robot That Has a Flair for Figures," *New York Times*, June 25, 1957, 31; "13 Year Old Improves Hollywood Robot: Marc's Creation Takes Its Own Pictures," *Chicago Tribune*, January 12, 1964, N2; "Robot Makes TV Debut," *Chicago Tribune*, January 12, 1964; "A Home-Made 'Otto-Tron,'" *Ebony*, June 1970, 65–68; "Prize Robot," *Jet*, May 1, 1958, 48.

60. "Can You Think Faster Than This Machine?," *Astounding Science Fiction*, May 1959, 3; Edmund C. Berkeley, "Simple Simon," *Scientific American*, November 1950, 40–43.

61. "New Toy," *Chicago Defender*, February 26, 1955, 16; advertisement in *Life*, November 22, 1954, 166.

62. Martha Weinman Lear, "5,000 Toys—and Most of Them Move," *New York Times*, December 10, 1961, SM17.

63. See several of the robot toys of the period in this collectors guide: Jim Bunte, Dave Hallman, and Heinz Mueller, *Vintage Toys: Robots and Space Toys* (Iola, WI: Antique Trader Books, 1999). On "Charley 'n Me," see Topper Toys, Charley 'n Me (Elizabeth, NJ: De Luxe Topper Corporation, 1967). On the larger world of toys in the period, see Gary Cross, *Kid's Stuff: Toys and the Changing World of American Childhood* (Cambridge, MA: Harvard University Pres, 1997), 147–87.

64. See *Gog*, directed by Herbert L. Strock ((United Artists, 1954); *Fail-Safe*, directed by Sidney Lumet (Columbia Pictures, 1964); *Dr. Strangelove or: How I Learned to Stop Worrying and Love the Bomb*, directed by Stanley Kubrick (Columbia Pictures, 1964); Harlan Ellison, "I Have No Mouth, and I Must Scream," in *Machines That Think: The Best Science Fiction Stories about Robots and Computers*, ed. Isaac Asimov and Patricia S. Warrick, et al. (New York: Holt, Rinehart and Winston, 1984), 233–50, originally published in *If*, March 1967, 24–36; and *Colossus: The Forbin Project*, directed by Joseph Sargent (Universal Pictures, 1970).

65. *The Terminator*, directed by James Cameron (Orion Pictures, 1984).

66. *Terminator 2: Judgment Day*, directed by James Cameron (TriStar Pictures, 1991).

67. *Terminator 2*, dir. Cameron.

68. *Terminator 2*, dir. Cameron.

Chapter 9

1. "Revival of R.U.R. with New Prologue," *New York Times*, May 7, 1950, 163. For more on Wiener's moth robot, see Jessica Riskin, *The Restless Clock: A History of the Centuries-Long Argument over What Makes Living Things Tick* (Chicago: University of Chicago Press, 2016), 324–26.

2. *Calling All Jobs: An Introduction to the Automatic Machine Age* (New York: National Association of Manufacturers, 1957), 21. For further contextualiza-

tion see Bix, *Inventing Ourselves Out of Jobs?*, 24. For more on this theme, see David P. Julyk, "'The Trouble with Machines Is People': The Computer as Icon in Post-War America, 1946–1970," PhD diss., University of Michigan, 2008.

3. For more on the rise of postindustrial visions and their connection to computers and automation, see Howard Brick, *Transcending Capitalism: Visions of a New Society in Modern American Thought* (Ithaca, NY: Cornell University Press, 2006), 152–85; Jefferson Cowie, *Stayin' Alive: The 1970s and the Last Days of the Working Class* (New York: New Press, 2012); Daniel T. Rodgers, *Age of Fracture* (New York: Belknap Press, 2012); and David Nye, *America's Assembly Line* (Cambridge, MA: MIT Press, 2013), 187–216.

4. Intercontinental Manufacturing Company, "Have You Ever Seen a Robot Santa Claus?," advertisement, *Aviation Week*, December 21, 1953, 48.

5. Mrs. Arthur L. Gardner, "Auxiliary Activities," *National Rural Letter Carrier*, December 20, 1952, 12; "Robot 'Santa' Aiding Needy Children," *New York Times*, December 14, 1954, 26; "Robot Santa," *Evening Times*, December 21, 1961, 2.

6. See *Galaxy*, December 1960, cover.

7. *Santa Claus Conquers the Martians*, directed by Nicholas Webster (Jalor Productions, 1964).

8. Carl Dreher, *Automation: What It Is, How It Works, Who Can Use It* (New York: Norton, 1957), 28–29.

9. John Diebold, *Automation: Its Impact on Business and Labor*, Planning Pamphlet no. 106 (Washington, DC: National Planning Association, 1959), 3.

10. "Mechanical Brain," *New York Times*, October 31, 1945, 17.

11. John Diebold, "Automation: Will It Steal Your Job?" *Los Angeles Times*, June 26, 1955, H7.

12. Victor Cohn, "1999: Our Hopeful Future: 'Man's Work Will Be Done by Robot,'" *San Mateo Times*, January 20, 1955, 1.

13. Carleton Beals, "Cybernetics," *Rotarian*, September, 1953, 14–16, 56–57, 16.

14. A. H. Raskin, "Automation Puts Industry on Eve of Fantastic Robot Era," *New York Times*, April 8, 1955, 14.

15. "Chivalrous Robot," *Life*, May 28, 1956, 126.

16. "Marvelous Mobot Will Do Work Too Hot for Man," *Popular Science*, September 1960, 82–83.

17. "Robot with a Super Deft Touch," *Life*, February 17, 1961, 55.

18. George C. Devol Jr., "Patent for Programmed Article Transfer," US patent 2988237, filed December 10, 1954, granted June 13, 1961.

19. "Unimate: The First Industrial Robot," *A Tribute to Joseph Engelberger: The Father of Robotics* (website), accessed March 6, 2019, http://www.robotics.org/joseph-engelberger/unimate.cfm.

20. Alden P. Armagnac, "New Factory Worker: Teachable Robot Can Remember 200 Commands," *Popular Science*, August 1962, 79.

21. Kasia Cieplak-Mayr von Baldegg, "Unimate Robot on Johnny Carson's Tonight Show (1966)," *Atlantic*, August 16, 2011, http://www.theatlantic.com/technology/archive/2011/08/unimate-robot-on-johnny-carsons-tonight-show-1966/469779/.

22. "Unimate: The First Industrial Robot"; "Unimate" Robot Hall of Fame, Carnegie Mellon University, 2003, http://www.robothalloffame.org/inductees/03inductees/unimate.html.

23. Wayne Johnson and Howard Gingold, "Automation: Portend of a Second Industrial Revolution," *Los Angeles Times*, September 9, 1957, B5.

24. See Philip K. Dick, "The Last of the Masters," *Orbit*, November 1954, 32–57; Philip K. Dick, "Autofac," *Galaxy Science Fiction*, November 1955, 70–95; Philip K. Dick, "To Serve the Master," *Imagination*, February 1956, 78–89. Nye also discusses these stories in *America's Assembly Line*, 158.

25. *Desk Set*, directed by Walter Lang (Twentieth Century Fox, 1957); *The Jetsons*, Season 1, episode 10, "Uniblab," aired November 25, 1962, on ABC; On *Desk Set*, see Bix, *Inventing Ourselves Out of Jobs*?, 251–52.

26. Howard Brick, "Optimism of the Mind: Imagining Postindustrial Society in the 1960s and 1970s," *American Quarterly* 44, no. 3 (September 1992): 348–80.

27. David Riesman, "Leisure and Work in a Post-Industrial Society," in *Mass Leisure*, ed. Eric Larrabee and Rolf Meyerson (Glencoe, IL: Free Press, 1958), 379–81. On the origins and development of the term, see Brick, *Transcending Capitalism*, 194–95.

28. Daniel Bell, *Work and Its Discontents: The Cult of Efficiency in America* (Boston: Beacon Press, 1956), 3, 21, 49, 55–56, 32, 38.

29. Bell, *Work and Its Discontents*, 48; Daniel Bell, *The Coming of Post-Industrial Society: A Venture in Social Forecasting* (New York: Basic Books, 1973), 127. For more on Bell and postindustrialism, see Brick, *Transcending Capitalism*, 197–98; and Howard Brick, *The Age of Contradiction: American Thought and Culture in the 1960s* (New York: Twayne Publishers, 1998), 54–57.

30. Donald N. Michael, *Cybernation: The Silent Conquest* (Santa Barbara, CA: Center for the Study of Democratic Institutions, 1982 [1962]), 9, 17.

31. "The Triple Revolution," *Liberation* (April 1964), 9–15; Full text can be accessed at https://www.marxists.org/history/etol/newspape/isr/vol25/no03/adhoc.html.

32. Brick, *The Age of Contradiction*, 54–56; Brick, "Optimism of the Mind," 348–80.

33. William M. Freeman, "Business Greets Push-Button Age," *New York Times*, January 5, 1943, 81; International Union, United Automobile, Aero-

space, and Agricultural Implement Workers of America, Education Department, *Automation: A Report to the UAW-CIO Economic and Collective Bargaining Conference Held in Detroit, Michigan, the 12th and 13th of November 1954* (Detroit: UAW-CIO Education Department, 1955), 2, 37. On the UAW and automation, see Kevin Boyle, *The UAW and the Heyday of American Liberalism* (Ithaca, NY: Cornell University, 1995).

34. Thomas F. Jackson, *From Civil Rights to Human Rights: Martin Luther King, Jr., and the Struggle for Economic Justice* (Philadelphia: University of Pennsylvania Press, 2007), 195–203

35. Quoted in Nat Hentoff, "James Baldwin Gets 'Older and Sadder,'" *New York Times*, April 11, 1965, X3.

36. "Vote Mobilization for the 1970s," *Ebony*, July 1971, 84–86.

37. Alex Poinsett, "How to Get a Job," *Ebony*, May 1964, 79–88.

38. Alex Poinsett, "Poverty amidst Plenty," *Ebony*, August 1965, 104–12.

39. "Unemployment among Youth: The Explosive Statistic!," *Ebony*, August 1967, 129.

40. Cowie, *Stayin' Alive*, 45–46.

41. Quoted in Studs Terkel, *Working: People Talk about What They Do All Day and How They Feel about What They Do* (New York: New Press, 2013), 191.

42. "Revolt of the Robots," *New York Times*, March 7, 1971, 38.

43. Otto Friedrich, Christopher Redman, and Janice C. Simpson, "The Robot Revolution for Good or Ill, It Is Already Transforming the Way the World Works," *Time*, December 8, 1980, 72. *Christian Science Monitor* article reprinted as David T. Cook, "Robots Populate Japanese Industry," *Sun*, January 21, 1982. For further analysis of the article, see Bix, *Inventing Ourselves Out of Jobs?*, 278–79.

44. Arthur Koestler, *The Lotus and the Robot* (New York: MacMillan, 1961), 165–88. On the cliché, see Robert Taylor, "Looking at Whimsical Pictures from Japan," *Boston Globe*, August 24, 1983, 1; and Tom Raithel, "Regimented Worker Image No Longer Fits," *Evansville Courier*, November 10, 1997, A3.

45. Styx, "Mr. Roboto," *Kilroy Was Here* (A&M, 1983), available at: https://www.youtube.com/watch?v=uc6f_2nPSX8.

46. Quoted in William Chapman, "Japanese Authorities Back Image-Making Effort on U.S. Cable TV," *Washington Post*, April 26, 1984. On the larger "yellow peril" in this rhetoric, see Nye, *America's Assembly Line*, 207.

47. *Futurama*, season 2, episode 4, "X-mas Story," aired December 19, 1999, on Fox.

48. On contemporary fears of robots, see Matt Thompson and Karen Yuan, "What If a Robot Wrote This Article," *Atlantic*, May 14, 2018, https://www.theatlantic.com/membership/archive/2018/05/what-if-a-robot-wrote-this-article/560327/; and Steve Lohr, "This Week in Tech: Are Robots Coming for

Your Job? Eventually, Yes," *New York Times*, September 21, 2018, https://www.nytimes.com/2018/09/21/technology/artificial-intelligence-jobs.html.

49. Betty Friedan, *The Feminine Mystique* (New York: Norton, 1963), 369–70.

50. An exception here is Valerie Solanas's call for women "to overthrow the government, eliminate the money system, institute complete automation and destroy the male sex," in her self-published *SCUM Manifesto*. This is discussed in Kathi Weeks, *The Problem with Work: Feminism, Marxism, Antiwork Politics, and Postwork Imaginaries* (Durham, NC: Duke University Press, 2011), 217; and Scott Selisker, *Human Programming: Brainwashing, Automatons, and American Unfreedom* (Minneapolis: University of Minnesota, 2016), 92–94.

51. "Vigorelli Super Robot," *Life*, May 2, 1955, 164.

52. Homko Lawn Mover, Western Tool & Stamping Company, advertisement, *Life*, April 5, 1954, 145.

53. "Automation Invades Garden: Gadgets for Most Chores," *Chicago Defender*, July 26, 1958, 16.

54. "McCulloch's Push-Button Paradise," *Life*, May 7, 1956, 71–78. For more on the obsession of push-button homes, see Elizabeth Fratterigo, *Playboy and the Making of the Good Life in Modern America* (New York: Oxford University Press, 2010), 89.

55. Letter to the editor, *Life*, May 28, 1956, 24.

56. *The Jetsons*, season 1, episode 1, "Rosey the Robot," aired September 23, 1962, on ABC.

57. Quoted in Eddy Gilmore, "Prof. Thring Makes the Thing," *Port Angeles Evening News*, February 2, 1967, 2.

58. Quoted in "Atom Chairman Sees Easy Life for 21st Century Housewife," *Chicago Defender*, February 11, 1967, 22.

59. Friedan, *Feminine Mystique*, 57; O. O. Binder, "You'll Own 'Slaves' by 1965," *Mechanix Illustrated*, January 1957, 62–65; Harry Carp, "Owns Own Slave," "Dear Editor," *Mechanix Illustrated*, March 1957, 30; Robert K. Plumb, "Science Envisions a Robot-Run Home," *New York Times*, January 31, 1956. See also Cowan, *More Work for Mother*, 196–201.

60. Friedan, *Feminine Mystique*, 255.

61. Friedan, 287.

62. Ira Levin, *The Stepford Wives* (New York: Random House, 1972); *The Stepford Wives*, directed by Bryan Forbes (Palomar Pictures, 1975); For further analysis, see Selisker, *Human Programming*, 94–95; Anna Krugovoy Silver, "The Cyborg Mystique: The Stepford Wives and Second Wave Feminism," *Women's Studies Quarterly*, Spring-Summer 2002, 60–76. On male intimacy after World War II, see John Ibson, *The Mourning After: Loss and Longing among Midcentury American Men* (Chicago: University of Chicago Press, 2018). For more on *The Stepford Wives* and Friedan, see Selisker, 94–95.

63. Marshall McLuhan, *The Mechanical Bride: Folklore of Industrial Man* (Berkeley: Ginko Press, 1951), 98–101; Levin, *Stepford Wives*, 34.

64. *My Living Doll*, season 1, episode 1, "Boy Meets Girl," aired September 27, 1964, on CBS. In this, the show mimicked the treatment of powerful women in *I Dream of Jeannie* and *Bewitched*. See Susan J. Douglas, *Where the Girls Are: Growing Up Female with the Mass Media* (Pittsburgh: Three Rivers Press, 1995), 123–38.

65. *A.I. Artificial Intelligence*, directed by Steven Spielberg (Warner Brothers, 2001); *Dr. Goldfoot and the Bikini Machine*, directed by Norman Taurog (American International Pictures, 1965); *Weird Science*, directed by John Hughes (Universal Pictures, 1985); *Buffy the Vampire Slayer*, season 5, episode 15, "I Was Made to Love You," aired February 20, 2001; *Buffy the Vampire Slayer*, season 5, episode 18, "Intervention," aired April 24, 2001.

66. Hines quoted in Peter Svensson, "It's a Life-Size Rubber Doll You Can Relate To," *Star Beacon*, January 22, 2010, C14; "The Real Side of Owning a RealDoll," CNET, August 10, 2017, https://www.cnet.com/pictures/realdolls-sex-doll-abyss-creations-owners-in-their-own-words/; FAQ, True Companion website, accessed December 10, 2018: http://www.truecompanion.com/shop/faq. For more analysis, see Christina Brown, "Sex Robots, Representation, and the Female Experience," *American Papers* 37, (2018-2019), 111–13.

67. Linda Strikwerda, "Legal and Moral Implications of Child Sex Robots," in *Robot Sex: Social and Ethical Implications*, ed. John Danaher and Neil McArthur (Cambridge, MA: MIT Press, 2017), 133–50.

68. Jack Williamson, "With Folded Hands," in *Men and Machines: Ten Stories of Science Fiction*, ed. Robert Silverberg, (New York: Willside Press, 1968), 184–240, 186, 217.

69. Williamson, "With Folded Hands," 231.

70. Williamson, 193, 206.

71. Norbert Wiener, *The Human Use of Human Beings: Cybernetics and Society* (Boston: Houghton Mifflin, 1950), 16; Brick, *Age of Contradiction*, 131–32.

72. McLuhan, *The Mechanical Bride*, 21.

73. Brick, *Age of Contradiction*, 131–36. The original argument for the counterculture's connection to technocracy comes from Theodore Roszak, *The Making of a Counterculture: Reflections on the Technocratic Society and Its Youthful Opposition* (Berkeley: University of California Press, 1968); Selisker, *Human Programming*, 70.

74. John Sinclair, "Rock and Roll Is a Weapon of Cultural Revolution," in *Guitar Army: Rock and Revolution with the MC5 and White Panther Party* (Los Angeles: Process, 2007), 97.

75. Andrew Kirk, "'Machines of Loving Grace': Alternative Technology, Environment, and the Counterculture," in *Imagine Nation: The American Counterculture of the 1960s & '70s*, ed. P. Braunstein and Michael William Doyle

(New York: Routledge, 2002), 353–78; R. John Williams, *The Buddha in the Machine: Art, Technology, and the Meeting of East and West* (New Haven, CT: Yale University Press, 2014), 179–80; Richard Brautigan, "All Watched Over by Machines of Loving Grace," *Digger Archives*, accessed March 6, 2016. http://www.diggers.org/digpaps68/allwat68.html.

76. *Silent Running*, directed by Douglas Trumball (Universal Pictures, 1972).

77. *Silent Running*, dir. Trumball.

78. On the conservative ascendency, see Jonathan M. Schoenwald, *A Time for Choosing: The Rise of Modern American Conservatism* (New York: Oxford University Press, 2001); Robert O. Self, *All in the Family: The Realignment of American Democracy since the 1960s* (New York: Hill and Wang, 2012); and Rick Perlstein, *Nixonland: The Rise of a President and the Fracturing of America* (New York: Scribner, 2008).

Chapter 10

1. New Science/Innovations," *Popular Mechanics*, April 1984, 61; Graydon E. Carter, "People," *Time*, November 15, 1982, 96.

2. William Grimes, "Barbara Goldsmith, Author of 'Little Gloria,' Dies at 85," *New York Times*, June 28, 2016, https://www.nytimes.com/2016/06/29/books/barbara-goldsmith-author-of-little-gloria-dies-at-85.html.

3. Barbara Goldsmith, "The Meaning of Celebrity," *New York Times*, December 4, 1983, SM75.

4. Jean Baudrillard, *Simulacra and Simulation*, trans. Sheila Glaser (Ann Arbor: University of Michigan Press, 1994), 1–2.

5. Quoted in Goldsmith, "The Meaning of Celebrity;" Daniel Boorstin, *The Image: A Guide to Pseudo-Events in America* (New York: Vintage Books, 1962), 240.

6. For the history of postmodern thinking, see James Livingston, *The World Turned Inside Out: American Thought and Culture at the End of the 20th Century* (New York: Rowman & Littlefield, 2010); David Steigerwald, *The Sixties and the End of Modern America* (New York: St. Martin's Press, 1995), 154–98; and Daniel T. Rodgers, *Age of Fracture* (Cambridge, MA: Belknap Press, 2011).

7. For more on Warhol and camp, see Roger Cook, "Andy Warhol, Capitalism, Culture, and Camp," *Space & Culture* 6, no. 1 (February 2003): 66–76; and David Serlin, *Replaceable You: Engineering the Body in Postwar America* (Chicago: University of Chicago Press, 2004), 191–200.

8. On potential links between Butler and behaviorism, see Carrie L. Hull, "Poststructuralism, Behaviorism, and the Problem of Hate Speech," *Philosophy & Social Criticism* 29, no. 5 (September 2003): 517–35, https://doi.org/10.1177/01914537030295002.

9. *RoboCop*, directed by Paul Verhoeven (Orion Pictures, 1987).

10. For a history of the development of the cyborg and posthumanism, see

N. Katherine Hayles, *How We Became Posthuman: Virtual Bodies in Cybernetics, Literature, and Informatics* (Chicago: University of Chicago Press, 1999).

11. On "the science of robots," see Ronald R. Kline, *The Cybernetics Moment: Or, Why We Call Our Age the Information Age* (Baltimore, MD: Johns Hopkins University Press, 2015), 68–101, esp. 88; Norbert Wiener, *Cybernetics: Or Control and Communication in the Animal and the Machine* (Cambridge, MA: MIT Press, 1948); Norbert Wiener, *The Human Use of Human Beings: Cybernetics and Society* (Boston: Houghton Mifflin, 1950), 8–9.

12. Wiener, *Cybernetics*, 58, 11. For more on the roots of cybernetics, see David A. Mindell, *Between Human and Machine: Feedback, Control, and Computing before Cybernetics* (Baltimore, MD: Johns Hopkins University Press, 2003); and Andrew Pickering, *The Cybernetic Brain: Sketches of Another Future* (Chicago: University of Chicago Press, 2010).

13. Wiener, *The Human Use of Human Beings: Cybernetics and Society*, 2nd ed. (Boston: Houghton Mifflin, 1954), 38. For more on Palomilla's connection to cybernetics, see Jessica Riskin, *The Restless Clock: A History of the Centuries-Long Argument over What Makes Living Things Tick* (Chicago: University of Chicago Press, 2016), 314–20.

14. Walter Ross Ashby, *Design for a Brain* (London: Chapman and Hall, 1948), 382–83, 10; Pickering, *Cybernetic Brain*, 1–3.

15. V. H. Brix, *You Are a Computer: Cybernetics in Everyday Life* (New York: Emerson Books, 1967).

16. Alan Turing, "Computing Machinery and Intelligence," *Mind* 59, no. 236 (October 1, 1950): 433–60. For further analysis, see Pickering, *Cybernetic Brain*, 328; Kline, *Cybernetics Moment*, 153–65; and Riskin, *Restless Clock*, 334–35.

17. John McCarthy, Marvin L. Minsky, et al., "A Proposal for the Dartmouth Summer Research Project on Artificial Intelligence," August 31, 1955, *AI Magazine*, Winter 2006, 27, 4, 12. See also Kline, *Cybernetics Moment*, 153–65.

18. Marvin Minsky, "Steps toward Artificial Intelligence," *Proceedings of the IRE* 41, no. 1 (January 1961): 8–30. On the fantasy of the unique human mind and AI, see Riskin, *Restless Clock*, 343–45.

19. Kline, *Cybernetics Moment*, 153–65; John von Neumann, *The Computer and the Brain* (New Haven, CT: Yale University Press, 1958); Norbert Wiener, "Chess-Playing Automata, the Turk, Mephisto, and Ajeeb," 1949.

20. Isaac Asimov, *The Caves of Steel* (New York: Bantam Books, 1991), 25, 27.

21. Asimov, *Caves of Steel*, 220–22.

22. James Sheldon and William Claxton, "I Sing the Body Electric," *The Twilight Zone*, season 3, episode 35 (May 18, 1962).

23. *2001: A Space Odyssey*, directed by Stanley Kubrick (Metro Goldwyn Mayer and Stanley Kubrick Productions, 1968); Carol Fry, "From Technology

to Transcendence: Humanity's Evolutionary Journey in '2001, a Space Odyssey,' *Extrapolation* 44, no. 3 (November, 2002): 331–43.

24. Philip K. Dick, *Do Androids Dream of Electric Sheep?* (New York: Doubleday, 1968). See also *Blade Runner*, directed by Ridley Scott (Ladd Company, 1982).

25. C. Wright Mills, *White Collar: The American Middle Classes* (New York: Oxford University Press, 1953), 235.

26. Mills, *White Collar*, 238.

27. On this point, see Benjamin Alpers, *Dictators, Democracy, and American Public Culture: Envisioning the Totalitarian Enemy, 1920s–1950s* (Chapel Hill: University of North Carolina Press, 2003), 250–302.

28. Malvina Reynolds, "Menace of Mediocrity: Robot Trend," *Washington Post*, May 21, 1947, 8.

29. See, for instance, "Eichmann Paints a Robot Portrait," *New York Times*, June 26, 1961, 8; and Homer Bigart, "Eichmann Tells Ransom Motive," *New York Times*, July 6, 1961, 8.

30. Lewis Mumford, *The Myth of the Machine: The Pentagon of Power* (New York: Harcourt Brace Jovanovich, 1964), 277. On the larger issue of the authoritarian personality, see Robert Genter, *Late Modernism, Art, Culture, and Politics in Cold War America* (Philadelphia: University of Pennsylvania Press, 2010), chap. 1 and 2.

31. See Kurt Vonnegut Jr., *Player Piano* (New York: Charles Scribner's Sons, 1952); Allen Ginsberg, "Howl," in "Howl and Other Poems, (San Francisco: City Lights Books, 1955), 18; Herbert Marcuse, *One-Dimensional Man: Studies in the Ideology of Advanced Industrial Society* (Boston: Beacon Press, 1964); and *The Graduate*, directed by Mike Nichols, (Embassy Pictures and United Artists,1967).

32. Marcuse, *One-Dimensional Man*, 10. For further discussion of the connection between Marcuse and postindustrialism, see J. Jesse Ramirez, "Marcuse among the Technocrats: America, Automation, and Postcapitalist Utopias, 1900–1941," *Amerikastudien / American Studies* 57, no. 1 (2012): 31–50; Andrew Feenberg, "Post-Industrial Discourses," *Theory and Society* 19, no. 6 (December 1990): 709–37; and Douglas Kellner, "Marcuse and the Quest for Radical Subjectivity," *Social Thought & Research* 22, no. 1/2 (1999): 1–24.

33. Quoted in Jason W. Stevens, *God-Fearing and Free: A Spiritual History of America's Cold War* (Cambridge, MA: Harvard University Press, 2010), 176.

34. For more on amusement parks and artificiality, see Gary Cross, *Packaged Pleasures: How Technology and Marketing Revolutionized Desire* (Chicago: University of Chicago Press, 2014), 207–40, 267-68.

35. "'Grandma' Flies First Class to N.Y., Fair," *Los Angeles Times*, February 20, 1964.

36. *Disneyland Goes to the Fair*, directed by Hamilton Luske (Walt Disney Productions, 1964).

37. Jack Smith, "Fun or a Reasonable Facsimilie of It," *Los Angeles Times*, September 5, 1966, D1.

38. Quoted in John M. Findlay, *Magic Lands: Western Cityscpaes and American Culture After 1940*, (Berkeley: University of California Press,1992), 70. Findlay notes that the quote may be apocryphal but that its fits with Disney's larger ideology.

39. Umberto Eco, *Travels in Hyperreality* (New York: Harcourt, 2014), 43, 44, 45, 47.

40. Maltz, *Psycho-Cybernetics* (New York: Prentice-Hall, 1960), x, 9.

41. See *Westworld*, directed by Michael Crichton (Metro-Goldwyn Mayer, 1973).

42. Marley Jackson, "The Jackson 5 Dancing Machine Live at Soul Train," August 2, 2014, video, 3:49, https://www.youtube.com/watch?v=CuyOGuXiGAk. For further analysis, see Megan Pugh, American *Dancing: From the Cakewalk to the Moonwalk* (New Haven, CT: Yale University Press, 2015), 271.

43. Pugh, *American Dancing*, 269–75.

44. See *Michael Jackson's Moonwalker*, by Emerald Software and Keypunch Software (US Gold, 1990).

45. Small-town newspapers around the country reported on robot dances at homecomings and other dances throughout the 1970s and 1980s. See, for instance, "Columbia Street Wins Talent Event," *Cumberland News*, August 6, 1977, 20; and "H-F Homecoming Slated," *Chicago Heights Star*, October 16, 1977, 3.

46. Jefferson Cowie, *Stayin' Alive: The 1970s and the Last Days of the Working Class* (New York: New Press, 2012), 345–48.

47. Jack Hamilton, *Just around Midnight: Rock and Roll and the Racial Imagination* (Cambridge, MA: Harvard University Press, 2016), 4. See also Styx, "Mr. Roboto," *Kilroy Was Here* (A&M, 1983). Video available at Styx, "Styx—Mr. Roboto (Relaid Audio)," August 10, 2017, YouTube video, 5:34, https://www.youtube.com/watch?v=uc6f_2nPSX8.

48. Manfred E. Clynes and Nathan S. Kline, "Cyborgs and Space," *Astronautics*, September 1960, 26–27, 74–76, 27; For further analysis, see Chris Hables Gray, *Cyborg Citizen: Politics in the Posthuman Age* (New York: Routledge, 2001), 10.

49. Paul Edwards, *The Closed World: Computers and the Politics of Discourse in Cold War America* (Cambridge: MIT Press, 1997), 271–74.

50. See Martin Caidin, *Cyborg: A Novel* (New York: Arbor House, 1972); *The Six-Million Dollar Man*, premiere January 18, 1974, on ABC; *The Bionic Woman*, premiere January 14, 1976, on ABC; Russ Manning, *Mangus, Robot Fighter* (Poughkeepsie, NY: Gold Key Comics, 1963). For more on these "posthuman bodies," see Scott Jeffery, *The Posthuman Body in Superhero Comics: Human, Superhuman, Transhuman, Post/Human* (New York: Palgrave MacMillan, 2016), esp. 144–46, on Ultron.

51. See John Markoff, *What the Dormouse Said: How the Sixties Counter-*

culture Shaped the Personal Computer Industry (New York: Penguin, 2005); John Markoff, *Machines of Loving Grace: The Quest for Common Ground Between Humans and Robots* (New York: HarperCollins, 2015); Fred Turner, *From Counterculture to Cyberculture: Stewart Brand, the Whole Earth Network, and the Rise of Digital Utopianism* (Chicago: University of Chicago Press, 2010), 103–40.

52. For more on the history of the computer, see Nathan Ensmenger, *Computers, Programmers, and the Politics of Technical Expertise* (Cambridge, MA: MIT Press, 2010). The endurance of this distinction into contemporary America is central to John Markoff's analysis. John Markoff, *Machines of Loving Grace*.

53. *THX 1138*, directed by George Lucas (American Zoetrope and Warner Brothers, 1971).

54. *Star Wars: Episode IV—A New Hope*, directed by George Lucas (Lucasfilm, Twentieth Century Fox, 1977).

55. Daniel T. Rodgers, *Age of Fracture* (Cambridge, MA: Belknap Press, 2011), 144–79.

56. Donna Haraway, *Simians, Cyborgs, and Women: The Reinvention of Nature* (New York: Routledge, 1990). For further discussion of Haraway, see Livingston, *World Turned Inside Out*, 100–101; Denise Handlarski, "ProCreation—Haraway's 'Regeneration' and the Postcolonial Cyborg Body," *Women's Studies* 39, no. 2 (January 2010): 73–99; Chris Hables Gray, *Cyborg Citizen: Politics in the Posthuman Age* (New York: Routledge, 2001), 9–20; and Anne Balsamo, *Technologies of the Gendered Body: Reading Cyborg Women* (Durham, NC: Duke University Press, 1996), 17–40.

57. On women and cyborg imagery in music, see Ann Powers, *Good Booty: Love and Sex, Black and White, Body and Soul in American Music* (New York: Deu St., 2017), 299–341; and Steven Shaviro, "Supa Dupa Fly: Black Women as Cyborgs in Hiphop Videos," *Quarterly Review of Film and Video* 22, no. 2 (2005): 169–79.

58. See Isaac Asimov, *Foundation* (New York: Gnome Press, 1951); Isaac Asimov, *Foundation and Empire* (New York: Gnome Press, 1952); and Isaac Asimov, *Second Foundation* (New York: Gnome Press, 1953).

59. See Isaac Asimov, *Foundation's Edge* (New York: Doubleday, 1982); Isaac Asimov, *Foundation and Earth* (New York: Doubleday, 1986); Isaac Asimov, *Prelude to the Foundation* (New York: Doubleday, 1988); and Isaac Asimov, *Forward the Foundation* (New York: Doubleday, 1993).

60. Ralph Waldo Emerson, "The Over-Soul," in *The Essential Writings of Ralph Waldo Emerson* (New York: Random House, 2009), 237.

61. On the roots of the utopianism of Silicon Valley culture, see Turner, *From Counterculture to Cyberculture*.

62. Since the late 1980s, there has been a growing literature on power and control in digital culture. See Gilles Deleuze, "Postscript on Control Societies,"

in *Negotiations*, trans. Martin Joughin (New York: Columbia University Press, 1995); James R. Beniger, *The Control Revolution: Technological and Economic Origins of the Information Society* (Cambridge, MA: Harvard University Press, 1989); Wendy Hui Kyong Chun, *Control and Freedom: Power and Paranoia in the Age of Fiber Optics* (Cambridge, MA: MIT Press, 2008); Seb Franklin, *Control: Digitality as Cultural Logic* (Cambridge, MA: MIT Press, 2015); John Cheney-Lippold, *We Are Data: Algorithms and the Making of Our Digital Selves* (New York: NYU Press, 2018); and Shoshana Zuuboff, *The Age of Surveillance Capitalism: The Fight for a Human Future at the New Frontier of Power* (New York: PublicAffairs, 2018).

63. Josephmaggio13, "Motorola Droid 2 for Verizon Commercial (HD)," August 14, 2010, video, 32 seconds, https://www.youtube.com/watch?v=q8bSLMcerCc.

64. Gray, *Cyborg Citizen*, 161–63.

Epilogue

1. Karel Čapek, *R.U.R. (Rossum's Universal Robots)*, trans. Paul Selver (Garden City, NJ: Doubleday, 1923), 16–17.

2. See Janelle Monáe, "Janelle Monae—Many Moons (Official Short Film)," video, 6:31, https://www.youtube.com/watch?v=LHgbzNHVg0c&list=RDLHgbzNHVg0c&start_radio=1.

3. James Montgomery, "Jermaine Jackson Brings 'Smile,' Tears to Michael Jackson Memorial," MTV News, July 7, 2009, http://www.mtv.com/news/1615435/jermaine-jackson-brings-smile-tears-to-michael-jackson-memorial/.

4. Janelle Monáe, *Dirty Computer*, Wondaland, Bad Boy, and Atlantic Records, April 27, 2018.

5. Graham Lee Brewer, "The Problems and Potential in HBO's Westworld," *High Country News*, June 15, 2018, https://www.hcn.org/articles/indian-country-news-the-problems-and-potential-in-westworlds-ghost-nation.

6. See *Westworld*, season 2, episode 8, "Kiksuya," aired June 10, 2018, on HBO.

7. See David E. Nye, *America as Second Creation: Technology and Narratives of New Beginnings* (Cambridge, MA: MIT Press, 2003); R. W. B. Lewis, *The American Adam: Innocence, Tragedy, and Tradition in the Nineteenth Century* (Chicago: University of Chicago Press, 1959).

8. See *Westworld*, season 2, episode 10, "The Passenger," aired June 24, 2018, on HBO.

9. *Westworld*, "Kiksuya."

Index

Ad Hoc Committee on the Triple Revolution, 253
Adam Link series (Binder), 176–86, 210, 220
African Americans: activism of, 61, 100, 254–55, 297; and automation, 254–55; cultural resistance of, 111, 134, 286–87, 291, 311n12. *See also* black manhood, depictions of; and black womanhood, depictions of
Ah-Sin (Twain and Hart), 54
Ajeeb (the Mystifying Chess Automaton), 45–50
Alien (film), 291
alienation, 5, 121, 206–7, 253–55, 266–68, 281–82. *See also* industrial labor, critiques of; workers, mechanization of
Allen, Ethan, 29
"All Watched over by Machines of Loving Grace" (Brautigan), 268
Amazing Stories, 99
Amazon.com, 11
Anderson, Sherwood, 147
Andy Warhol's Overexposed: A No-Man Show (play), 271–72
animatronics, 271–73, 282–87
Arch-android, The (album), 298
Arminianism, 26
artificial intelligence, 2–3, 11, 229, 275–76
artificiality, 1, 15–17, 33, 51, 65–66, 68, 76–77, 110–13, 164, 167, 262–63, 270–73, 279–88, 295, 297–98
Artificial Mother: A Marital Fantasy, The (Putnam), 65–68, 165
"Artist of the Beautiful, The" (Hawthorne, N.), 16
Art of Loving, The (Fromm), 207
Ashby, Walter Ross, 274–75, 283–88
"Asians," generic caricatures of, 1, 10, 45–52, 54, 164, 190–93

Asimov, Isaac, 4, 186, 209–10, 212–13, 219–24, 231, 239, 248, 250, 291–93; fan response to, 223, 248; fiction, 209–10, 220, 276–78, 291–93; Three Laws of Robotics, 221–22, 232–34, 277
atomic bomb, cultural response, 206, 218, 227–30, 235, 250–51, 267
authenticity, 7–8, 15–17, 48, 65, 76–77, 110, 272–73, 279–88
"Autofac" (Dick), 250
"Automatic Bridget" (Fielding), 62
Automatic Toy Works, 52–53
automation, 5, 211–13, 243–45; advocacy of, 226–30, 243, 251–53; in Europe, 3–4, 23, 32; home, 243, 258–61, 283; industrial, 242–49; military, 218, 224, 226–30, 244; opposition to, 238–43, 249–57
automaton: meaning, 6–7, 30–31, 44–45, 47; and philosophy/science, 23–27, 78–84, 276; stage 19–20, 23, 30–34, 45–50, 77
Automaton Chess Player, The (The Turk), 32, 39, 47–50
Automaton Regiment, The (Brewerton), 85
automats, 108
Autry, Gene, 189–90
Avengers: Age of Ultron (film), 289

Babbage, Charles, 39, 89
Baldwin, Hanson, 227–28
Baldwin, James, 254
Bates, Harry, 218
Baudrillard, Jean, 272
Baum, L. Frank, 91–95
Bayliss, William, 166
Beard, Charles, 138
behaviorism, 165, 167–68, 170–71, 174–75, 181, 185–86, 262, 274–75, 359
Behaviorism (Watson), 167–68
Belcher, John W., 57
Bell, Daniel, 251–53
Bellamy, Edward, 68, 108
Bellamy, Elizabeth, 63–64, 68–69
Berkeley, Edmund C., 235
Berman, Louis, 166–67
Beyoncé, 291
Bierce, Ambrose, 71–74, 95
Binder, Eando, 165, 176–86, 220–21, 234
Binder, Otto, 165, 261
Bionic Woman, The (television series), 289
black manhood, depictions of, 52, 54, 131–35, 152, 154–55
black womanhood, depictions of, 52–54, 99, 140, 152, 159, 286–87, 297–99
Blade Runner (film), 280
Bourne, Randolph, 113, 159
Boyle, Robert, 23
Bradbury, Ray, 278–79
Brautigan, Richard, 268
Brave New World (Huxley), 164
Breuer, Miles J., 148–49
Brewer, Graham Lee, 299
Brewerton, G. Douglas, 85, 95
Brooks, Van Wyck, 113, 159
Broun, Heywood, 115
Brown-Sequard, Edouard, 166
Bryner, Gary, 255
Buck, Pearl S., 225
Buck Rogers (film), 193–94
Buffy the Vampire Slayer (television series), 8, 264
bureaucratization, 6, 206, 268, 281–82
Bush, Vannevar, 170, 226
Butler, Judith, 273, 291
Butler, Octavia, 8
Butler, Samuel, 118, 193

Calvinism, 25–26, 74, 219
Cameron, James, 238–39
Campbell, John W., 164, 221
Čapek, Karel, 5, 103–4, 114, 117–18, 150, 185, 187, 194, 296
Captain Jinks, Hero (Crosby), 86–87
Captain Video and His Video Rangers (television series), 231
Carlyle, Thomas, 6, 75–76, 89
Catholicism, 52, 310n9
Caves of Steel, The (Asimov), 276–78, 280
Cazzamali, Ferdinando, 169
Centennial Exposition (Philadelphia), 43–44
Century of Progress Exposition, 153
Chaplin, Charlie, 197, 200–201, 298
Chase, Stuart, 121, 126–29, 135, 137, 148, 172, 184, 195, 205, 282, 335
children, 34, 38, 57, 61, 67, 80, 161, 167, 189, 192–93, 259; and automatons/robots, 7, 23, 11, 30, 32, 34, 38, 50–54, 63, 65–66, 68, 94, 125, 158, 173, 203, 209–10, 219, 231, 234–37, 239–40, 278–79, 287; innocence of, 231, 234, 237; and leisure, 110–11
Christianity, evangelical, 4, 74, 115
Christmas, 50–51, 243–45, 259
City, The (film), 160–62
Civil War, 71, 79, 84–86
Clynes, Manfred, 288
Coates, Ta-Nehisi, 230
Coblentz, Stanton A., 203–4
Cohn, Victor, 245–46
Cold War, 210–13, 258; anxieties, 207, 210, 218, 230, 238; cultural responses to, 207, 215–19, 231–40, 261, 288–89, 311n12; military technologies of, 226–30; and military training, 224–26; and moral innocence, 230–34, 237
colonization, 49, 59–60, 73, 86–88
Colossus: The Forbin Project (film), 238
Coming of Post-Industrial Society, The (Bell), 252
commercialization, 4–6, 34, 49–55, 74, 91, 107–11, 113, 293, 296
Computer and the Brain, The (von Neumann), 276
computers: and automation, 245–46, 252; cultural impact, 211–13, 218, 273–80, 283, 289–94; fictional, 63–64, 147–49, 174–77, 181, 233–34, 238–40, 251, 279; meaning, 211; responses to, 223, 226–27, 252, 267–68, 275–76, 289, 298; technology, 73–74, 142, 144, 164–66, 170–71, 211, 226–27, 229, 264, 289
Comte de Buffon, Georges-Louis LeClerc, 21, 28
Congress of Industrial Organizations, 205
consciousness, 7, 9, 22, 26–31, 72–75, 81–83, 94, 167–68, 171, 211, 232, 275, 282, 299
consumerism, 11, 44–45, 49–50, 54, 61, 74, 89, 91, 100, 107–10, 127, 136, 139, 143, 151, 159, 162, 165, 174–76, 184–85, 223, 243, 253–54, 258–61, 265–72, 282–83, 289–90, 297; critiques of, 34, 51–52, 56, 98, 212, 250, 262–63, 271–72, 283–84
Corbin, John, 114, 118–19
Counterculture, The, 243, 267–70
Coxe, Tench, 36–38
Crichton, Michael, 9, 264, 285–88
Crosby, Ernest Howard, 86
crowds, fear of, 190, 193, 197, 206, 215–18
cultural lag, 123–24

cultural reinvigoration , 111, 113–14, 121, 125, 159
Cummings, Ray, 172
cybernetics, 211–12, 239–40, 268, 274–78
Cybernetics (Wiener), 211–212, 274
Cyborg (Caidin), 289
cyborgs, 3, 211–12, 239–40, 273–74, 288–94

"Dancing Machine" (song), 286–87
"Defenders, The" (Dick), 231
Darwin, Charles, 73
Day the Earth Stood Still, The (film), 215–18, 231, 240
Debs, Eugene, 90
deism, 25–26
Delaney, Samuel, 8
del Rey, Lester, 165, 173–76, 184–86, 220, 296, 342–43
democracy, 4, 76, 100, 111, 120–21, 134, 160–62, 182, 192, 205, 207, 268, 295
Der Fuehrer's Face (short film), 200, 202
Descartes, Rene, 23
Design for a Brain (Ross), 275
Desk Set (film), 251
determinism, 78–79, 82–83, 111, 171, 175–76
Devo, 287
Devol, George, 210, 240
Dewey, George, 198
Dexter, J. D., 77
Dick, Philip K., 213, 231, 250–51, 280
dictatorship, 99–100, 128, 182, 184, 188–90, 192–94, 197–98; by machines, 148–50, 184
Diderot, Denis, 28
digital identities, 273–74, 300–301
disenchantment, 45, 49
Disney: Disneyland, 283–84; films, 154–56, 199–202; Walt, 282–83, 287–88, 362
Dissertation on Liberty and Necessity, Pleasure, and Pain, A (Franklin), 26
Do Androids Dream of Electric Sheep? (Dick), 280
Dr. Goldfoot and the Bikini Machine (film), 264
Dr. Strangelove (film), 238
Drederick, Zadoc, 41, 42, 44–45, 55
drones, 11, 227–30, 293
Drucker, Peter, 194, 196–97, 201, 206

Eckert, J. Presper, 226
Eco, Umberto, 283
Eden Musee, 45–46
Edison, Thomas, 52, 91, 107–8, 293–94
Education for Death (film), 199–201
Eichmann, Adolf, 281
Einstein, Albert, 111
Electric Lady, The (album), 298
Ellis, Edward S., 57–59, 61
Ellison, Harlan, 238
"Ely's Automatic Housemaid" (Bellamy, El.), 63, 95
Emerson, Ralph Waldo, 14, 16, 293
Emerson, Reverend Joseph, 35
emotions: as dividing line, 10, 20–22, 30, 148–50, 163–64, 199–200, 255, 263, 279–80, 287–88; mechanization of, 173–74, 263, 279–80, 297–99; understandings of, 21–30 75, 81, 149, 166–68, 174, 283. *See also* sensibility/sentiment
Encyclopédie (Diderot), 28
endocrinology, 165–67, 175
End of Economic Man, The (Drucker), 196–97
Engelbart, Douglas, 289

Engelberger, Joseph, 210, 248
ENIAC, 226–27
Enton, Harry, 60–61
Erewhon (Butler), 118, 193
Escape from Freedom (Fromm), 195–96
Essay Concerning Human Understanding, An (Locke), 28
Evans, Oliver, 36
evolution, 81–82, 111
"Exile of Time, The" (Cumming), 172

Fail-Safe (film), 238
Faith for Living (Mumford), 194–95, 203
Falconi, Signor, 19–22, 31–34, 38, 40, 54, 298, 300–301
"Farewell to the Master" (Bates), 218
farmer activism, 64–65, 108
fascism, conceptions of, 188–89, 194–203
feedback (technology), 165–66, 169–71, 211, 227, 241–42, 245–49, 348n8
"Feminine Metamorphosis, The" (Keller), 164
Feminine Mystique, The (Friedan), 258, 261–62
feminism, 8, 176, 243, 258, 261–63, 291, 297–98, 357n50; fears of, 67, 164, 172–73, 176, 260–61, 344
Fielding, Howard, 62–63
Flanders, Ralph E., 100, 121, 145, 147–48
Flash Gordon Conquers the Universe (film serial), 190, 192
Forbidden Planet (film), 231–33
Ford, Henry, 5, 109, 128
Fordism, 109–10, 119
Foundation series (Asimov), 291–93
Frank, Waldo, 113–14
Frank Reade series (Enton), 60–61
Frankenstein (Shelley), 4, 114
Franklin, Benjamin, 24, 26, 30–31
Freeman, Damita Jo, 286–87
free will, 6, 24–30, 34–36, 72–76, 82–84, 165–71, 175–76, 187, 220–22
Freud, Sigmund, 112, 114
Friedan, Betty 258, 261–62, 264
Fromm, Erich, 194–96, 201, 206–7
Frontier, the. *See* Western iconography
Fuehrer, Sherwood, 235–36
Fuller, Alice W., 65, 67–69
Fuqua, Robert, 176–78, 181–83
Futrelle, Jacques, 89
Futurama (television series), 1–3, 5, 8–9, 257, 264
Future of Industrial Man, The (Drucker), 196–97

Galbraith, John Kenneth, 210
Galvani, Luigi, 168–69
Garden of Allah, The (Hitchen), 49
General Electric, 156, 246, 256
General Motors, 156, 162, 255
Gernsback, Hugo, 99, 134, 140, 338
Gilman, Charlotte Perkins, 68
Ginsberg, Allen, 212, 282
Glands Regulating Personality, The (Berman), 166–67
Gog (film), 238
Goldman, Emma, 90
Goldsmith, Barbara, 272
Goodwin's Automaton Toys, 52
Gort (aka Gnut), 215–19, 238, 240
Government control of innovation, 120–21, 127–28, 136–37, 143–46, 159–62, 216, 268
Graham, Billy, 219, 282
Great Depression, 98–101, 131–33, 143–46, 150, 220
Great Dictator, The (film), 197, 200

Great War. *See* World War I, cultural impact
Green, William, 144–45
Greene, Frank L., 87

HAL 9000, 279
Hamilton, Alexander, 35–36
Haraway, Donna, 291
Harte, Bret, 54
Hawthorne, Julian, 14–17, 165
Hawthorne, Nathaniel, 14, 16, 312
Hazard, Rowland G., 82
Heisenberg, Werner, 112
"Helen O'Loy" (del Rey), 173–76, 220, 296
Helmholtz, Hermann von, 79, 80
Hemingway, Ernest, 204
Hensel, H. Struve, 225–26
Here Comes Tobor (failed television series), 231
Hines, Douglass, 264
Hitchens, Robert, 49
Hitler's Children (film), 201
Hobbes, Thomas, 24–25, 28
Hoffmann, E. T. A., 112, 165
home, the: fictional robots in, 11, 61–69, 136–40, 158–59, 174, 258–61, 265–66; mechanization of, 67–69, 108–9, 152, 172, 242–43, 258–61, 283
Houdini, Harry, 122–23
Huge Hunter; or, The Steam Man of the Prairies, The (Ellis), 57–61, 64, 71, 298
Hughes, John, 264
Hughes, Ted, 234
Hughes Aircraft, 246–47
Humanoids, The (Williamson), 265–66
Human Use of Human Beings, The (Wiener), 266–67, 274–75, 288
Huxley, Aldous, 164, 252
Huxley, Thomas Henry, 73, 81–84
hyperreality, 272, 283–84

Ickes, Harold, 197
Ideal Toy Company, 235, 237
"I Have No Mouth and I Must Scream" (Ellison), 238
immigrants, fears of, 52, 100–101, 114–15, 119–20, 157–58, 219–20
immigration, 84, 184–85
imperialism, 49, 59–60, 73, 86–88
Indian, automaton, 4, 19–22, 30–34, 54, 298, 300–301
individualism, celebration of, 14, 16, 59–61, 73–76, 148, 151, 156–57, 179–80, 204, 225, 269; loss of, 5–6, 16, 54–55, 75–78, 87, 97, 113–14, 124, 128, 188, 194–202, 206, 212, 258, 266, 281–82; political implications of, 4, 7–8, 30, 34–36, 64–65; selfish, 2, 11, 73, 86, 99, 243, 265
individualization, 6, 10–11, 74–75, 110–11, 161, 262–65, 286–87
individuals, nature of, 23–30, 78–84, 88–89, 91–95, 165–69, 285
industrial labor, critiques of, 77–78, 89–91, 116, 117, 118, 121, 124–25, 127–29, 144–46, 147–50, 168, 195–97, 200, 206, 282, 296. *See also* alienation; workers, dehumanization of
Industrial Workers of the World, 115–17
industrialization, 36–40, 43–45, 52, 56–57, 61–62, 74, 78, 84–85, 89–90, 92–93, 108–9, 118–20, 135; defenses of, 37–40, 134–35, 151–52, 157–62, 205
Invisible Boy, The, 233–34
iRobot, 229–30
I, Robot (Asimov), 210, 222

"I, Robot" (Binder), 176–79, 181, 234
Iron Man (character), 289
Irony of American History, The, 230
Iron Man, The: A Children's Story in Five Nights (Hughes, T.), 234
Iron Man in Industry, The (Pound), 120–21, 167
"I Sing the Body Electric" (television episode), 278–80

Jackson, Michael, 273, 286–87
Jaquet-Droz, Henri-Louis and Pierre, 23, 32–33
James, William, 82–84, 112
Japan, robots in, 235, 255–56, 310–11
Japanese, caricatures of, 192, 199, 202, 230, 256–57, 265, 346n29
Jefferson, Thomas, 21, 24, 34–36
Jentsch, Ernst, 112, 114
Jetsons, The (television series), 251, 259–60
Johnson, Edwin C., 225
Johnson-Reed Act, 114–15

Kaempffert, Waldemar, 136–37, 166–67, 171
Kames, Lord Henry Homes, 24, 35
Keller, David H., 97–100, 111, 128, 140, 148, 164, 172–73
Kempelen, Wolfgang von, 31, 34
Khrushchev, Nikita, 254
"Kiksuya" (television episode), 299
King Jr., Martin Luther, 254
Kinsey, Alfred , 212
Kintner, S. M., 132
Kline, Nathan, 288
Knuuti, Rosa, 117–18
Korean War, 228
Krim, Norman , 170–71
Kubrick, Stanley, 238, 279

labor activism, 88–89, 108–9, 115–17, 143–46, 205, 207, 255–57
La Mettrie, Julien Offray de, 24
Lang, Fritz, 124–26, 165
"Last of the Masters" (Dick), 250
Le Guin, Ursula, 8
League for Industrial Democracy, 254
leisure, 30–31, 51, 110–11; critiques of, 34, 37, 99–100, 111, 121–22, 148, 251–53, 281; identity and, 110–11, 116, 142, 238, 296; promotion of, 37, 49, 51, 95, 136–37, 142–43, 151, 161, 212, 242–46, 258–60, 277, 283
Leviathan (Hobbes), 24–25
Levin, Ira, 8, 262–64
"Liar" (Asimov), 222
liberalism, 22, 28–30, 75–76, 156–57, 196–97; critiques of, 75–76, 291
Lindsay, Malvina, 281
Link, Stanley, 192–93
Locke, John, 28–29, 51, 75–76
Loeb, Harold, 146
Looking Backward (Bellamy, Ed.), 68, 108
Lordstown, Ohio, strike, 255
Lost in Space (television series), 234
Lucas, George, 290–91

Maelzel, Johann Nepomuk, 39, 47–50
magic, as power source, 13–17, 45–47, 51–52, 91–93
Maltz, Maxwell, 285, 287–88
Man a Machine (La Mettrie), 24
Mangus, Robot Fighter (Manning), 289
manhood, 9, 74, 172,
"Man That Was Used Up, The" (Poe), 77
"Many Moons" (song), 297–98
Marcuse, Herbert, 282

Marx, Karl, 90
masculinity. *See* manhood
mass media, 189, 193–94, 206, 281–82
Master Mystery, The (film serial), 122–23
materialism (philosophy), 23–29, 35–36, 72–75, 78–84, 91, 119, 164–71, 184–86, 296
Mather, Cotton, 25
Mauchly, John , 226
McCormick, Cyrus, 39
McCulloch, Robert, 259, 274
McDougall, William, 168, 185
McGeehan, John E. Justice, 194, 202
McLuhan, Marshall, 262, 267
Mechanical Bride, The (McLuhan), 262, 267
mechanization of, 68–69, 75–76, 77–78, 85–90, 92–93, 194–98. *See also* black manhood, depictions of; white manhood
Melville, Herman, 14, 79
Men and Machines (Chase), 121, 127–28, 172, 195
Men Must Act (Mumford), 194–95
Mental Philosophy (Upham), 74–75
Merrick, F. A., 151–52
Methodism, 26–27
Metropolis (film), 124–26, 165
Metropolis: The Chase Suite (album), 297–98
Michael Jackson's Moonwalker (game), 287
Michael, Donald, 252–53
Mickey's Mechanical Man (short film), 154
Middleton Family at the Fair, The (film), 157–59
Mills, C. Wright, 5, 212, 272, 281
Minsky, Marvin, 275–76
Modern Inventions (short film), 155–56
Modern Times (film), 200, 298
Monáe, Janelle, 297–99
Moral Man in Immoral Society (Niebuhr), 204
Morris, William, 90
Morse, Wayne Senator, 225
Motorola, 293
"Moxon's Master" (Bierce), 71–73
"Mr. Corndropper's Hired Man" (Stannard), 64–65, 95, 296
"Mr. Roboto" (Song), 256, 288
"Mullenville Mystery, The" (Hawthorne, J), 14, 16, 112–13, 165
Mumford, Lewis, 113, 159–62, 188, 194–96, 201, 203, 206, 267, 281
My Living Doll (television series), 263
Mysterious Doctor Satan (film serial), 190

Naked Sun, The (Asimov), 276, 280
NASA, 288
Native Americans, representations of, 19–21, 250, 299–301
Neumann, John von, 276
neuroscience, 168–69, 274–75
"New Age and the New Man, The" (Flanders), 145
New Deal, 146, 161
New Left, 243, 267–68
New Thought, 91, 285
Newton, Isaac 21, 23
Niebuhr, Reinhold, 206, 219, 230–31, 237
Nixon, Richard, 258, 261
North, Edmund, 216–18
Nott, Eliphalet, 34, 38

Obama, Barack, 8, 230
O'Brien, Fitz-James, 51–52

"Occurrence at Owl Creek Bridge, An" (Bierce), 71
Ogburn, William F., 123–24, 147, 172
Olivaw, R. Daneel (character), 276–78, 280, 292
Orientalism, 48–49, 54
Outer Limits, The (television series), 234
Ozma of Oz (Baum), 93

Palmer, Frederick, 199
Palomilla (robot), 241–42, 275
Paradise and Iron (Breuer), 148–49
"Paradise of Bachelors and The Tartarus of Maids, The" (Melville), 78
passions, the. *See* emotions
Peale, Charles Willson, 33
Peale, Norman Vincent, 285
Pegler, Westbrook, 199
Peirce, Charles Sanders, 112
Perew, Phillip Louis, 57
Perhaps Women (Anderson), 147
Phantom Empire, The (film serial), 189–90
Pillsbury, Harry Nelson, 48
Pitts, Walter, 274
planning, economic, 127, 144–46, 160–62
Player Piano (Vonnegut), 250, 267
Poe, Edgar Allan, 48, 77
Poinsett, Alex, 254
postindustrialism, 161, 212, 251–53
Pound, Arthur, 120–22, 124, 167
Principles of Scientific Management, The (Taylor), 89
Product Integraph, 170
Protestantism, 81, 115, 195–96, 310–11n9
Psychocybernetics (Maltz), 285
"Psychophonic Nurse, The" (Keller), 140
Puritanism, 4, 22, 25, 74, 292
Putnam, George Haven, 65–68, 165

quantum mechanics, 111

"Reason" (Asimov), 220–21
Reason: The Only Oracle of Man (Allen), 29
regimentation, 85–88, 92–93, 140, 194–98, 224–25, 237
republicanism 29–40, 43, 52, 98, 180
Revolt against Civilization, The (Stoddard), 119
"Revolt of the Pedestrians" (Keller), 173
Revolution, American, 22, 25, 29, 30, 40
"Rex" (Vincent), 149–50, 164
Rickenbacker, Eddie, 224
Rising Tide of Color against White World-Supremacy, The (Stoddard), 119, 192
"Robbie" as "Strange Playfellow" (Asimov), 209–10, 223
Robby the Robot, 232–35
RoboCop (film), 273, 289
robot: dance, 273, 286–87; definitions, 4–7, 103, 105–7, 114, 115–16, 119, 126–27, 132–34, 140–42, 163, 188, 193–94, 211, 281–82, 295
robotics, industry, 211, 243–49
Roddenberry, Gene, 210, 223
Romney, Mitt, 8
Roosevelt, Franklin Delano, 146, 186, 188, 202, 225
Roosevelt, Theodore, 88
Rosa, E. B., 80
responses, 115–19, 188

R.U.R. (Rossum's Universal Robots) (Čapek), 5, 103–6, 114–16, 119, 121, 187–88, 204–5, 296
Ruskin, John, 90
Rustin, Bayard, 254

Sandburg, Carl, 115
"Sand-man, The" (Hoffmann), 112, 165
Santa Claus, 50, 243–45, 257
Santa Claus Conquers the Martians (film), 245
Schlumberger, William, 48
science fiction (genre), 59–60, 99
scientific management, 88–89
Scott, Howard, 145–46
Seaborg, Glenn, 260
Semi-Automatic Ground Environment (SAGE), 227
Senarens, Luis P., 60–61, 91
sensibility/sentiment, 16, 21–22, 26–31, 35–38, 75, 92, 98, 113, 125, 143, 201, 293
Seward, William, 39–40
sex with robots, 1–2, 9–10, 65, 67, 134, 139–40, 143, 153–54, 174, 176, 262–65, 280, 296
Shannon, Claude, 276
Shedd, John G., 88
Shelley, Mary, 99
"Sign of the Times" (Carlyle), 75
Silent Running (film), 268–70, 296
Silicon Valley, ideology of, 270, 293
Sinclair, John, 268
Six Million Dollar Man, The (television series), 289
slavery, as fantasy, 8–10, 39–40, 43–44, 61–62, 69, 94, 100, 131–36, 142, 151–53, 156, 162, 220–22, 243, 246, 261, 265–66, 296–97, 298, 300–301
Sloan, M. S., 152
Smith, Adam, 75
Social Change with Respect to Culture and Original Nature (Ogburn), 123–24
soldiers: as citizens, 85–86, 180, 224; mechanization of, 35, 85–88, 113, 200
Soul Train (television series), 286
Sousa, John Philip, 110, 144
Spencer, Herbert, 72–73, 168
Stannard, W. M., 64, 65
Star Trek: The Next Generation (film and television series), 280, 291, 294
Star Wars (film series), 290–91
Steam Man (Drederick's) 41–45, 55–57, 62
steam men, literary responses, 57–69, 296
Stepford Wives, The (Levin), 8, 262
Stern, Bernhard J., 220
Stoddard, Lothrop, 119, 192
Styx, 256, 288

Taming Our Machines (Flanders), 145, 147
Taylor, Frederick Winslow, 88–89, 116, 293–94
Technics and Civilization (Mumford), 159–60
Technocracy Movement, 145–46
technological determinism, 100, 120–23, 154, 160–62
technological unemployment: debate over, 98, 117, 127, 141–47, 151–52, 158, 161, 249–55, 257; in fiction, 97–98, 148, 197, 220–21, 239, 250–51, 257, 277
Terminator, The (film), 9, 238–39
Terminator 2: Judgment Day (film), 239–40

Tesla, Nicola, 6, 83–84
Theater Guild, 103–5, 116
Theory of the Leisure Class (Veblen), 122
thermodynamics, 79–81, 274
Thomas, Phillips, 131–33
Thoreau, Henry David, 14
"Threat of the Robot, The" (Keller), 97–99, 111, 148
THX 1138 (film), 290
Tin Woodman of Oz, The (Baum), 93
Tiny Tim and the Mechanical Men (Link), 192–93
Tobor the Great (film), 231
"To Serve the Master" (Dick), 250–51
totalitarianism, 194–203, 206, 281–82
Tousey, Frank, 60
toys, 50–54, 235–37
Transcendentalism, 14, 16, 291–93
True Companion, 264–65, 296
Truman, Harry S., 224
Turner, Sir Alfred, 198–99
Twain, Mark, 54
"Twilight" (Campbell), 164
Twilight Zone, The (television series) 278–79
2001: A Space Odyssey (film), 279

uncanny, the, 112–13
Undersea Kingdom (film serial), 190–93
Unimate, 210, 248–49, 255
Unimation, 248–49
United Auto Workers, 241, 253–54
Universal Military Training, 224–26
Upham, Thomas Cogswell, 74

Vaucanson, Jacques de, 23, 32, 33
Veblen, Thorstein, 121, 126
Verne, Jules, 60, 99
Vietnam War, 229
Vincent, Harl (Harold Vincent Schoepflin), 149–50, 164
vitalism, 26–30, 74–76, 81, 92–93, 137, 168, 171, 277–78, 296–97
Vonnegut, Kurt, 250, 267

Walker, Timothy, 76
Warhol, Andy, 271–73, 300
Washington, Charles "Robot," 286
Watson, John B., 167–68, 171, 184–85
Webster, Daniel, 39–40
Weiner, Tim, 229–30
Weird Science (film), 264
Wells, H. G., 60, 99, 126
Wensley, Roy 135–38
Wesley, John, 26–27
Western iconography, 4, 9, 59–61, 128, 148, 189–93, 204, 224, 255–56, 269, 277, 298–301
Westinghouse, George, 88
Westinghouse Electric Company, 128–29, 132–43, 149–53, 156–59, 161–62, 176, 220, 248; engineering culture, 151–52; and unions, 146
Westinghouse Robots, 133–35; Elektro, 133, 156–58, 162, 220; fictional responses, 140, 149–50, 154–55; journalist responses, 136–43, 170; Katrina von Televox, 151–53; Mr. Televox, 128–29, 136–43, 152; Rastus Robot, 132–35, 151–52, 296; Telelux, 142–43; Willie Vocalite, 151–54, 157
Westworld (film), 9, 264, 285–86, 298–99
Westworld (television series), 4, 298–301
Whedon, Joss, 8, 264

white manhood: conceptions of, 60, 124, 152, 190–92, 238–40; restoration of, 8–10, 88, 148–49, 285–86; threats to, 77–78, 99, 119–20, 147–49, 154–56, 185, 189–94, 205, 285–86, 288, 344n64

whiteness and machinery, 42–43, 52, 54–58, 128, 134, 137–43, 280

white womanhood: depictions of, 9, 23, 68, 172–76, 181, 201, 238–40, 262–65; threats to, 20, 78, 99, 171–73, 216–17

Whitney, Eli, 31, 38

Whytt, Robert, 27, 28

Wiener, Norbert, 211, 241–43, 266–67, 270, 274–76, 288

Wilcox, Ella Wheeler, 81

Williamson, Jack, 265–66

Willkie, Wendell, 198

Wise, Robert, 215–18

womanhood, 9, 68–69, 172, 175, 185; as mechanical, 30, 32–35, 37–39; mechanization of, 61–69, 78, 90, 173–76, 258–68. *See also* black womanhood, depictions of; white womanhood

"Woman Manufactured to Order, A" (Fuller), 65

Wonderful Wizard of Oz, The (Baum), 91

"Wondersmith, The" (O'Brien), 51

work: ethic, 30–31, 37–40, 75, 98–100, 104, 250–53, 261, 266, 296; hours, 109; organization of, 36–38, 84–85, 108–11, 205–7, 250–53

Work and Its Discontents (Bell), 251

workers, dehumanization of, 5, 11, 38, 77–78, 88–92, 105–7, 115–20, 124–25, 127, 149–50, 168, 255

World's Fair, 1939–1940, 156–62

World's Fair, 1964, 282

World War I, cultural impact, 97–100, 103–7, 114

World War II: cultural impact, 204–7, 215–19; propaganda during, 199–204; technologies of, 203, 224, 226

Wyndham, John, 179

You and Machines (Ogburn), 147, 172

You Are a Computer (Brix), 375

Zuckerberg, Mark, 293